Aromatic Basmati Rice

Aromatic Basmati rice, also known as 'the scented pearl' by its connoisseurs, has been grown in a specific geo-climatic region of India and Pakistan for centuries. It is appreciated by consumers in more than 100 countries around the globe. Due to its specific aroma and grain quality traits, Basmati rice also fetches a much higher price in the international market compared to other non-aromatic rices, benefiting its producer farmers and also traders. Timely information on Basmati rice acreage and production is a necessity for planners, policy makers, and traders alike. Satellite remote sensing and other geoinformatics tools and techniques are presently being used in Basmati rice area estimation and production forecast. Unfortunately, there is no publication available that deals with the various aspects of Basmati rice, including its cultivation practices, area mapping, monitoring crop growth, and forecasting production. The present book may be the first of its kind, which deals not only with the above aspects of aromatic Basmati rice, but also the global trade and issues arising from climate change and its likely impact on future production.

Specifically, the book covers:

- Basmati rice—Historical and cultural perspective
- Varieties and cultivation practices
- Science of rice aroma
- Environmental factors influencing Basmati aroma and grain quality
- Remote sensing in Basmati area mapping
- Yield prediction using machine learning techniques
- Hyperspectral data analysis in Basmati discrimination
- Traditional methods of area and yield estimation
- Global trade in Basmati rice
- Influence of climate change on Basmati rice production

In essence, the book covers both agronomical and geo-spatial technological aspects involved in aromatic Basmati rice cultivation, mapping, and monitoring. It is hoped that the book will be useful for the researchers, geospatial technologists, agronomists, environmentalists, farmers, traders, consumers, and all those interested in this wonderful grain gifted to us by Mother Nature.

Aromatic Basmati Rice

Science, Mapping, and Monitoring

Manik H. Kalubarme
Dharmesh Verma
Ritesh Sharma
A.N. Singh

CRC Press
Taylor & Francis Group
Boca Raton London New York

CRC Press is an imprint of the
Taylor & Francis Group, an **informa** business

Designed cover image: Manik H. Kalubarme, Dharmesh Verma, Ritesh Sharma, and A.N. Singh

First edition published 2026
by CRC Press
2385 NW Executive Center Drive, Suite 320, Boca Raton FL 33431

and by CRC Press
4 Park Square, Milton Park, Abingdon, Oxon, OX14 4RN

CRC Press is an imprint of Taylor & Francis Group, LLC

ISBN: 978-1-032-65937-4 (hbk)
ISBN: 978-1-032-65938-1 (pbk)
ISBN: 978-1-032-65939-8 (ebk)

DOI: 10.1201/9781032659398

Typeset in Times
by SPi Technologies India Pvt Ltd (Straive)

Contents

Forewords

FOREWORD BY GEORGE JOSEPH

I am delighted to write a foreword for the book entitled *Aromatic Basmati Rice: Science, Mapping and Monitoring*, which explores various aspects of discriminating, mapping, and monitoring Basmati rice varieties using remote sensing data. Rice plays a vital role in global food security, particularly in Asia, where nearly 90% of its global production and consumption takes place. Aromatic rice, such as Basmati, is prized internationally for its unique sensory attributes, nutty aroma, and excellent cooking qualities, which command it a premium price in the international market. India and Pakistan are the two countries growing Basmati, of which India leads both in its production and export. Indian annual Basmati rice production varies between 8 to 11 million metric tonnes, and its contribution to global export is between 65–70%, the remaining 30–35% being shared by Pakistan.

Remote sensing data from satellites have been widely used by researchers in mapping and monitoring rice cultivation. The Space Applications Centre of ISRO initiated studies on crop acreage and production estimation using remote sensing data in the late 1980s, which was later expanded into a national-level programme called Forecasting Agricultural Output, using Space, Agro-meteorology, and Land-based observations (FASAL) in 2006, with the objective of providing multiple pre-harvest acreage and production forecasts for rice and other crops at the national, state, and district levels. The programme, presently carried out by the Ministry of Agriculture and Farmers' Welfare of the Government of India, is being used operationally for rice crop acreage, crop condition assessment and yield forecasting in India.

A feasibility study on Basmati rice discrimination and mapping was also initiated by ISRO during the late 1980s, using IRS LISS-II data covering Karnal district in Haryana state of India. The differences in phenological stages of different rice varieties on a given date of data acquisition resulted in differences in their spectral signatures on the satellite data, enabling the discrimination of Basmati rice from other high-yielding rice varieties. However, at that time, only traditional, low-yielding Basmati rice varieties were grown, due to which its acreage, production and export were limited in extent. In the later years, the development of new high-yielding, short-duration Basmati varieties led to a significant increase in Basmati rice area and production in India. During the early years of the new Millennium, due to the increased quantum of Basmati production and export, the exporters' organisations like All India Rice Exporters Association and Basmati Export Development Foundation (BEDF) realised the necessity of remote sensing technology, and started using it to obtain timely information on Basmati rice acreage and production.

This book provides information on the analysis of remotely sensed satellite data from several rice-growing seasons, covering the Basmati-growing states of the north and north-west India. Case studies using a variety of satellite data, from IRS to

Sentinel, and different sensors operating in both microwave and optical (multispectral to hyperspectral) ranges have been described in the book, which have been used in mapping Basmati on an operational basis. Analysis of multitemporal Sentinel-1 SAR and Sentinel-2 MSI data using Random Forest classifier in the Google Earth Engine cloud computing platform has shown significant differences between early, long- and short-duration Basmati varieties. PRISMA hyperspectral narrow-band data analysis has also been used to discriminate aromatic Basmati from other high-yielding rice varieties. Similarly, different methods in Basmati rice yield prediction and production estimation, from the historical weather and yield data trend analysis to the use of empirical and machine learning models, Moderate Multiple Linear Regression, Random Forest, and XGBOOST, have been described and used. Apart from area mapping and production monitoring, the book also provides information on the historical and cultural aspects of Basmati, its cultivation practices, the aroma of rice, and the influence of environmental factors, as well as global trade and the impact of climate change on Basmati production in the future.

All the authors have pursued distinguished research careers in reputed institutions dedicated to advancing remote sensing technologies and their practical applications. Much of this book reflects their own extensive work in using remote sensing for agricultural applications. The carefully selected references provided at the end of each chapter will be of particular benefit to readers seeking a deeper understanding of the subject. I extend my sincere congratulations to the authors for this timely and informative publication. I am confident that the book will serve as a valuable resource not only for researchers, students, and professionals in remote sensing and geospatial technologies, but also for agricultural planners, environmentalists, traders, and all others interested in Basmati rice.

Dr. George Joseph
Former Director, Space Applications Centre (ISRO), Ahmedabad
Former Director, Centre for Space Science and Technology
Education in Asia and the Pacific, Dehradun
Former Chairman, Lunar Mission Study Task Force, ISRO
Former Editor-in-Chief, Journal of the Indian Society
of Remote Sensing
1999 Padma Bhushan Awardee

FOREWORD BY VIRENDRA PAL SINGH

Among the different aromatic rices grown in the Indian sub-continent, Basmati is considered a rice *par excellence* due to its unique fragrance, cooking and eating qualities, and has found wider acceptance all over the world. Its special quality is attributed to a unique combination of agroclimatic conditions, besides the inherent genetics governing these features. Grown in a specific contiguous geographical region of India and Pakistan since centuries, Basmati rice is consumed today in more than a hundred countries globally. The increased importance of Basmati rice has led to its increased acreage and production in both growing countries. It is interesting to note that during the past twenty years, Basmati rice production has quadrupled in India, whereas it has doubled in Pakistan; and consequently, the domestic consumption and export have also increased significantly. This has been made possible due to the development of several new high-yielding, disease- and pest-resistant varieties by our plant breeders in recent years using new biotechnological tools, and better management practices followed by the farmers. Considering the increasing global importance of this wonderful grain, publication of the present book, *Aromatic Basmati Rice: Science, Mapping and Monitoring*, is both valuable and timely.

In this book, the authors have addressed not only the agronomical, environmental, and geo-spatial technological aspects involved in aromatic Basmati rice cultivation, mapping the acreage, and monitoring production, but also the scientific basis of its aroma, global trade, and the future impact of climate change on Basmati rice production and sustainable approaches for its mitigation. The use of geospatial technology for determining the acreage and production of Basmati is relatively new, more so the discrimination between the different varieties—early, short-duration, and long-duration. The use of different remote sensing satellites and sensors in the discrimination of different Basmati varieties and production estimation methods has been succinctly dealt with. The authors have also studied the long-term trends in Basmati cultivation, showing that the long-duration Basmati varieties are increasingly being replaced by the short-duration ones. This transition is driven by factors such as climate change, reduced irrigation needs, and the potential for post-rice crop diversification.

I congratulate and extend my appreciation to the authors for writing this book and sharing their knowledge and long experience in this area of work. I hope that this book will be of value not only to the researchers, geospatial technologists, agronomists, and environmentalists but also to Basmati farmers, traders, and consumers alike.

Virendra Pal Singh
Former Senior Agronomist, International Rice Research Institute,
Manila, Philippines
Former Regional Co-ordinator, South Asia, World Agroforestry Centre,
Nairobi, Kenya

Preface

"In a grain of rice, the universe resides"

— *A Japanese proverb*

Rice, one of the three most important food crops in the world, forms the staple food of 2.7 billion people and is grown in various parts of the world, from 39°S (Australia) to 50°N latitude (China). Realising its global importance, the Food and Agriculture Organisation (FAO) of the United Nations had declared 1966 as *the Year of Rice*, and recently, the UN General Assembly also declared 2004 as *the International Year of Rice*. Aromatic rice constitutes a small but important subgroup of rice, prized for its special grain quality. Among the aromatic rices grown in different countries, the two most traded globally are Basmati and Jasmine. While Jasmine rice is grown primarily in Thailand, Basmati rice is grown in a specific geo-climatic region of India and Pakistan. Grown for centuries in this geographical region, Basmati is also called 'the scented pearl' by its connoisseurs, and is enjoyed by consumers spread over more than a hundred countries, including throughout the Middle East, Europe, and the United States of America. Basmati rice has a special significance not only amongst its producer farmers and traders in India and Pakistan, but also to its consumers spread globally.

Though Basmati rice is prized globally for its unique aroma and other grain characteristics, there is no single publication available dealing with different aspects of Basmati rice, including its cultivation practices, area mapping, monitoring crop growth, and forecasting production. The present book may be the first of its kind, which deals not only with the above aspects of aromatic Basmati rice, but also the global trade and issues arising from climate change and its likely impact on future production. In essence, the book covers both agronomical and geo-spatial technological aspects involved in aromatic Basmati rice cultivation, mapping, and monitoring.

The first chapter of the book traces the geographical origin and history of rice cultivation, evolution of aromatic rice and varieties grown in different countries, and subsequently describes Basmati rice, its history, quality traits, geographical dependence, varietal improvement in India and Pakistan, and cultural and religious importance. The second chapter describes the currently grown aromatic Basmati rice varieties, their area, production, and cultivation practices, apart from the total rice area and production in major rice-growing countries of the world. The science of rice aroma, its biochemical and genetic perspectives, and the environmental factors influencing Basmati rice aroma and other grain quality traits have been discussed in the third chapter.

Remote sensing and other geo-information technologies have been used in rice area mapping for quite some time now. However, its use in Basmati rice varietal discrimination and mapping has been done only to a limited extent. It may be interesting to mention here that one of the authors (Manik H. Kalubarme) conducted the first study on discrimination and area estimation of Basmati rice using Indian Remote Sensing Satellite (IRS-1A) data in Karnal district of Haryana, India, as early as 1989. The technology has taken a great leap since then, and hence the fourth chapter, apart from providing the basic information on remote sensing, describes the satellite data analysis in Basmati rice discrimination, mapping, and acreage estimation. Two case studies, one describing the use of the technology in an operational project covering the entire Basmati growing area of India, and another on the use of high-resolution Sentinel-1 and -2 data in the discrimination and mapping of Basmati varieties of different growth durations, have been described in detail.

The fifth chapter reviews different yield prediction models being used, and also presents a case study evaluating the use of different empirical and machine learning models in pre-harvest yield estimation of Basmati rice. Weather parameters serve as the primary drivers of yield variability in this agroecologically uniform Indo-Gangetic Plains, having an assured irrigation facility across the entire Basmati-growing area. The use of hyperspectral data in the discrimination and area mapping of Basmati rice has not been attempted yet. The sixth chapter, therefore, shows an example of the use of the data from the PRISMA satellite in Basmati mapping.

Apart from the remotely sensed data-based crop area and yield estimation method, which has become nearly operational for crops grown on a large area, traditional methods are still being used in the case of crops grown on a small scale. The seventh chapter describes the traditional method used in crop area and yield estimation in India. Basmati rice forms a major agricultural export commodity from India and Pakistan. Global trade in aromatic Basmati rice has been described in the eighth chapter. The ninth and the last chapter describes the influence of climate change and its likely impact on future Basmati rice production, and adaptation and mitigation measures to combat it.

We hope that the book will be useful for the researchers, geospatial technologists, agronomists, environmentalists, farmers, traders, consumers, and all those interested in this wonderful grain gifted to us by Mother Nature. We wholeheartedly welcome suggestions and criticisms from the readers, which will help us to carry this publication towards perfection in future.

We are thankful to the two eminent scientists, Dr. George Joseph and Dr. V. P. Singh, who wrote the foreword of our book and encouraged us greatly. We also acknowledge Dr. Ajai, former Group Director, SAC (ISRO), Mr A. K. Mittal, the founder President, and Mr Satish Goel, currently President of AIREA, Mr Abhishek Dev, Chairman, APEDA, and Dr. Tarun Bajaj, Director, BEDF, for their encouragement and support. The consistent cooperation and encouragement received from our spouses, Mrs Anjali Kalubarme, Mrs Meenu Verma, Mrs Neerja Sharma, and Mrs Sanjulata Singh, respectively, are specially acknowledged.

Manik H. Kalubarme
Dharmesh Verma
Ritesh Sharma
A.N. Singh

About the Authors

Manik H. Kalubarme is a postgraduate in agriculture (agronomy) from G.B. Pant University of Agriculture and Technology, Pantnagar, and Ph.D. in Environmental Science, Uttarakhand Technical University, Dehradun. He joined the Space Applications Centre, Indian Space Research Organisation, in 1976 and contributed to various projects related to remote sensing applications for agriculture and environment for more than 34 years. His major contributions were related to the development of procedures for crop acreage and production estimation using remote sensing satellite data for various crops at the district and state levels. He has also contributed to different application projects like land use/land cover monitoring, soil degradation studies, urban land use monitoring, and so on, using multi-temporal remote sensing satellite data. He has also developed procedures for Basmati rice discrimination based on phenological stage differences using multi-temporal remote sensing satellite data, and was also involved in the BEDF (APEDA)–sponsored project on Varietal discrimination, acreage, and yield estimation of Basmati aromatic rice in North-Western India. He has also received many awards, including a Team Excellence Award from ISRO for successfully implementing remote sensing applications for the crop acreage and production estimation project. He has published over 90 research papers in peer-reviewed journals and contributed to several book chapters.

Dharmesh Verma completed his master's in soil science from G.B. Pant University of Agriculture and Technology, Pantnagar, and his doctorate in agriculture chemistry from C.C.S. University, Meerut (UP). He is a domain expert in remote sensing and GIS applications for natural resources with more than 25 years of experience. He joined as a scientist at the Remote Sensing Applications Centre—Uttar Pradesh, and worked on remote sensing projects in soil mapping, crop suitability, crop mapping, and yield forecasting. He has worked on World Bank–aided projects on sodic land mapping at the cadastral and regional levels; irrigation impact evaluation using remote sensing and GIS in irrigation schemes of the Andhra Pradesh Economic Restructuring Project. He has also worked with AgriNet Solutions, a remote sensing geospatial company, originated by UPL Ltd., a plant protection global multinational group of companies, for a decade. He worked as Chair for Pillar-1 of the ASP of the UNFAO's Global Soil Partnership program for two years. He also worked as a Global Head—Technical for a Dubai-based fertilizer company for a short period. He and his team worked for Basmati rice acreage and yield estimation for a decade for APEDA, Govt. of India. He has many publications, including research articles and books, to his credit.

Ritesh Sharma holds a Ph.D. in Plant Breeding & Genetics from Dr. B.R. Ambedkar University, Agra. Currently, he is the Joint Director at Basmati Export Development Foundation of APEDA, Ministry of Commerce, Govt. of India. He is involved in the maintenance of Basmati germplasm and breeding and authentic seed production. He has developed several new technologies, including Pata Practice and Basmati Paddy Thresher (Patented). He is actively involved in providing extension services to the stakeholders of Basmati-growing states. He has been nominated as a member of the national committees, as a two-member committee for registration of rice mills for the export of rice to the USA and China as APEDA representative, and also nominated as a member of the variety identification committee by DDG, ICAR. He was a member of the team for the audit of the public sector of Basmati-growing states by the European Committee, and also a member of the Chinese delegation. He is also actively providing expertise to different state governments, including Jammu and Kashmir, on different issues related to Basmati, including doubling the farmers' incomes by 2022. He has also published several books/chapters and many research papers.

A.N. Singh is a postgraduate in soil science from G.B. Pant University of Agriculture and Technology, Pantnagar, and Ph.D. from the University of Allahabad. He was a Visiting Scientist at LARS, Purdue University, USA, and a Post-doctoral Fellow at the International Rice Research Institute (IRRI), Philippines. He joined NRSC, Hyderabad, in 1975 and initiated satellite remote sensing applications to soil mapping in the North-Eastern, Western Coast, and Indo-Gangetic Plains regions of India. Later, he joined the Remote Sensing Applications Centre—Uttar Pradesh (RSAC-UP), and carried out various national mapping projects, including the salt-affected soils mapping, wasteland mapping, and Natural Resources Census. He was the director of RSAC-UP from 2001 to 2008, and his notable contribution at this centre was the use of geo-informatics in the execution and monitoring of large-scale sodic land reclamation in Uttar Pradesh. Later, he joined IRRI as a GIS consultant and was responsible for the targeted dissemination of stress-tolerant rice varieties in South Asia. He has published more than 65 research papers in peer-reviewed journals and several book chapters, including chapters in two international books on aromatic rice. He was also involved in Basmati rice mapping and production estimation for BEDF (APEDA), Ministry of Commerce. He is an active member of the Indian Society of Remote Sensing and remained vice-president of the society. He also remained the vice-president of the Indian Society of Soil Survey and Land Use Planning.

Abbreviations

Chapter 2: Aromatic Basmati Rice Varieties, Area, Production, and Cultivation Practices

2-AP	2-Acetyl-1-Pyrroline
BADH	Betaine Aldehyde Dehydrogenase
CF-SPME	Cold-Fibre Solid Phase Micro Extraction
CRISPR	Clustered Regularly Interspaced Short Palindromic Repeats
DRR	Directorate of Rice Research, Hyderabad, India
IARI	Indian Agricultural Research Institute, New Delhi
Fgr	Fragrant-Related Gene
GABA	Y-Aminobutyric Acid
GABald	Y-Aminobutyraldehyde
GC-FID	Gas Chromatography-Flame Ionisation detector
GC-MS	Gas Chromatography-Mass Spectrometry
IUPAC	International Union of Pure and Applied Chemistry (naming scheme for organic compounds)
KDML	Khao Dawk Mali, a Jasmine rice variety
LC-MS	Liquid Chromatography-Mass Spectrometry
MAS	Marker Added Selection
OAC	Odour Active Compound
RAPD	random Amplified Polymorphic DNA
RFLP	restriction Fragrant Polymorphism
SPME	Solid Phase Micro Extraction
VOC	Volatile Organic Compounds

Chapter 3: Science of Rice Aroma and Environmental Factors Influencing Basmati Aroma and Other Grain Quality Traits

BADH2	Betaine Aldehyde Dehydrogenase 2
QTL	Quantitative Trait Loci
2-AP	2-Acetyl-1-Pyrroline
Ppb	Parts Per Billion
fgr	Recessive Fragrance Gene
RFLP	Restriction Fragment Length Polymorphism
GABald	γ-Aminobutyraldehyde
SNP	Single-Nucleotide Polymorphisms
VOC	Volatile Organic Compounds
GC	Gas Chromatography (GC)
OAC	Odour-Active Compounds
GABA	Gamma-Aminobutyric Acid
FID	Flame Ionisation Detection
MS	Mass Spectrometry
PGPR	Plant Growth-Promoting Rhizobacteria
QTL	quantitative Trait Loci

Chapter 4: Remote Sensing Applications for Aromatic Basmati Rice Discrimination and Mapping

AUC	Area Under Curve
NIR	Near Infrared
LAI	Leaf Area Index
NDVI	Normalised Difference Vegetation Index
UAV	Unmanned Aerial Vehicles
GPS	Global Positioning System
IFOV	Instantaneous Field of View
SWIR	Short Wave Infrared
MIR	Medium Infrared
MSS	Multi-Spectral Scanner
ROC	Receiving Operating Characteristics
SAR	Synthetic Aperture Radar
TM	Thematic Mapper
VNIR	Visible and Near Infra-Red

Chapter 5: Basmati Rice Yield Prediction Using Machine Learning Techniques

RMSE	Root Mean Squared Error
MAPE	Mean Absolute Percentage Error
LSTM	Long Short-Term Memory
CNN	Convolutional Neural Network
PAR	Photosynthetically Active Radiation
GDD	Growing Degree Days
LACIE	Large Area Crop Inventory Experiment
VIs	Vegetation Indices
CCE	Crop Cutting Experiment
LR	Linear Regression
RF	Random Forest
SVR	Support Vector Regression
LAI	Leaf Area Index
EVI	Enhanced Vegetation Index
PVI	Perpendicular Vegetation Index
GAI	Green Area Index
VCI	Vegetation Condition Index
FPAR	Fraction of Photosynthetically Active Radiation
XGBoost	Extreme Gradient Boosting

Chapter 6: PRISMA Hyperspectral Data Analysis for Basmati Rice Discrimination

DESIS	Deutsches Zentrum fur Luftund Raumfahrt (DLR) Earth Sensing Imaging Spectrometer
PRISMA	PRecursore IperSpettrale della Missione Applicativa
EnMAP	Environmental Mapping and Analysis Program

EO-1	Earth Observing-1
NASA	National Aeronautics and Space Administration
ESA	European Space Agency
VNIR	Visible Near Infrared
SWIR	Short Wave Infrared
ISS	International Space Station
ALOS	Advanced Land Observing Satellite
HySIS	Hyperspectral Imaging Satellite
SAVI	Soil-Adjusted Vegetation Index
TBVIs	Two-Band Vegetation Indices
OSAVI	Optimised Soil Adjusted Vegetation Index
TCARI	Transformed Chlorophyll Absorption Reflectance Index
DD	Double Difference
DPI	Double Peak Index
REIP	Red-Edge Inflection Point
Clred-edge	Chlorophyll Red-Edge Index
CARI	Chlorophyll Absorption Reflectance Index
GEE	Google Earth Engine
GPS	Global Positioning System
HDF5	Hierarchical Data Format version 5
QGIS	Quantum Geographic Information System
LULC	Land Use and Land Cover
EVD	Eigenvalue Decomposition
LSWI	Land Surface Water Index
APAR	Absorbed Photosynthetically Active Radiation
LISS IV	Linear Imaging Self-Scanning Sensor-IV

Chapter 7: Traditional Methods of Crop Area and Yield Estimation in India

CCE	Crop Cutting Experiments
GCES	General Crop Estimation Surveys
TRS	Timely Reporting Scheme
ICS	Improvement in Crop Statistics
ICAR	Indian Council of Agricultural Research
IASRI	Indian Agricultural Statistics Research Institute
CSO	Central Statistical Organisation
NSSO	National Sample Survey Organisation
CES	Crop Estimation Surveys

Temporarily Settled or Land Record States

Permanently Settled States (Non-Land Record States)

SASA	State Agricultural Statistics Authority
DES	Directorate of Economics and Statistics
BEDF	Basmati Export Development Foundation
APEDA	Agricultural and Processed Food Products Export Development Authority

Chapter 8: Global Trade in Aromatic Basmati Rice

GDP	Gross Domestic Product
CGR	Compound Growth Rate
DGFT	Directorate General of Foreign Trade
GATT	General Agreement on Tariffs and Trade
WTO	World Trade Organisation
CAGR	Compound Annual Growth Rate
MEP	Minimum Export Price

Chapter 9: Influence of Climate Change on Aromatic Basmati Rice

CERES-Rice	Crop Environment Resource Synthesis-Rice (simulation model)
CSA	Climate Smart Agriculture
DSR	Direct Seeded Rice
DSSAT	Decision Support System for Agrotechnology Transfer
GCISC	Global Change Impact Studies Centre (Pakistan)
GCM	General Circulation Model
GHG	Greenhouse gases
$GtCO_2$-eq	Giga Tons of Carbon Dioxide Equivalent
ICAR	Indian Council of Agricultural Research
IMD	India Meteorological Department
IMPACT	International Model for Policy Analysis of Agricultural Commodity and Trade
IPCC	Intergovernmental Panel on Climate Change
$MtCO_2$-eq	Million tons of Carbon dioxide equivalent
NAPCC	National Action Plan on Climate Change (of India)
NCCP	National Climate Change Policy (of Pakistan)
NMSA	National Mission on Sustainable Agriculture
NRSC/ISRO	National Remote Sensing Centre/ Indian Space Research Organisation
PMD	Pakistan Meteorological Department
ppm	Parts Per Million (concentration)
RCP	Representative Concentration Pathways (of GHGs)
SRES	Special Report on Emission Scenario
SSP	Shared Socio-economic Pathways
TPR	Transplanted Rice
TROPOMI	TROPOspheric Monitoring Instrument (a sensor on board Sentinel-5P satellite)
UNFCCC	United Nations Framework Convention on Climate Change
VIIRS	Visible Infrared Imaging Radiometer Suite (sensor onboard Suomi NPP satellite)

Glossary

Chapter 1: Aromatic Basmati Rice: Importance, Historical and Cultural Perspective

Geographical indication (GI) product A geographical indication (GI) is a sign used on products that have a specific geographical origin and possess qualities or a reputation that are due to that origin.

Oryza sativa Oryza is the rice genus, and sativa in Latin means 'sown'. Oryza sativa, also known as Asian rice.

Ecological diversification It refers to the process where species evolve to occupy different ecological niches, leading to increased biodiversity within a community or ecosystem.

Genetic groups Genetic groups of different plant species that share a significant amount of DNA, indicating shared ancestry and geographic origins.

Polyphyletic origin Polyphyletic origin refers to a grouping of organisms where the members do not share a single, most recent common ancestor. Instead, they have evolved from two or more distinct ancestral lineages.

Taxonomy Taxonomy is the science of classifying and naming living organisms.

Morphology of Basmati Morphology, in general, refers to the study of form and structure. Morphologically, Basmati is an annual grass with a slender stem, long and slender grains, and a terminal panicle.

Traditional Basmati rice varieties Traditional Basmati rice varieties are the original, naturally aged Basmati rice types, prized for their long, slender grains, delicate aroma, and fluffy texture when cooked.

Evolved Basmati rice varieties Evolved Basmati rice varieties are those developed through breeding programs, often involving hybridisation, using traditional Basmati varieties as one of the parents.

Glycaemic Index It is a value that indicates the relative ability of carbohydrate foods to increase the level of glucose in the blood.

APEDA Agricultural and Processed Food Products Export Development Authority

Indian folklore Indian folklore refers to the traditional beliefs, stories, customs, and arts of the people of India, passed down through generations, often orally.

Chapter 2: Aromatic Basmati Rice Varieties, Area, Production, and Cultivation Practices

Kharif season The *kharif* season, also known as the monsoon season, is a major cropping season in India, typically spanning from June to October. The major crops cultivated during the *kharif* season in India are paddy, maize, soybean, cotton, pulses, and groundnut.

Transplanting It refers to the process of moving the rice plants from seedbed to the field.

Crop phenology Crop phenology is the process of crop growth and yield formation, which is largely driven by climatic conditions.

Rice phenology refers to the study of the timing of key developmental stages in rice plants, such as germination, tillering, flowering, and maturity, in response to environmental factors.

Chapter 3: Science of Rice Aroma and Environmental Factors Influencing Basmati Aroma and Other Grain Quality Traits

Allele Allele is the term used to describe 'different versions of a gene.'

Cultivar Cultivar is a term referring to a plant produced in cultivation. It is a kind of cultivated plant that people have selected for desired traits and which retains those traits when propagated.

Gene A gene is the basic unit of heredity, a segment of DNA that carries instructions for building and operating an organism. Genes determine an organism's genotype, whereas alleles determine its phenotype.

Glutinous rice Glutinous rice has a different starch composition than normal rice. It is also known as sticky rice or sweet rice, a type of rice characterised by its high amylopectin and low amylose starch content, which makes it sticky when cooked.

Introgression Introgression is the transfer of genetic information from one species to another as a result of hybridisation between them and repeated backcrossing.

Polygenetic origin Polygenetic origin refers to how species evolved through genetic changes.

Polyphyletic origin Polyphyletic origin refers to a plant or animal related to, or derived from, more than one ancestral stock.

Quantitative trait loci (QTL) These are specific regions within a genome that influence the variation of a quantitative trait, which is a characteristic that varies continuously, like height or weight.

2-Acetyl-1-pyrroline (2-AP) It is a volatile organic compound known for its potent, "popcorn-like" aroma, particularly in fragrant rice varieties like Basmati and Jasmine.

Recessive fragrance gene *(fgr)* A recessive fragrance gene, often found in plants like rice, requires two copies (one from each parent) for the fragrant trait to be expressed.

Restriction Fragment Length Polymorphism (RFLP) It is a technique used to detect variations in DNA sequences by analysing the lengths of DNA fragments produced after digestion with restriction enzymes.

γ-aminobutyraldehyde (GABald) It is a four-carbon amino acid and a key intermediate in the biosynthesis of γ-aminobutyric acid (GABA).

Single-nucleotide polymorphisms (SNPs) These are the most common type of genetic variation in humans, where a single nucleotide (A, T, C, or G) in a DNA sequence differs between individuals or paired chromosomes.

Volatile organic compounds (VOC) These are a group of chemicals that easily evaporate at room temperature and are found in many products and processes.

Gas chromatography (GC) is an analytical technique used to separate and identify volatile and semi-volatile compounds in a mixture.

Odor-active compounds (OAC) These are also known as aroma compounds or odorants, and are volatile chemical compounds that humans and animals perceive through the sense of smell.

Flame ionisation detection (FID) is a widely used detector in gas chromatography (GC) that detects organic compounds by measuring the ions produced when those compounds are burned in a hydrogen flame.

Mass spectrometry (MS) is an analytical technique that identifies and quantifies molecules by measuring their mass-to-charge ratio.

Plant growth-promoting rhizobacteria (PGPR) are beneficial bacteria that live in the rhizosphere (the soil near plant roots) and enhance plant growth and health through various mechanisms.

Chapter 4: Remote Sensing Applications for Aromatic Basmati Rice Discrimination and Mapping

Area under curve Refers to a quantitative metric that measures the overall ability of a binary classifier to distinguish between positive and negative classes, calculated as the area under the ROC curve.

Electromagnetic spectrum The electromagnetic spectrum encompasses the entire range of electromagnetic radiation, organised by frequency and wavelength. It includes various forms of radiation, such as radio waves, microwaves, infrared, visible light, ultraviolet, X-rays, and gamma rays.

Backscatter radiation It refers to radiation that is scattered back toward its source after interacting with a material.

Dielectric constant It is a measure of a material's ability to store electrical energy in an electric field, relative to a vacuum.

Microwave remote sensing Microwave remote sensing involves using sensors that operate in the microwave portion of the electromagnetic spectrum to gather information about the Earth's surface.

Active microwave Active microwave remote sensing involves using radar systems that transmit microwave signals and analyse the reflected (backscattered) signals to gather information about the Earth's surface.

Passive microwave Passive microwave remote sensing involves measuring naturally emitted microwave radiation from the Earth's surface, unlike Synthetic Aperture Radar (SAR).

Synthetic Aperture Radar (SAR) It is a radar imaging technique that uses the motion of a radar antenna to create high-resolution images of a target area. It's an active sensor, meaning it generates its own microwave signals and analyses the reflected waves.

Optical remote sensing It is a technique that uses sensors to detect and measure reflected or emitted electromagnetic radiation from the Earth's surface to gather information about it.

Infrared radiation (reflective, middle, thermal) Infrared (IR) radiation is classified into near-, mid, and far-infrared based on wavelength, each with different properties and applications.

Reflective infrared This typically refers to near-infrared, which is mainly reflected sunlight and is used in applications like remote controls and night vision.

Mid-infrared is This is used in thermal imaging, spectroscopy, and gas detection.

Thermal infrared It is often associated with far-infrared, is emitted by objects as heat, and is used in thermal imaging to measure temperature.

Multispectral It refers to imaging or data collection across multiple bands of the electromagnetic spectrum, encompassing both visible and invisible light.

Receiving Operating Characteristics (ROC) It refers to a graphical plot of the True Positive Rate (Sensitivity) against the False Positive Rate (Specificity). It is used to evaluate the performance of a binary classification model, such as Random Forest or SVM.

Spectral bands These are specific ranges of wavelengths in the electromagnetic spectrum that sensors can detect and use to gather information about the Earth's surface.

Spectral resolution It refers to the ability of a sensor to distinguish between different wavelengths of electromagnetic radiation.

Spatial resolution It refers to the ability of an imaging system to distinguish between closely spaced objects or the smallest-sized objects.

Temporal resolution Temporal resolution is the frequency at which a sensor captures data over the same geographic area.

Panchromatic It refers to something sensitive to all colours of visible light. Panchromatic imagery is a single-band image that captures the overall brightness of a scene across the visible spectrum.

Leaf Area Index (LAI) It is a dimensionless quantity that represents the total one-sided leaf area per unit of ground area.

Unmanned aerial vehicle (UAV) It is commonly known as a drone, an aircraft without a pilot on board, operated remotely or autonomously.

Normalised Difference Vegetation Index (NDVI) NDVI uses the difference between near-infrared (NIR) and red reflectance divided by their sum.

Spectral reflectance (of a material) It refers to how different materials reflect light across various wavelengths of the electromagnetic spectrum.

Wavelength It's a fundamental property of waves, including electromagnetic waves (like light and radio waves) and sound waves, and is denoted by the Greek letter lambda (λ). Wavelength is inversely proportional to frequency; a higher frequency wave has a shorter wavelength, and vice versa.

Chapter 5: Basmati Rice Yield Prediction Using Machine Learning Techniques

Moderate Multiple Linear Regression (MMLR) is form of multiple linear regression that includes interaction terms between one or more predictors and a moderator variable to test for moderation effects.

Random Forest (RF) It is a popular supervised machine learning algorithm used for both classification and regression tasks.

Extreme gradient boosting (XGBoost) It stands for eXtreme Gradient Boosting, which is a powerful and efficient machine learning algorithm known for its speed and performance.

Root Mean Squared Error (RMSE) It is a commonly used metric in machine learning and statistics to evaluate the accuracy of predictive models.

Mean Absolute Percentage Error (MAPE) It is a statistical measure used to assess the accuracy of forecasting models.

Long short-term memory (LSTM) networks These are a type of recurrent neural network (RNN) designed to learn and remember long-term dependencies in sequential data, overcoming the limitations of standard RNNs in handling long sequences.

Convolutional neural networks (CNNs) These are a specialised type of artificial neural network, particularly effective for processing data with a grid-like structure, like images.

Environmental factors These factors encompass all the external elements that can influence living organisms and ecosystems. These factors can be broadly categorised as abiotic (non-living) and biotic (living).

Photosynthetically active radiation (PAR) It refers to the specific range of solar radiation, between 400 and 700 nanometers, that plants utilise for photosynthesis.

Growing degree days (GDD) It is a measure of heat accumulation used to predict the development of plants.

Large Area Crop Inventory Experiment (LACIE) It was a pioneering program from 1974 to 1978, focused on using satellite remote sensing to estimate wheat production over large areas, specifically the U.S. Great Plains and the Soviet Union.

Vegetation Indices (VIs) These are calculated using remotely sensed data with different band combinations to help quantify and analyse vegetation cover and its health.

Crop cutting experiments (CCEs) These are a method used to estimate crop yields by harvesting and weighing a small, randomly selected portion of a field.

Linear regression (LR) It is a statistical method used to model the relationship between a dependent variable and one or more independent variables by fitting a linear equation to the observed data.

Random Forest Regression (RFR) is a powerful ensemble learning method that uses multiple decision trees to make predictions, particularly for continuous numerical outputs.

Support vector regression (SVR) is a machine learning algorithm used for regression tasks, extending the principles of support vector machines (SVMs) to handle continuous numerical outputs.

Leaf Area Index (LAI) is a dimensionless value that quantifies the amount of leaf area in a vegetation canopy per unit of ground surface area.

Enhanced Vegetation Index (EVI) is a remote sensing index that measures vegetation greenness and health, designed to be more sensitive and accurate than the Normalised Difference Vegetation Index (NDVI).

Perpendicular Vegetation Index (PVI) is a remote sensing index used to assess vegetation health and cover by measuring the perpendicular distance of a pixel from a 'soil line' on a scatterplot of near-infrared (NIR) and red reflectance.

Green Area Index (GAI) is a measure of the green plant area (primarily leaves and stems) within a plant canopy, expressed as the ratio of green area to the ground area it covers.

Vegetation Condition Index (VCI) is a remote sensing–based indicator that assesses the health of vegetation by comparing the current Normalised Difference Vegetation Index (NDVI) to its historical range of values.

Fraction of Photosynthetically Active Radiation (FPAR) is the proportion of incoming solar radiation within the PAR (Photosynthetically Active Radiation) range (400–700 nm) that is absorbed by vegetation for photosynthesis.

Chapter 6: PRISMA Hyperspectral Data Analysis for Basmati Rice Discrimination

PRISMA The PRISMA (PRecursore IperSpettrale della Missione Applicativa) satellite is an Italian Space Agency (ASI) mission focused on Earth observation, specifically utilising hyperspectral imaging technology.

Hyperspectral data Hyperspectral data refers to data sets that cover a wide range of wavelengths in the electromagnetic spectrum with multiple channels.

Narrow-band data Narrowband data refers to information transmitted or received within a narrow range of frequencies.

Principal Component Analysis (PCA) It is a dimensionality reduction technique used to transform a dataset with many correlated variables into a smaller set of uncorrelated variables called principal components, while retaining most of the original data's variance.

Selective Principal Component Analysis (SPCA) It is a method that focuses on analysing specific subsets of variables, spectral bands in remote sensing, to extract meaningful information.

Sentinel-2 MSI It is a sensor on board the Sentinel-2 satellites, designed for high-resolution, wide-swath, multi-spectral imaging

Hyperspectral Indices These are mathematical combinations of reflectance values from multiple narrow spectral bands within a hyperspectral image.

Red-Edge Indices These are spectral indices that leverage the steep change in reflectance between the red and near-infrared (NIR) portions of the electromagnetic spectrum, known as the red edge, to assess vegetation health and characteristics.

Ratio Vegetation Index (RVI) Also known as the Simple Ratio (SR), this is a vegetation index calculated by dividing the near-infrared (NIR) reflectance by the red reflectance of a multispectral image.

Green Normalised Difference Vegetation Index (GNDVI) It is a vegetation index used to assess the health and photosynthetic activity of plants, particularly in later stages of development.

Chlorophyll Green Index (Clgreen) It is also known as the Green Chlorophyll Index (GCI), a vegetation index used to estimate chlorophyll content in leaves, particularly for various plant species.

Chlorophyll Red-Edge Index (Clred-edge) This vegetation index is used to estimate chlorophyll content in plants, which leverages the unique spectral reflectance properties of plants in the red-edge and near-infrared (NIR) bands to assess plant health and photosynthetic activity.

Plant Senescence Reflectance Index (PSRI) It is a satellite-derived indicator used to detect plant senescence, or aging, and is sensitive to changes in chlorophyll concentration as plants mature.

Normalised Difference Red Edge Index (NDRE) It is a remote sensing vegetation index that measures chlorophyll content in plants, particularly useful for assessing crop health and nitrogen status, especially in later stages of plant development.

NIR/RED ratio It is a simple vegetation index used to estimate the amount of vegetation and its health.

SWIR/Green ratio It is a spectral ratio that compares the reflectance values in the shortwave infrared (SWIR) and green bands of the electromagnetic spectrum.

Land Surface Water Index (LSWI) It is a satellite-derived vegetation index that indicates the amount of liquid water in vegetation and its soil background.

Spectral Angle Mapper (SAM) algorithm It is a method used in remote sensing to determine the spectral similarity between image spectra and reference spectra (also known as endmembers).

Geographic coordinates They define a point on the Earth's surface using latitude and longitude.

Indian Hyperspectral Imaging Satellite (HySIS) It is an Earth observation minisatellite developed by the Indian Space Research Organisation (ISRO).

Phenological stages These refer to the recurring natural events in the life cycle of plants, particularly those influenced by seasonal changes.

Soil-Adjusted Vegetation Index (SAVI) It is a remote sensing index used to minimise the influence of soil brightness on vegetation measurements, particularly in areas with low vegetation cover.

Optimised Soil Adjusted Vegetation Index (OSAVI) It is a vegetation index used in remote sensing to minimise the influence of soil brightness on vegetation cover estimations.

Two-Band Vegetation Indices (TBVIs) These are a type of vegetation index (VI) that uses the reflectance values from two spectral bands to assess vegetation health and characteristics.

Transformed Chlorophyll Absorption Reflectance Index (TCARI) It is a vegetation index used to estimate chlorophyll content in vegetation, particularly when soil background effects are significant.

Double Difference (DD) Index This index, often used with hyperspectral data, is a vegetation index that helps assess chlorophyll content and other plant characteristics. It is useful in the red-edge region, where chlorophyll reflectance exhibits significant variation.

Double Peak Index (DPI) This index, calculated from hyperspectral data, is a vegetation index that can be used to detect subtle changes in plant health, particularly related to chlorophyll fluorescence.

Red-Edge Inflection Point (REIP) This spectral feature is used in remote sensing to assess plant health and chlorophyll content. It refers to the wavelength where the rate of change in reflectance transitions from decreasing (red absorption) to increasing (near-infrared reflectance) in vegetation spectra.

Chlorophyll Absorption Reflectance Index (CARI) This index is used to estimate chlorophyll content in vegetation, particularly in scenarios with varying leaf area index (LAI) and background reflectance.

GPS measurements These measurements refer to the techniques and methods used to determine locations and distances using signals from GPS satellites.

Google Earth images Google Earth uses a combination of satellite and aerial imagery to create its global view, digitally stitching together images from various sources to provide a detailed and interactive map.

Flag leaf to ear head emergence stages In rice, the flag leaf to ear head emergence (also known as heading) stages are crucial periods in the reproductive phase.

***Kharif* season** The *kharif* season, also known as the monsoon season, is a major cropping season in India, typically lasting from June to October.

Hierarchical Data Format version 5 (HDF5) It is an open-source file format designed to store and manage large, complex, and heterogeneous datasets.

PRISMA Level 1 product These products are Top-of-Atmosphere (TOA) spectral radiance data products, which have been radiometrically corrected and calibrated.

GeoTIFF format It is a public domain raster format that embeds georeferencing information into standard TIFF image files.

Land Use and Land Cover (LULC) It refers to the physical characteristics of the Earth's surface (land cover) and how humans utilise that land (land use).

Eigenvalue Decomposition (EVD) It is also known as eigen-decomposition and is a mathematical technique that decomposes a square matrix into its constituent eigenvectors and eigenvalues.

Covariance Matrix It is a square matrix that represents the covariances between each pair of elements in a set of random variables.

Absorbed photosynthetically active radiation (APAR) It is the amount of solar radiation in the 400–700 nm range that is absorbed by vegetation for photosynthesis.

Spectral separability It refers to the ability to distinguish between different land cover types based on their unique spectral reflectance or emittance patterns across various wavelengths.

End-member selection It involves choosing representative spectral signatures of specific materials (endmembers) to classify image pixels based on their spectral similarity.

Chapter 7: Traditional Methods of Crop Area and Yield Estimation in India

Crop Cutting Experiments (CCE) These are a method used to estimate crop yields by harvesting and measuring the produce from a small, randomly selected area within a field.

General Crop Estimation Survey (GCES) This is a scientifically designed, nation-wide survey conducted to estimate crop yields and production for major crops across India.

Timely Reporting Scheme (TRS) It is a centrally sponsored scheme focused on improving the timeliness and accuracy of agricultural statistics, particularly for area and production estimates of major crops.

Improvement in Crop Statistics (ICS) This scheme aims to enhance the accuracy and reliability of agricultural statistics, particularly regarding crop area and yield, through rigorous supervision and quality checks.

Cadastral maps These are also known as land ownership maps or Bhu Naksha, and are detailed maps that depict the boundaries and ownership details of individual land parcels within a specific area.

Revenue administration It refers to the system and processes involved in managing and collecting revenue, primarily through taxation and other means, to fund public services and operations.

e-Crop Estimation Surveys (e-CES) It is a digital system used for planning, managing, and reporting crop cutting experiments for yield estimation.

Patwari A Patwari, also known as a Talati in some regions, is a village-level administrative official primarily responsible for maintaining land records and revenue collection in rural areas of India.

Tehsil/taluka It is a local administrative unit in India.

Sampling It is a statistical method used to select a subset of individuals (a sample) from a larger group (a population) to study and draw conclusions about the entire population.

***Annawari* method** It was a traditional Indian system for estimating crop yields, particularly during the pre-independence era, based on a 16-anna scale.

Randomly selected plots It refers to a process where specific plots are chosen from a larger group in a way that each plot has an equal chance of being selected.

Rabi crops These are winter crops that are sown in India between October and December, and harvested in the spring, typically from March to April.

Sampling design This is a crucial step in research methodology that outlines how a researcher will select a subset of individuals or items from a larger population to study.

Multistage stratified random sampling It is a complex sampling technique that combines stratified sampling and multistage sampling.

Stratum In stratified random sampling, a population is divided into subgroups called strata, based on shared characteristics.

Basmati Export Development Foundation (BEDF) It is a society established by the Agricultural and Processed Food Products Export Development Authority (APEDA) to strengthen the Basmati rice export supply chain.

Agricultural and Processed Food Products Export Development Authority (APEDA) It is a statutory body established in 1986 under the Ministry of Commerce and Industry, Government of India.

Geotagged photographs These contain embedded location data, typically latitude and longitude coordinates, within their metadata.

Chapter 8: Global Trade in Aromatic Basmati Rice

Global trade Also known as international trade, it involves the exchange of goods and services between countries.

Trend analysis It is a method used to identify patterns and make predictions about future outcomes by examining historical data over a period of time.

Geographical region It is a specific area on Earth distinguished by its unique characteristics, including climate, geological features, and other environmental factors.

Agro-climatic factors These are the key elements of climate that significantly influence agricultural practices and crop production. These factors include temperature, rainfall, humidity, wind, and solar radiation.

Fragrance and flavour These are both sensory experiences created by volatile chemical compounds, but they are perceived through different senses.

Irrigation infrastructure This refers to the man-made systems and facilities used to provide water to crops and agricultural land, supplementing natural rainfall.

Commercial cultivation This is the large-scale production of crops and livestock for the primary purpose of selling them for profit.

Gross Domestic Product (GDP) is a fundamental measure of a country's economic activity, representing the total market value of all final goods and services produced within a country's borders during a specific period, usually a year.

Geographical Indication (GI) status It is a sign used on products that have a specific geographical origin and possess qualities or a reputation that are essentially due to that origin.

Compound Annual Growth Rate (CAGR) It represents the mean annual growth rate of an investment over a specified time period, assuming profits are reinvested each year.

Green Revolution It was a period of agricultural development, primarily in the mid-20th century, focused on increasing food production through the adoption of modern techniques and technologies.

Directorate General of Foreign Trade (DGFT) It is the key government body in India responsible for regulating and promoting international trade.

Minimum Export Price (MEP) It is a trade policy tool used by governments to set a minimum price below which a commodity cannot be exported.

Chapter 9: Influence of Climate Change on Aromatic Basmati Rice

Adaptation Adaptation refers to the adjustment by plants, animals, or humans in response to the existing or forecasted adverse situation.

Forecast/Prediction When a projection is branded as 'most likely', it becomes a forecast or prediction. A forccast is often obtained using deterministic models, or possibly a set of these, the outputs of which can enable some level of confidence to be attached to projections.

General Circulation Model (GCM) GCMs are complex computer models that simulate the Earth's climate system, including atmospheric and oceanic circulation, radiative transfer, and the water cycle.

GHG Emission GHG emission refers to the emission of gases arising from various anthropogenic sources, responsible for global warming and climate change such as CO_2 from fossil fuel combustion, industrial processes, and land use change; methane (CH_4); nitrous oxide (N_2O); and fluorinated gases comprising hydrofluorocarbons (HFCs), perfluorocarbons (PFCs), sulphur hexafluoride (SF_6), as well as nitrogen trifluoride (NF_3).

Pathway Pathways are strategies and actions aimed at mitigating and adapting to the impacts of a changing climate, while also promoting sustainable development.

Projection A projection can be regarded as any description of the future and the pathway leading to it. However, in climate projections referred to by IPCC, it indicates model-derived estimates of future climate.

Representative Concentration Pathways (RCPs) These are scenarios that project future greenhouse gas concentrations and climate change impacts, helping to assess potential risks and inform mitigation strategies.

Scenario A scenario is a coherent, internally consistent, and plausible description of a possible future state of the world. It is not a forecast; rather, each scenario is one alternative image of how the future can unfold.

Sentinel-5P Launched under the Copernicus programme, Sentinel-5P is a satellite mission dedicated to atmospheric monitoring that provides high-resolution data on atmospheric trace gases, which are helpful in understanding climate change.

Simulation A simulation study is a computer-based experiment where a model of a real-world situation is created and then subjected to repeated trials to analyse its behaviour and predict outcomes. These studies are used when direct experimentation is impractical, costly, or potentially dangerous.

Shared Socio-economic Pathways (SSPs) SSPs are alternative projections of socio-economic development that may influence future GHG emissions.

Suomi NPP satellite Suomi National Polar-orbiting Partnership (Suomi NPP) is a satellite launched by NASA-NOAA in 2011, to collect data on land, ocean, and air. It has sensors that can observe active fires, ocean colour, sea surface temperature, and other surface features.

1 Aromatic Basmati Rice
Importance, Historical and Cultural Perspective

1.1 IMPORTANCE OF RICE

Feeding more than half the world's population, rice plays a vital role in global food security, contributing significantly to the daily energy intake of billions of people. In Asia, which accounts for nearly 90% of global rice production and consumption, rice holds a central place in the daily diet and cultural practices. Rice holds deep cultural and traditional significance across much of Asia. In 2009, it accounted for nearly 29% of the average dietary calorie intake in the region (GRiSP, 2013). That same year, rice contributed about 19% of global per capita energy and 13% of per capita protein. Beyond Asia, rice is also the staple grain in many tropical regions of Latin America and the Caribbean, serving as the main daily energy source. Its role in daily meals has steadily expanded in Africa, where it is now a routine part of the diet. Although rice is not a staple food in Europe, North America, and Australia, its per capita consumption has been increasing over the years. The total demand for rice continues to rise in Asia, Africa, and Latin America. Due to population and economic growth, global rice consumption has been rising steadily. A recent study by the CGIAR Initiative on Foresight has shown that global rice consumption, which remained at 450 million tons in 2010, is likely to increase by about 18% to 505 million tons in 2050, mainly due to population growth (Valera and Pede, 2023).

1.2 GEOGRAPHICAL ORIGIN AND HISTORY OF RICE CULTIVATION

Researchers have debated the birthplace of rice cultivation for many years. The rice genus *Oryza* is believed to have emerged around 14 million years ago in the biogeographic region of Malaysia, which includes the Malay Peninsula, the Philippines, New Guinea, Borneo, and parts of the Indonesian archipelago (GRiSP 2013). The name *Oryza* has its roots in Greek, signifying 'of the eastern world'. Genus *Oryza* has 27 species, but only two species are cultivated and used as a human food source: *Oryza sativa* and *Oryza glaberrima* (*sativa* in Latin means 'sown', and *glaberrima* means 'smooth' and free of hairs). *Oryza sativa*, also called Asian rice, originated from its wild perennial relative, *Oryza rufipogon*, through *Oryza nivara*, a wild annual species extensively found in India and many parts of Southeast Asia. From its early beginnings in different parts of Asia, *Oryza sativa* spread to other regions of the world and is now grown on every continent except Antarctica. *Oryza glaberrima*, also called African rice, originated from its wild perennial progenitor, *Oryza longistaminata*,

DOI: 10.1201/9781032659398-1

through its wild annual species *Oryza brevigulata* in West Africa. Rice is widely cultivated in the floodplains of the Niger and Sokoto river basins, where it thrives under both upland conditions and in deepwater environments. However, due to the low yield of these varieties, Asian kinds of rice are slowly replacing them, and currently, these are grown in only 20% of the rice area in West Africa. The two cultivated rice species—*Oryza sativa* in Asia and *Oryza glaberrima* in Africa—are considered classic examples of parallel evolution in domesticated crops across distinct geographic regions (Khush, 1997).

Domestication of *Oryza sativa*, the dominantly grown rice species, probably began about 9,000–10,000 years ago in different parts of Asia. As noted by Callaway (2014), the domestication of rice marked a turning point in human history, encouraging the shift from a nomadic lifestyle to settled farming and eventually laying the foundation for the rise of cities, empires, and civilisations. Through generations of selective cultivation, early farmers gradually shaped wild rice into a more robust and upright plant that yielded more grain, exhibited less dormancy, and adapted to a variety of growing environments. It was also transformed from its earlier grain-shattering (grain tumbling to the ground after maturity) type to a non-shattering type. A recent rice genome study has found the *sh4* allele to be responsible for the non-shattering of grains in rice (Vaughan et al., 2008a). In Asia, rice domestication is believed to have taken place independently at multiple locations across a wide region stretching from the lowlands near the eastern Himalayas in India through upper Myanmar, northern Thailand, Laos, and Vietnam to southern and southwestern China (Ramiah, 1937; Chang, 1976). Within this geographic arc, rice was likely cultivated in forest clearings using shifting agricultural practices. Early cultivation probably relied on dry direct seeding methods, without the use of standing water (Khush, 2000). The agricultural techniques of puddling, converting soil into a muddy consistency, and transplanting young rice seedlings are believed to have originated in China. Over time, these practices became central to rice cultivation and continue to be widely used. Drawing from archaeological findings, Oka (1988) proposed that rice farming traditions in Asia were firmly established approximately 7,000 years ago in regions such as northern India, Thailand, and eastern China.

India is widely regarded as among the earliest centres for the cultivation of *Oryza sativa* (Khush, 2000). Archaeological discoveries from the Neolithic site in the Belan Valley have yielded carbonised rice grains dated to around 6750 BC, marking some of the earliest evidence of rice domestication in the region (Sharma and Manda, 1980). Additional findings from Mohenjo-Daro, located in present-day Pakistan, include rice remains dating to approximately 2500 BC (Andrus and Mohammed, 1958). In China, archaeological evidence of rice grain and farming implements dating back to about 8000 years BP was discovered in the Lower Yangtze River valley (GRiSP 2013). Carbonised rice grains about 7,040 years old have been found in Tongxiang County of Zhejiang Province. The second-earliest site, about 6,960 years old, has been found in the Hemdu relic in Yayao County of Zhejiang Province (Khush, 2000). In Thailand, evidence of early rice cultivation was uncovered by Solheim (1972) at Non Nok Tha in the Korat region, where pottery fragments imprinted with rice grains and husks were dated to around 4000 BC. Even earlier plant remains, dating back to approximately 10,000 BC, were found in Spirit Cave near the Thailand–Myanmar border, suggesting a long history of plant use in the region (GRiSP, 2013).

The domestication of *Oryza glaberrima* or African rice, probably started about 1500 BC in tropical West Africa (Carpenter, 1978). Genomic studies comparing cultivated African rice and its wild ancestors indicate a single domestication event, likely centred around the Niger River region in West Africa. Interestingly, these rice varieties share mutations in genes associated with traits like reduced seed shattering, similar to those seen in domesticated Asian rice, along with other key domestication characteristics. However, control of shattering in African and Asian rices has been found to be of independent domestication (Wang et al., 2014). African rice varieties are characterised by small, pear-shaped grains with red bran and seed coats ranging from olive to black. Their panicles tend to be upright with simple branching and short, rounded ligules. However, compared to Asian rice, they exhibit some less favourable traits, such as a tendency for easy seed shattering, brittle grains that are challenging to mill, and generally lower yields. Due to these reasons, *O. sativa* is slowly replacing *O. glaberrima* not only in West Africa but in other parts of Africa also *Oryza glaberrima* varieties are generally classified into two main ecotypes: upland and deepwater types (GRiSP, 2013). In the seasonally flooded lowlands of the Niger and Sokoto river basins, rice is cultivated as a rainfed wetland crop, either through broadcasting or transplanting methods. Deepwater rice systems are primarily found in regions such as the Niger Valley near Mopti in Mali, Birnin Kebbi in Nigeria, and parts of northern Guinea. Along the coastal zones of Guinea-Bissau, Guinea, Sierra Leone, and Senegal, African rice is grown in scattered, localised areas. Upland rice cultivation, on the other hand, is typically practiced within bush fallow or slash-and-burn farming systems.

1.3 ECOLOGICAL DIVERSIFICATION IN ASIAN RICE

As early cultivars of *Oryza sativa* spread through the movement of farmers, traders, and settlers into diverse environments, ranging from dry uplands and seasonal deepwater zones to tidal wetlands, the crop underwent repeated cycles of hybridisation, adaptation, and selection. This ecological and geographical diversification eventually gave rise to thousands of rice varieties. Over time, isolation and environmental pressures led to the emergence of two primary ecogeographic groups: (i) *indica*, suited to the warm, humid climates of the tropics and subtropics, and (ii) *japonica*, adapted to cooler temperate zones and higher-altitude tropical regions. Indica was domesticated in India, whereas japonica was domesticated in China, and hence japonicas are also known as sinicas or keng (Chang, 1976). Within the two major groups, five subgroups have been identified based on recent DNA studies (GRiSP, 2013). The *indica* group is further classified into two subgroups: *indica* proper and *aus*. The *aus* subgroup consists of varieties traditionally cultivated during the *aus* season (March to July) in regions such as Bangladesh and northeastern India. These varieties are notable for carrying several stress-resilient genes that are not commonly found in other rice groups. Japonica group has been classified more specifically as temperate and tropical japonica. In addition to the two main groups, the Basmati or aromatic cultivars, primarily cultivated in northwestern India and Pakistan, are considered a distinct lineage that evolved from the *japonica* group. Tropical japonicas were also known as javanicas earlier. The indica group of rice is grown in most countries of the world. Indica rice is predominantly cultivated across a number of countries, particularly in Asia, including China, India, Indonesia, Bangladesh, Vietnam, Myanmar, Thailand, the Philippines, Cambodia, Laos, Malaysia, and Pakistan. It is also grown in parts of South America, notably in Brazil. Temperate japonica is grown in China, Japan, South Korea, Taiwan, Egypt, Italy, Russia, and in California state of USA, where there are distinct seasonal changes in temperature. Tropical japonica rices are grown in Indonesia, Madagascar, and in the southern states (Arkansas, Louisiana, Mississippi, and Texas) of the USA. In terms of global production, indica rice production during 2017 was 417 million tons, whereas japonica was 71 million tons (Koizumi and Furuhashi, 2020). China produces both japonica and indica; however, in terms of its total production, indica is two-thirds, and japonica is one-third. Indicas are medium- to long-grained, with their kernel 4–5 times longer than their width. Japonicas are round-grained and sticky in texture (glutinous), which makes them easier to pick up with chopsticks.

1.4 GENETIC GROUPS SUGGEST THE POLYPHYLETIC ORIGIN OF ASIAN RICE

Glaszmann (1987) conducted a comprehensive isozyme-based analysis of 1,688 traditional *Oryza sativa* accessions collected from diverse regions across Asia to investigate allelic variation. The study identified six genetically distinct groups, designated as Groups I through VI. Among these, Groups I and VI were predominant, while Groups II and V represented minor clusters, and Groups III and IV were identified as satellite groups with region-specific adaptation. Group I, found extensively across tropical Asia, including India, Bangladesh, Indonesia, and China, was later shown to correspond with the *indica* varietal group (Sharma et al., 2021). Group VI, distributed primarily in temperate regions and high-altitude areas of the tropics, is aligned with *japonica* types. Group II matched the *aus* rices of Bangladesh and eastern India, typically cultivated between March–April and June–July. Groups III and IV were geographically restricted, representing floating rice varieties, Aswina from India and Rayadas from Bangladesh, respectively. Group V comprised aromatic rice types from the Indian subcontinent, including the renowned Basmati varieties. The distinctiveness of Groups II through V from the primary *indica* and *japonica* groups suggests alternative domestication routes or diversification events (Figure 1.1).

Further support for the genetic groupings within cultivated rice was provided by Garris et al. (2005), who analysed 234 *Oryza sativa* accessions utilising 169 SSR markers and two chloroplast DNA loci, identifying five major genetic groups: aus, aromatic, indica, tropical japonica, and temperate japonica. These findings reinforce the complex evolutionary history and diversification of rice. Observations on hybrid performance show that sterility predominantly arises in F_1 progeny from inter-group crosses, while intra-group hybrids generally retain fertility. For example, hybrids between Group I accessions and those from other isozyme-defined groups often exhibit partial or complete sterility, whereas tropical and temperate japonica crosses

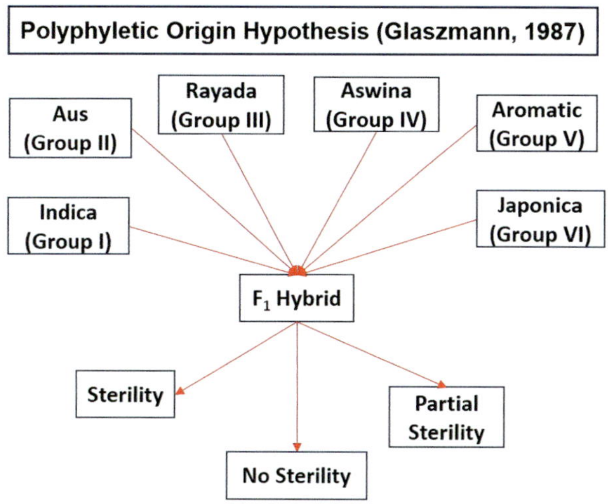

FIGURE 1.1 Polyphyletic origin of rice.

tend to be fertile (Khush, 2000). Such reproductive barriers, together with isozyme variation and morphological classifications, lend support to the polyphyletic origin hypothesis. According to Glaszmann (1987), the six genetic groups of *O. sativa* may have independently originated through separate domestication events from distinct *O. nivara* populations, likely occurring in different geographic regions and timeframes, explaining the extensive genetic and phenotypic diversity seen in cultivated rice today.

However, this view has been challenged by more recent molecular evidence favoring a monophyletic origin. Vaughan et al. (2008a, 2008b) proposed that the domestication process was initiated in the Yangtze River basin, where repeated cycles of introgression, human-mediated selection, and ecological adaptation led to the emergence of diverse rice lineages. They argue that the presence of an identical mutation in both *indica* and *japonica* at a key gene associated with seed shattering contradicts the notion of independent domestications. Instead, this shared mutation suggests a single domestication origin followed by wide dissemination and subsequent diversification across Asia (Figure 1.2).

1.5 TAXONOMY, HISTORICAL EVIDENCE, AND EVOLUTION OF AROMATIC RICES

Aromatic rices are a distinctive group of high-quality rice cultivars valued for their unique fragrance and superior grain quality. These are highly favoured by consumers globally because of their special aroma, soft texture, appealing taste, and other grain quality traits. Due to their specialty, they fetch much higher prices, generally double that of the non-aromatic types. All aromatic rice varieties possess recessive genes responsible for their characteristic fragrance. Multiple alleles associated with aroma have been identified, with the *badh2* allele being the most prevalent across these

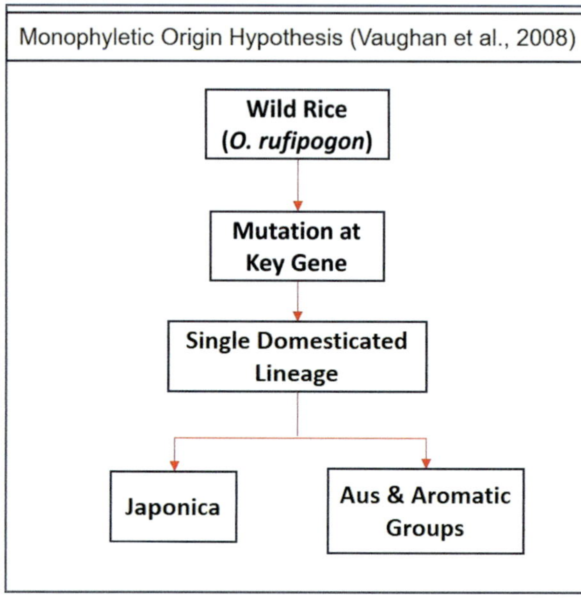

FIGURE 1.2 Monophyletic origin hypothesis.

varieties (Vamireddy et al., 2021). These rices are primarily cultivated and consumed in countries such as India, Pakistan, Thailand, Iran, Afghanistan, Indonesia, China, Vietnam, and the USA. The isozyme-based analysis by Glaszmann (1987), which included 44 aromatic rice cultivars from regions such as Afghanistan, Bangladesh, Cambodia, China, India, Indonesia, Iran, Iraq, Myanmar, the Philippines, Thailand, and Vietnam, revealed that most aromatic types clustered in Group V. However, a few were placed in Group I (indica) and Group VI (japonica), indicating genetic diversity within aromatic rices. Group V notably includes globally renowned Basmati varieties from India and Pakistan, along with several other long- and medium-grain fragrant rices—Ambemohar, Barah, Hansraj, Kataribhog, Lawangin, and Sadri rices of Iran. It also includes short-grain Bindli, Badshahbhog, Nama ThaLay, Prasadbhog, Tulsimanjri, and other aromatic rices. A country-wise list of aromatic rices classified under different groups by Glaszmann is given in Table 1.1. It can be seen that several long- and short-grain rices from India are included in this list.

Historical records indicate that aromatic rice cultivation in India dates back to ancient times. Sushruta (circa 400 BC), a renowned pioneer of Indian medicine and surgery, documented several rice cultivars, including Sugandhaka (Sugandhika), a fragrant *sali* (white) type, along with Swastika, a short-duration (60-day) variety, Kalama, and others (Kumar, 1988; Krishnamurthy, 1991; Nene, 2003). Charaka (circa 100 BC to 100 AD), a distinguished figure in ancient Indian medicine, mentions Gandhana (scented), apart from the other 29 rice cultivars. Kashyapa in his Kashyapiya Krishisukti (800–900 AD) has documented several aromatic cultivars with their characteristics, including Dirghasali (long), having sweet flavor; Palasa Vrihi, which is full of flavour; Kalama, somewhat hard and strong flavored; and

TABLE 1.1
Classification/ Grouping of Some Aromatic Rice Varieties According to Glaszmann (1987)

Group	Country of origin	Grain Type	Cultivar/Variety
I	Thailand		Jao Mali, Hawm Mali, Khao Dawk Mali, Khao Mali, Nahng Nuan, Som Hong
	Vietnam		Tam Xuan, Tam Xuan Hai Hau
	China		Zhao Xing 17
	Cambodia		Somali
V	India	Long-grain aromatic	Basmati 370, Basmati 5853, Basmati 5877
		Short-grain aromatic	Bindli, Dubraj, Jeeraga Samba, Kamini Bhog, Kali Munchh, Kamod, Ram Tulsi, Tulsi Manjri
	Bangladesh	Short-grain	Badshahbhog, Kala Nimak, Kalijira
	Myanmar		Nama Tha Lay, Ngakywe, PawsanHmwe, TaungpyanHmwe,
	China		Xiang Keng 3, Xiang Nuo 4
	Afghanistan	Long-grain	Barah, Lawangun
	Iran	Long-grain	Dom Siah, Moosa Tarum
	Iraq	Long-grain	Anbarboo
VI	Indonesia		Mentik Wangi, Rojolele, Sukanandi
	Philippines		Azucena, Milfore 2, Milagrosa

Source: **Glaszmann, J.C. 1987**. Isozymes and classification of Asian rice varieties. Theor. Appl. Genet. 74:21–30.

several other rices with their specific characteristics. Records from the 12th century by Someshwara Deva and the 13th century by Sarang Dhara have mentioned aromatic rices. Some of the cultivars, like Kalama, have continued to be mentioned for more than 1000 years (Nene, 2003). During the 16th-century Mughal rule, Abul Fazl Allami, in his book Ain-I-Akbari written around 1590 AD, provided details of rices along with their prices, which include aromatic Mushkin and Sukhdas (Jarret, 1927). Sukhdas cultivar was being used in the royal kitchen during Akbar's period. As the name suggests, Muskin had a musk-like aroma and small grain. Tavernier, a French traveler to India in 1654, also mentions Muskin as a popular variety grown in large areas in Surat (Gujarat, India). Aromatic rices grown in different areas of India from the 17th century onwards have been described by several authors, which include long-grain Basmati and several short-grain aromatic cultivars (Watt, 1891; Kumar, 1988; Ahuja et al., 1997; Ahuja et al., 2003; Nene, 2003; Shobha Rani and Singh, 2003). Every province in India has its own popular fragrant rice variety.

According to genome-wide diversity studies, aromatic rice appears to have developed in the Indian subcontinent through genetic mixing between wild and cultivated rice populations (Vamireddy et al., 2021). The primary centre of diversity for **Group V** aromatic rices is located in the Himalayan foothills, particularly across Uttar

Pradesh and Bihar in India, and the Tarai region of Nepal (Khush, 2000; Vamireddy et al., 2021). A significant number of aromatic rice varieties continue to be cultivated in this region. From this center of diversity, their cultivation expanded north-westward into Punjab (India), Pakistan, Afghanistan, and Iran; eastward into Bangladesh and Myanmar; and further south and southeast into other parts of India. According to Glaszmann's (1987) classification, the easternmost distribution of Group V aromatic rices extends to Myanmar while being absent in Southeast and East Asia (Khush, 2000). Some of the important aromatic varieties grown in these countries are given in Table 1.2. Basmati rice from India and Jasmine rice from Thailand dominate the international aromatic rice trade. Together, exports of these two premium varieties accounted for nearly two-thirds of global aromatic rice shipments between 2015 and 2017 (Bairagi et al., 2020). Other countries involved in the aromatic rice trade include Pakistan, exporting Basmati; Vietnam, Cambodia, and China trading Jasmine; and the USA trading Della rice.

TABLE 1.2
Some Important Aromatic Rice Varieties Grown in Different Countries

Country	Grain Type	Prominent Aromatic Variety	Growing Area
India	Long-grain Basmati	Basmati-370, Taraori Basmati, Karnal Local, Mushkan, Dehraduni Basmati, Pusa Basmati-1, Pusa-1121, Pusa Basmati-1509	Northwestern and northern states
	Medium- and short-grained	Sakkarchini, Lalmati, Bindli, Kamod, Tusimanjari, Kalimunchh, Badshahbhog, Kalanamak, Kaminibhog	Entire country
Pakistan	Long-grained Basmati type	Basmati-370, Basmati-385, Super Basmati, Punjab Basmati,	Punjab
Afghanistan	Long-grain	Bahra, Bala, Germa Bala, Lawangi, Sela Takhar	
Iran	Long-grain	Tarom, Sadri, Hassan Saraie, Domsiah, Hassani	
Thailand	Long-grained, non-glutinous (Jasmine)	Khao Dawk Mali (KDML-105), RD-15, Howm Klong Luang, Hawn Suponburi	North and North-eastern region
	Long-grained, glutinous (Jasmine)	RD-6	North and North-eastern region
Vietnam	Jasmine	Nang Thom Cho Dao, Tam Xoan	Mekong and Red River Delta
China	Jasmine	Xianggeng-3, Xiangyou-63, Zhe-9248, Ganwanxian-22	
Bangladesh	Medium- and short-grained	Chinigura, Kalijira, Kataribhog	Entire country
USA	Long grained	A-201, Della, Dellamont, Kasmati, Texmati	Texas, California

1.6 BASMATI AROMATIC RICE

Basmati rice, native to the Indian subcontinent and renowned for its distinctive quality attributes, is highly esteemed by rice enthusiasts worldwide (Siddiq et al., 2012). Basmati rice is characterised by a tall, traditionally low-yielding plant type that produces long, slender, and lustrous grains with a distinct aroma. Upon cooking, it exhibits a soft texture, a mildly sweet flavour, graceful curvature, and remarkable kernel elongation with minimal lateral expansion. The cooked grains are non-sticky and smooth, offering a buttery mouthfeel in contrast to the drier, coarser texture typical of non-Basmati varieties (Thakrar and Ahuja, 1993). These qualities of Basmati make it a specialty rice *par excellence*.

1.6.1 Basmati Rice in Historical Context

'Basmati' derives its name from two Sanskrit root words, *vas* meaning aroma, and *mayup* meaning inherent from the outset. While combining the two, *mayup* has changed to *mati*, and it became *Vasmati*. In general pronunciation, it has become Basmati. One of the earliest mentions of Basmati was made by the Punjabi poet Waris Shah in his epic poem *Heer Ranjha* in 1766 (Thakrar and Ahuja, 1993; Singh 2000; Siddiq et al. 2012). A reference to Basmati rice also appears in the 16th-century epic poem *Padmavat*, composed by Malik Muhammad Jayasi in 1540 (Ahuja et al. 2019). During the British period, Hamilton (1828) recorded Basmati as a species of rice of excellent flavor and greatly superior even to the finest Patna rice, growing in Bareilly, Uttar Pradesh. Basmati is additionally documented in *A Dictionary of the Economic Products of India*, authored by George Watt and published in 1891. The Agriculture Ledger (Anonymous 1911), in its section on Races of Rice in India, has used 22 names and spellings of Basmati growing in different areas. These are: Bansmati Lal, Bansmatti, Bansmutee, Bansmutty, Bashmati, Basmathi, Basmati, Bansmatee, Bansmati, Basmattee, Basmatti, Basmaty, Basmuti, Basmuttee, Basmutti, Basumati, Basubati, Baseemati, Dhan Basmeti, Safed Basmati, Salam bansmati, and Shada bansmati. Growing areas of Basmati have been shown extending from Kashmir, Punjab (divided now in India and Pakistan), United Provinces and Oudh, Bihar, and Bengal. Recognised as a 'heritage rice', Basmati is regarded as a traditional farmers' variety, cultivated and preserved for over 250 years by farming communities in the Punjab region (India and Pakistan), as well as in Haryana, Uttar Pradesh, and Jammu (Nene, 1998).

1.6.2 Morphological and Other Quality Traits of Basmati

Although morphologically similar to the indica group, Basmati rice is distinguished by its unique phenol reaction, isoenzyme profile, intermediate amylose content, opaque kernel, and alkali spreading value. These distinct traits place Basmati in an intermediate classification between indica and japonica types (Ahuja et al., 2008). To be officially recognised as Basmati, a variety must conform to the minimum quality standards prescribed by the Central Sub-Committee on Crop Standards, Notification and Release of Varieties under India's Seed Act, 1966, as well as the *Export of*

Basmati Rice (Quality Control and Inspection) Rules, 2003. The minimum standards for Basmati rice grain include parameters for primary characteristics such as aroma, kernel length (L), breadth (B), L/B ratio, cooked rice length, ratio of cooked rice length and precooked rice length, volume expansion after cooking, and the texture of cooked grain. Apart from these, the parameters for ancillary characteristics include grain amylose content (AC), alkali spreading value, and percent trice recovery on milling. The minimum standards for these quality traits are shown in Table 1.3.

Apart from the Indian minimum standards of grain quality traits, the Export of Basmati Rice (Quality Control and Inspection) Rules, 2003 also defines the genotypes (varieties) that can be called Basmati. Basmati rice varieties are broadly classified into two categories: (i) ***Traditional Basmati***, which are pure line selections derived directly from landraces and include six officially recognised types; and (ii) ***Evolved Basmati***, developed through breeding methods such as hybridisation, where at least one parent is a Traditional or pure line Basmati variety. The Traditional Basmati varieties notified in India include *Basmati-217*, *Basmati-370*, *Type 3* (Dehraduni Basmati), *Taraori Basmati* (Karnal Local), *Ranbir Basmati*, and *Basmati-386*. As of 2021, a total of 28 evolved Basmati varieties have been officially notified in India. A list of notified Traditional and Evolved Basmati varieties of India and approved Basmati varieties of Pakistan is provided in Table 1.4. In Pakistan, the number of approved Basmati varieties till 2021 is 14, which includes both traditional and evolved Basmati.

TABLE 1.3
Eligibility Criteria of Basmati Rice in India

Primary Quality Characteristics[a]		Ancillary Quality Characteristics[a]	
Parameters[b]	**Value**	**Parameters[b]**	**Value**
Minimum average precooked milled rice length (mm)	6.61	Amylose content range (%)	20–25
Average precooked milled rice breadth (mm)	<2.00	Alkali spreading value (ASV) range	4.0–7.0
Minimum length/breadth ratio of precooked milled rice (L/B ratio)	3.50	Minimum brown rice recovery (%)	76
Minimum average cooked rice length (mm)	12.00	Minimum milled rice recovery (%)	65
Minimum cooked rice length/ precooked rice length ratio	1.70	Minimum head rice recovery (%)	45
Average volume expansion ratio	>3.5		
Aroma	Present[c]		
Texture of cooked grain for high integrity (without bursting the surface), non-stickiness, tenderness, good taste, and good mouth feel	Present[c]		

[a] Eligibility standards of a rice variety to be notified as Basmati under the Seed Act 1966

[b] The grain sample for analysis will necessarily have to be 'aged' for three months at normal room temperature as milled grain.

[c] Qualitative sensory analysis as Panel Test to be carried out as per the Standard Protocol of the Directorate of Rice Research, Hyderabad, India.

TABLE 1.4
Traditional and Evolved Basmati Rice Varieties Notified/Approved in India and Pakistan

India		Pakistan	
Notified Variety[a]	Year of Notification	Approved Variety[c]	Year of Release
		Basmati-370	1933
Basmati-217[b]	1969	Basmati-385	1985
Basmati-370[b]	1973	Super Basmati	1996
Type 3 (Dehraduni Basmati)[b]	1978	Basmati-2000	2001
Punjab Basmati-1 (Bauni Basmati	1984	Shaheen Basmati	2000
Pusa Basmati-1	1989	Basmati-515	2011
Kasturi	1989	Pak-1121 Aromatic	2013
Haryana Basmati-1	1991		
Mahi Sugandha	1995	Kissan Basmati	2016
Taraori Basmati (HBC-19/ Karnal Local)[b]	1996	Punjab Basmati	2016
Ranbir Basmati[b]	1996	Chenab Basmati	2016
Basmati-386[b]	1997	Super Basmati-2019	2019
Improved Pusa Basmati-1 (Pusa-1460)	2007	Super Gold	2019
Pusa Basmati-1121	2008	Pak-2021 Aromatic	2021
Vallabh Basmati-22	2009	Nibge-Basmati-2020	2020
Pusa Basmati-6 (Pusa-1401)	2010		
Punjab Basmati-2	2012		
Basmati CSR-30	2012		
Malviya Basmati Dhan-10-9 (IET 21669)	2013		
Vallabh Basmati-21(IET 19493)	2013		
Pusa Basmati-1509 (IET 21960)	2013		
Basmati-564	2015		
Vallabh Basmati-23	2015		
Vallabh Basmati-24	2015		
Pusa Basmati-1609	2015		
Punjab Basmati-3	2016		
Pusa Basmati-1637	2016		
Pusa Basmati-1728	2016		
Pusa Basmati-1 (IET 21665)	2016		
Pusa Basmati-2 (IET 21953)	2016		
Pusa Basmati-1718	2017		
Punjab Basmati-4	2018		
Punjab Basmati-5	2018		
Haryana Basmati-2	2019		

(Continued)

TABLE 1.4 (Continued)

India		Pakistan	
Notified Variety[a]	Year of Notification	Approved Variety[c]	Year of Release
Pusa Basmati-1692	2021		
Pusa Basmati-1847	2022		
Pusa Basmati-1885	2022		
Pusa Basmati-1886	2022		
Pusa Basmati-1985	2022		
Pusa Basmati-1979	2022		
Pusa Basmati-1882	2022		
Punjab Basmati-7	2022		

[a] Notified under Indian Seed Act, 1966.
[b] Traditional Basmati in India, others are evolved.
[c] Approved Basmati Variety in Pakistan (Sabar and Sabir, 2020).

The morphological characteristics of the Basmati plant, its height, average yield, ripening period, and so on, are quite distinct as compared to dwarf high-yielding varieties (HYVs). Traditional Basmati varieties, such as Basmati-370, are characterised by tall plant stature, susceptibility to lodging, and sensitivity to both photoperiod and temperature, which collectively contribute to their low yield potential and restricted productivity (Singh et al., 2011). Among the evolved Basmati varieties, tall cultivars like Haryana Basmati-2, Pusa Basmati-1121, and Pusa Basmati-1 mature late, whereas the semi-dwarf cultivar Pusa Basmati-1509 matures earlier. Maximum days to maturity (131 & 135 days) and maturity days increase with the increase in N-doses. Pusa Basmati-1121 represents a breakthrough in Basmati rice breeding, combining high yield potential with key quality attributes derived from ancestral varieties like Basmati-370 and Type 3. Released in 2008 for commercial cultivation, this semi-dwarf Basmati rice variety grows to a height of 110–120 cm, matures in 140–145 days in Indian Basmati regions, and yields up to 5.5 tons per hectare (Singh et al. 2018).

1.6.3 Geographical Dependence of Quality Traits and GI for Basmati Rice

The distinctive quality characteristics of Basmati rice are predominantly expressed when cultivated in the northwestern Himalayan foothills of the Indian subcontinent (Siddiq et al., 2012). Due to this region-specific expression of its attributes, Basmati rice has been granted Geographical Indication (GI) status, linking its identity to a defined area. GI status recognises the value that geographical origin imparts to a product, reinforcing its uniqueness and market identity. GIs serve to safeguard traditional products from unauthorised use or misrepresentation, particularly when the product's distinct quality, reputation, or characteristics are intrinsically linked to its place of origin. Products like 'Basmati' rice, 'champagne' wine, and 'scotch' whisky fall under this category. Owing to its exceptional and location-specific attributes, Basmati rice, recognised as an Indian GI, commands global recognition and fetches a higher market value. This fragrant rice continues to be cultivated using traditional methods across select regions of the Indo-Gangetic plains, spanning seven northern and northwestern Indian states and the Punjab province of Pakistan (Mishra and Fatesaria 2022). In India, the Agricultural and Processed Food Products Export Development Authority (APEDA), a central government agency, submitted the GI application for Basmati rice in November 2008, identifying Delhi, Haryana, Himachal Pradesh, Punjab, Uttarakhand, and portions of Jammu and Kashmir and Uttar Pradesh as traditional Basmati-growing areas. In Pakistan, Basmati is cultivated primarily in the Punjab province, mainly on a broad strip of land between the Ravi and Chenab rivers, known as the 'Rachna Doab' (Akram 2009). The main growing districts in Pakistan are Gujranwala, Hafizabad, Sheikhupura, Sialkot, Narowal, Nankana Sahib, Chiniot, Gujrat, Mandi Bahauddin, and Kasur (Sabar and Sabir 2020).

1.6.4 Efforts in Basmati Rice Improvement

The scientific enhancement of Basmati rice through pure line selection began in 1921, following the establishment of the Rice Research Station at Nagina in present-day Uttar Pradesh (formerly the United Provinces of Agra and Oudh). The rice variety Type-3, known as Dehraduni Basmati, was developed here as a pure line selection and became a famous export-quality rice variety (https://www.svpuat.edu.in/research-stations.php). Another Rice Research Station was established in 1926 at Kala Shah Kaku in Undivided India (now in Punjab province of Pakistan) from where the rice variety Basmati-370 was released for commercial cultivation in 1933 (Singh 2000; Sabar and Sabir 2020). Basmati-370, a traditional tall-statured variety known for its low yield and vulnerability to pests and diseases, was long regarded as the benchmark for Basmati quality. It remained widely cultivated and traded across India and Pakistan until the 1980s. Traditional Basmati varieties are tall and suffer from lodging, are low-yielding, and are susceptible to various pests and diseases. Hence, attempts were made to develop dwarf, disease/pest-resistant varieties through selection, mutagenesis, and hybridisation. Efforts to develop a dwarf, high-yielding Basmati prototype met with limited success, hindered by environmental constraints, reproductive barriers between genetic groups, the polygenic nature of key traits, and the intricate inheritance patterns governing grain quality (Shobha Rani and Singh 2003). After more than 25 years of sustained research at the Indian Agricultural Research Institute (IARI), New Delhi, the world's first high-yielding, semi-dwarf Basmati variety, Pusa Basmati-1 was released in 1989. Combining key traditional Basmati traits such as superfine grains, characteristic aroma, soft texture, and notable elongation upon cooking, this variety also offered a yield advantage of 1–1.5 tons per hectare, gaining wide acceptance among both farmers and traders. It accounted for 40–60% of the Basmati rice exported every year from India up to 2007–08 (Siddiq et al. 2012). Later, several improved high-yielding Basmati rice varieties were developed with Pusa Basmati 1 serving as a key parental line in their breeding programs. The selective intermating of two sister lines of Pusa Basmati-1 led to the development and release of Pusa-1121 by IARI in 2003. Pusa-1121 had an extra-long kernel

(>8 mm), high kernel elongation on cooking (>2.5 mm)., was strongly aromatic, with yield potential slightly less than Pusa Basmati-1 but quite higher than Basmati-370 and Taraori Basmati (Shobha Rani and Singh, 2003). Further breeding improvements in the available Basmati varieties by incorporating several other desirable characteristics have led to the development of new rice varieties included in Table 1.4. Efforts on improvements of Basmati varieties in Pakistan continued at the Rice Research Institute (RRI), Kala Shah Kaku in Punjab. To overcome the low yield potential of Basmati-370, Basmati-385 was released from RRI in 1985. This variety matured early and has a yield potential of 5.5 t/ha. Later, Super Basmati was released in 1996. It has extra-long grain and matures earlier than Basmati-370. Super Basmati fetched a high price in the international market due to its superior grain quality and remained a major export earner for several years. Another variety, Basmati 2000, having high yield and good cooking quality, was released in the year 2001 by the same Institute. Till 2021, 14 Basmati rice varieties have been released in Pakistan (Sabar and Sabir 2020), as listed in Table 1.4.

Agronomic and grain characteristics of some of the important Basmati varieties being grown in India and Pakistan have been given in Table 1.5.

1.6.5 Major Factors Affecting Aroma and Flavor in Basmati Rice

The aroma in Basmati and other aromatic rices is due to the presence of certain chemical compounds, mainly 2-acetyl-1-pyrroline (2AP), and has been detected both in the uncooked grain and various parts of the plant (Buttery et al. 1983; Routray and Rayaguru 2018). The presence of this chemical compound in the vegetative parts leads to the development of aroma in the standing crop even at early stages of growth. The amount of 2-acetyl-1-pyrroline present in Basmati rice grain is 0.09 ppm, which is almost 12 times more than the non-Basmati rice. Aroma expression in rice is governed by a combination of genetic makeup and environmental conditions. Research has identified a single recessive allele (fgr) located on chromosome 8, linked to the betaine aldehyde dehydrogenase (BADH2) gene, as the primary genetic factor responsible for aroma in Basmati rice (Rai et al., 2015). Abiotic stresses, which include salinity, drought, extreme temperature, submergence, and ultraviolet irradiation, affect aroma adversely (Dutta et al. 2022). The unfavorable environmental conditions, abnormal temperatures (high >30°C, low <20°C), and strongly acidic or alkaline soil pH lead to a reduction in the aroma of Basmati varieties. The *badh2* gene expression level and concentration of 2AP have been found to be influenced by the temperature, which also regulates the phenotypic expression of aroma (Prodhan et al. 2017). The temperature range of 20°C–30°C has been found to be optimum for the aroma development and agronomic performance of Basmati aromatic rice. The temperature variation may be one of the reasons for variation in aroma quality at different geographic locations.

1.6.6 Basmati Rice in Indian Folklore

Basmati has been mentioned in the folklore of multiple Indian languages like Punjabi, Urdu, Hindi, Maithili, Nepalese, Kumaoni, Garhwali, Himachali, Dogri, and Awadhi.

TABLE 1.5
Agronomic and Grain Characteristics of Important Basmati Varieties

S. No.	Name of the Variety	Average Yield (q/ha)	Maturity (Days)	Salient Features	Plant Height (cm)	Amylose Content (%)
India						
1	Basmati-370	28	150	Tall, low yielding, long duration with very high aroma	165	21
2	CSR-30	33	140	Salt tolerant and susceptible to BLB.	139	23.3
3	Pusa Basmati-1	50	140	Semi-dwarf, high yielding and short duration	100	18.81
4	Pusa Basmati-6	55	140	Semi-dwarf, high yielding and short duration		18.5
5	Pusa Basmati-1121	45	145	Extra-long slender grain with cooked kernel length of 20–25 mm. Susceptible to BLB	120	22
6	Pusa Basmati-1509	60	115	Semi-dwarf, high yielding and short duration	92	22.44
7	Pusa Basmati-1692	52.6	115	Improved version of PB-1509, semi-dwarf, high yielding and short duration	92	24.47
8	Pusa Basmati-1637	42	130	Improved version of PB-1, resistant to blast disease	109	22.8
9	Pusa Basmati-1718	46.4	135	Improved version of PB-1121, resistant to BLB and blast	121	22.2
10	Pusa Basmati-1847	57.1	120	improved version of PB-1509, semi dwarf, non-shattering, short duration, resistant to BLB and blast	110	22.88
11	Pusa Basmati-1885	46.8	140	Improved version of PB-1121, semi dwarf, resistant to BLB and blast	120	21.76
12	Pusa Basmati-1886	44.9	143	Improved version of PB-6, resistant to BLB and blast	95	23.71
13	Punjab Basmati -7	47	126	Resistant to BLB	111	22–24
Pakistan						
1	Punjab Basmati	48.7	100–110	Resistant to lodging.	107	24.8
2	Super Basmati	35	120–125	Basmati-320/10486, milled grain length 7.4 mm, plant height 120 cm, strong aroma	120	24
3	Basmati-515	40.3	114	Longer grain than Super Basmati, better head rice recovery, mature 4–5 days earlier than Super.	132	23.45

(*Continued*)

Ahuja, S. C., U. Ahuja, and S. Ahuja. 2019. History and folklore of basmati rice. *Journal of Cereal Research* 11: 206–214.
Ahuja, S. C., D. V. S. Panwar, and U. Ahuja. 2003. Aromatic rice in Haryana. In *A Treatise on the Scented Rices of India*. ed. R. K. Singh and U. S. Singh, 271–286. Ludhiana, India: Kalyani Publishers.
Ahuja, U., S. C. Ahuja, and R. Thakrar. 1997. Rice in the folklores of India. *Asian Agri-History* 1: 321–330.
Ahuja, U., S. C. Ahuja, R. Thakrar, and N. S. Rani. 2008. Scented rices of India. *Asian Agri-History* 12: 267–283.
Akram, M. 2009. Aromatic rices of Pakistan- A review. *Pakistan Journal of Agriculture Research* 22: 154–160.
Andrus, J. and A. F. Mohammed. 1958. *The Economy of Pakistan*. Oxford: Oxford University Press.
Anonymous. 1911. Races of rice in India. *In The Agricultural Ledger* 1910, No. 1, Vol. XVI. Ed. The Reporter on Economic Products to the Government of India 60-61. Calcutta: Superintendent Government Printing.
Bairagi, S., M. Demont, M. C. Custodio, and J. Ynion. 2020. What drives consumer demand for rice fragrance? Evidence from South and Southeast Asia. *The British Food Journal (BFJ)* 122: 3473–3498. https://doi.org/10.1108/BFJ-01-2019-0025
Buttery, R. G., L. C. Ling, B. O. Juliano, and J. C. Turnbaugh. 1983. Cooked rice aroma and 2-acetyl-1-pyrrolidine. *Journal of Agricultural and Food Chemistry*, 31: 823–826.
Callaway, E. 2014. The birth of rice. *Nature* 514: 56–59.
Carpenter, A.J. 1978. The history of rice in Africa. In *Rice in Africa*. ed. I. W. Buddenhagen and G. J. Parsley, 3–10. London: Academic Press.
Chang, T. T. 1976. The origin, evolution, cultivation, dissemination, and diversification of Asian and African rices. *Euphytica* 25: 435–444.
De, M. 2014. Folk rice varieties and cultural heritage. *The Beats of Natural Sciences*. 7: 1–7.
Dutta, C., D. J. Nathand, D. Phyllei. 2022. Aromatic rice and factors affecting aroma in rice. *International Journal of Environment and Climate Change* 12: 1773–1779. Article no. IJECC.91077ISSN:2581-8627.
Garris, A. J., T.H. Tai, J. Coburn, S. Kresovich, and S. Mc Couch. 2005. Genetic structure and diversity in *Oryza sativa* L. *Genetics* 169: 1631–1638.
Glaszmann, J. C. 1987. Isozymes and classification of Asian rice varieties. *Theoretical and Applied Genetics* 74: 21–30.
GRiSP (Global Rice Science Partnership). 2013. *Rice Almanac*, 4th edition. Los Baños (Philippines): International Rice Research Institute.
Hamilton, W. 1828. *Hindostan, India Beyond Ganges*. London: The East India Gazetteer.
Jarret, H. S. 1927. *The Ain-I-Akbari of Abul Fazl Allami*. 2nd Edition. Vol. I. Calcutta: The Asiatic Society.
Khush, G.S. 1997. Origin, dispersal, cultivation, and variation of rice. *Plant Molecular Biology* 35: 25–34. https://doi.org/10.1023/A:1005810616885
Khush, G. S. 2000. Taxonomy and origin of rice. In *Aromatic Rices*. ed. R. K. Singh, U. S. Singh, and G. S. Khush, 6–14. New Delhi: Oxford & IBH Publishing Co.
Koizumi, T. and G. Furuhashi. 2020. Global rice market projections distinguishing Japonica and indica rice under climate change. *The Japan Agricultural Research Quarterly (JARQ)* 54: 63–91 https://jircas.go.jp
Krishnamurthy, K. H.. 1991. *Wealth of Susruta*. Coimbatore, India: International Institute of Ayurveda.
Kumar, T. 1988. *History of Rice in India*. New Delhi: Gyan Publishing House.
Mishra, A. and H. Fatesaria. 2022. Basmati rice: The ongoing domestic challenge. *Journal of Intellectual Property Rights* 27: 100–106.

Nene, Y. L. 1998. Basmati rice: A distinct variety (cultivar) of the Indian subcontinent. *Asian Agri-History* 2: 175–188.
Nene, Y. L. 2003. Basmati rice: A distinct variety (cultivar) of the Indian subcontinent. In *A Treatise on the Scented Rices of India*. ed. R. K. Singh and U. S. Singh, 7–21. Ludhiana (India): Kalyani Publishers.
Oka, H. I. 1988. *Origin of Cultivated Rice*. Tokyo: Japan Scientific Societies Press.
Prodhan, Z. H., G. Faruq, K. A. Rashid, and R. M. Taha. 2017. Effects of temperature on volatile profile and aroma quality in rice. *International Journal of Agricultural and Biological Engineering* 19: 1065–1072.
Rai, V. P., A. K. Singh, H. K. Jaiswal, S. P. Singh, R. P. Singh, and S. A. Waza. 2015. Evaluation of molecular markers linked to the fragrance and genetic diversity in Indian aromatic rice. *The Turkish Journal of Botany*. 39: 209–217.
Ramiah, K. 1937. *Rice in Madras- Popular Handbook*. Madras (India): Goverment Press.
Routray, W. and K. Rayaguru. 2018. 2-Acetyl-1-pyrroline: A key aroma component of aromatic rice and other food products. *Food Reviews International* 34: 539–565.
Sabar, M. and A. M. Sabir. 2020. *Production Technology of Rice*. Kala Shah Kaku (Pakistan): Rice Research Institute.
Sharma, A., A. Srivastava, S. Singh, et al. 2021. Aromatic rice of India: Its types and breeding strategies. *Integrative Advances in Rice Research*, https://doi.org/10.5772/intechopen.99232
Sharma, G. R. and D. Manda. 1980. Excavations of Mahagara 1977–1978. A neolithic settlement in Belan Valley. Archeology of the *Vindhyas and Belan Valley*, 6. Department of Ancient History, Culture and Archaeology, University of Allahabad, India.
Shobha Rani, N. and R. K. Singh. 2003. Efforts in aromatic rice improvement in India. In *A Treatise on the Scented Rices of India*. ed. R. K. Singh and U. S. Singh, 23–72. Ludhiana (India): Kalyani Publishers.
Siddiq, E. A., L. R. Vemireddy and J. Nagaraju. 2012. Basmati rices: Genetics, breeding and trade. *Agriculture Research* 1: 25–36. https://doi.org/10.1007/S40003-011-0011-5
Singh, A. K., S. Gopala Krishnan, V. P. Singh et al. 2011. Marker-assisted selection: A paradigm shift in basmati breeding. *The Indian Journal of Genetics* 71: 1–9.
Singh, V., A. K. Singh, T. Mohapatra, S. Gopala Krishnan, and R.K. Ellur 2018. Pusa Basmati 1121 – a rice variety with exceptional kernel elongation and volume expansion after cooking. *Rice* 11: 19. Published online 9 Apr 2018. https://doi.org/10.1186/s12284-018-0213-6
Singh, V. P. 2000. Basmati rice of India. In *Aromatic Rices*. ed. R. K. Singh, U. S. Singh, and G. S. Khush, 135–154. New Delhi: Oxford & IBH Publishing Co.
Solheim, W. G. II. 1972. An earlier agricultural revolution. *Scientific American* 226: 3441.
Thakrar, R. and S. C. Ahuja. 1993. Potential and prospects for the export of Basmati rice. In *New Frontiers in Rice Research, Hyderabad*. ed. K. Murlidharan and E. Siddiq, 382–387. Hyderabad (India): Directorate of Rice Research.
Upadhyay, K. D. 1991. *Bhojpuri Lok-Sanskriti*. Prayag, Allahabad (India): Hindi Sahitya Sammelan.
Valera, H. and V. Pede. 2023. What do we know about the future of rice in relation to food system transformation? CGIAR Initiative on Foresight.
Vamireddy, L. R., B. Tanti, L. Lahkar, and Z.M. Shandilya. 2021. Aromatic rices: Evolution, genetics and improvement through conventional breeding and biotechnological methods. In *Molecular Breeding for Abiotic Stress Tolerance and Nutritional Quality*. ed. M. A. Hossain, L. Hassan, K. N. Iftekharudaula, A. Kumar, R. Henry, 341–357. New York: John Wiley & Sons Ltd.
Vaughan, D. A., B.R. Lu, and N. Tomooka. 2008a. The evolving story of rice cultivation. *Plant Science* 174: 394–408.

Vaughan, D. A., B. R. Lu, and N. Tomooka. 2008b. Was Asian rice (*Oryza sativa*) domesticated more than once? *Rice* 1: 16–24. https://doi.org/10.1007/s12284-008-9000-0

Wang, M., Y. Yu, G. Haberer et al. 2014. The genome sequence of African rice and evidence for independent domestication. *Nature Genetics* 46:982–988.

Watt, G. 1891. *A Dictionary of the Economic Products of India*. Vol. V. New Delhi: Cosmo Publications. (Reprint 1972).

2 Aromatic Basmati Rice Varieties, Area, Production, and Cultivation Practices

2.1 RICE CULTIVATION IN INDIA AND THE TOP 10 COUNTRIES OF THE WORLD

Rice (*Oryza Sativa*), is an important crop, and half of the world's population depends on it; it is also one of the staple food crops for more than 70% of the Indian population. In India, it is cultivated on about 150 million hectares producing about 132 million metric tons during the 2022–23 cropping season. Globally, Asia dominates the world in rice production as it accounts for about 90% of the world's rice area and 92% of production. The total production of Asia in the year 2021, according to FAOSTAT, was about 708 million tonnes. India is the second largest producer of rice, 195.4 million tonnes preceded by China, 214.4 million tons, during the year 2021. Rice covers 11.5% of the world's available land area, and India is the second largest rice producing country after China, which accounts for 17.95% of the world's total rice production. FAO declared 1966 as the Year of Rice, and again in 2002, the United Nations General Assembly (UNGA) declared 2004 as the International Year of Rice. In India, rice is grown mainly in two seasons, *Kharif* (about 85%) and Rabi (about 15%), and a major share is cultivated during the *kharif* season (June to October). The sowing of *kharif* rice starts in June–July and is harvested in October–November. Rice (*kharif*) accounts for 37% of the total *kharif* sown area and 70% of *kharif* food grain production in the country (MNCFC/FASAL/2017/01).

The total rice acreage, production, and productivity in India during the last 47 years, from cropping seasons of 1975–76 to 2023–24, are given in Table 2.1 and Figure 2.1 (Source: Agricultural Statistics at a Glance 2021, Directorate of Economics & Statistics, DA&FW). The rice acreage increased from 39.48 million ha in 1975–76 to 46.38 million ha in 2020–21, and rice production increased from 48.74 million tons in 19875–76 to 130.29 million tons in 2021–22. The area under rice has not increased drastically; however, rice production has increased significantly, mainly due to the improvement in rice productivity over the period of 46 years, mainly due to adaptation of newly released high-yielding varieties and drastic improvement in the irrigation sector. Rice yield has increased from 1235 kg/ha in 1975–76 to 2882 kg/ha during 2023–24 (Table 2.1).

DOI: 10.1201/9781032659398-2

TABLE 2.1
Total Rice Acreage, Production, and Productivity in India

Sr. No.	Year	Area (Million Ha)	Production (Million Tons)	Yield (kg/ha)
1	1975–76	39.48	48.74	1235
2	1979–80	39.42	42.33	1074
3	1984–85	41.16	58.34	1417
4	1989–90	42.17	73.57	1745
5	1994–95	42.81	81.81	1911
6	1999–00	45.16	89.68	1986
7	2004–05	41.91	83.13	1984
8	2009–10	41.92	89.09	2125
9	2010–11	42.86	95.98	2239
10	2011–12	44.01	105.30	2393
11	2012–13	42.75	105.23	2461
12	2013–14	44.14	106.65	2416
13	2014–15	44.11	105.48	2391
14	2015–16	43.50	104.41	2400
15	2016–17	43.77	109.70	2494
16	2017–18	43.77	112.76	2576
17	2018–19	44.46	116.48	2638
18	2019–20	43.66	118.87	2722
19	2020–21	45.07	122.27	2713
20	2021–22	46.28	129.47	2798
21	2022–23	47.83	135.76	2838
22	2023–24	47.83	137.83	2882

Source: Agricultural Statistics at a Glance 2024–25, Directorate of Economics & Statistics, DA & FW.

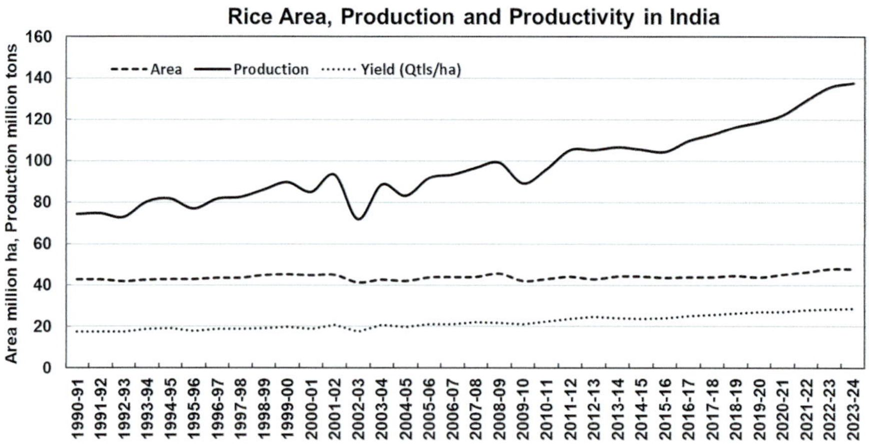

FIGURE 2.1 Total rice acreage and production in India during 1980–81 to 2023–24.

2.2 RICE PRODUCTION IN THE TOP 10 COUNTRIES OF THE WORLD

Rice, being the main staple food for more than half of the world's population (Maclean et al., 2002) and has been studied for ages. During 2021, the global rice production was 787 million tonnes, which is the world's third most-produced agricultural crop after sugarcane and maize. In Asia, 95% of the world's rice is grown (Bhattacharjee et al., 2002), which includes both aromatic and non-aromatic cultivars. India stands second after China in terms of total rice production among the top ten rice-producing countries. The rice production during the *kharif* season of 2021–22 in the top ten countries of the world is given in Figure 2.2.

2.3 AROMATIC BASMATI RICE CULTIVATION IN INDIA

Rice is mainly cultivated under two main categories, fragrant and non-fragrant. In the case of fragrant rice, India dominates the trade among its Basmati rice, followed by Pakistan, China, Indonesia, Bangladesh, and so on. Locally known as "scented pearls," Basmati rice is endemic to the Indian subcontinent, where it has been cultivated by farmers for over 250 years (Nene, 1998; Singh and Singh, 2009; Ahuja et al., 2019). Traditionally, Basmati rice is a crop of the northwest Himalayas in India and it is cultivated in six states and 78 districts. The major states and number of districts in each state where Basmati rice is cultivated are Punjab (22 districts), Haryana (20 districts), Uttar Pradesh (27 districts), Jammu and Kashmir (3 districts), Himachal Pradesh (2 districts) and Uttarakhand (4 districts). In Haryana, the total rice area during the 2023 crop season was 1520 thousand hectares, out of which the total Basmati rice area was 812 thousand hectares. Kaithal district has the maximum transplanted area under rice of 168.7 thousand hectares, followed by Jind district with 165.3 thousand hectares. The total Basmati rice area was high in Jind district (111.27 thousand hectares), out of which the maximum area under Basmati rice varieties was PB-1121, that is, 35.7 thousand hectares, followed by Sonepat in which total Basmati rice area was 86.94 thousand hectares out of which the maximum area under Basmati rice varieties was PB-1718 , that is, 60.31 thousand hectares. In India, with the evolution of Basmati varieties, several high-yielding, disease-tolerant, salt-tolerant, and short-duration varieties have been released during the last decade. Since the farmers need to grow wheat (in some districts, potato) immediately after the paddy, they prefer to grow short-duration evolved Basmati varieties in the rice-wheat system of the Indo-Gangetic Plains (IGP) of North India. Therefore, the short-duration varieties, namely, PB-1509, PB-1692, and PB-1847, have replaced much of the area (48% in 2023) under other long-duration varieties, including both traditional and evolved varieties. Other recently released disease-tolerant and high-yielding varieties, namely, PB-1718, PB-1886, PB-1885, PB-1, and PB-6, are also being adopted by the progressive farmers of the Basmati growing regions. The acreage and production statistics of aromatic Basmati rice growing states of India from *kharif* seasons 2005 to 2023 are given in Tables 2.2 and 2.3, respectively.

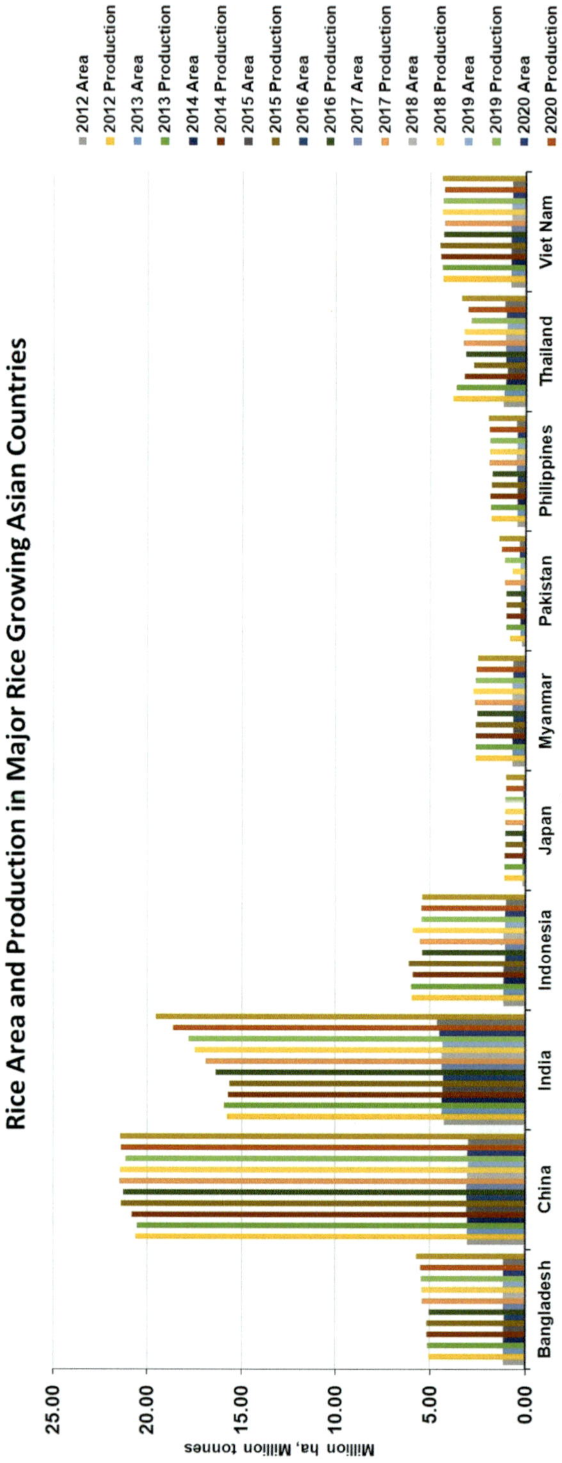

FIGURE 2.2 Rice area and production in major rice-growing Asian countries.

TABLE 2.2
Aromatic Basmati Rice Acreage ('000 ha) in Major Basmati Growing States in India

	Acreage ('000 ha) in Different States						
Year	Haryana	Punjab	Uttar Pradesh	Uttara-khand	Jammu & Kashmir	Himachal Pradesh	Total
2005	221.5	139.6	271.1	8.9	37.6	NA	**678.7**
2006	177.7	171.0	195.4	8.4	17.1	NA	**569.6**
2007	458.4	234.5	231.1	5.9	31.1	NA	**961.0**
2008	658.5	648.7	283.3	13.3	33.4	NA	**1637.2**
2009	610.7	792.7	254.3	17.5	32.9	NA	**1708.1**
2010	816.6	738.9	377.8	53.7	33.0	18.7	**2038.7**
2011	705.1	600.7	428.3	23.5	38.0	5.6	**1801.2**
2012	594.1	573.1	366.6	16.9	36.9	2.85	**1590.5**
2013	711.0	590.0	319.0	18.3	37.3	1.0	**1677.0**
2014	833.0	858.0	354.4	20.3	68.5	0.45	**2135.0**
2015	833.0	864.0	340.0	15.8	62.9	2.2	**2119.0**
2016	720.0	549.0	266.0	15.6	62.3	7.0	**1686.0**
2017	652.0	562.0	256.0	14.8	61.5	7.02	**1554.0**
2018	634.0	547.0	251.0	14.6	62.4	7.07	**1515.0**
2019	NA	NA	NA	NA	NA	NA	NA
2020	NA	NA	NA	NA	NA	NA	NA
2021	NA	NA	NA	NA	NA	NA	NA
2022	NA	NA	NA	NA	NA	NA	NA
2023	812.4	874.3	679.5	14.5	57.8	1.6	**2442.1**

Source: Basmati Export Development Foundation (APEDA), New Delhi.

2.4 AROMATIC RICE VARIETIES IN INDIA

Aromatic or scented rice has long been highly regarded in Indian society not only because of its excellent quality but also because it has been considered auspicious. The aromatic rice varieties in the states of West Bengal, Orissa, Chhattisgarh, Bihar, and North East region are very short, fine-grained, and highly scented. Farmers mainly grow these varieties for their own consumption and ceremonial purposes, and they do not have a well-developed market (Shamim, et al., 2013). There are nearly 300 indigenous collections of aromatic rice varieties in India, which are classified into different types based on kernel length: **Small grained (≤5.51 mm); medium grained (5.51–6.60 mm); long grained (6.61–7.51 mm), and very long grained (>7.51 mm)**. Most of the aromatic rices of India are short to long grained and only a few of them are very long grained (Basmati types). In general, there are two types of aromatic rice: Basmati and non-Basmati types. This differentiation is done on the basis of kernel dimensions (kernel length, kernel breadth, and ratios of length/

TABLE 2.3
Aromatic Basmati Rice Production ('000 tons) in Major Basmati Growing States in India

	Production ('000 tons) in Different States						
Year	Haryana	Punjab	Uttar Pradesh	Uttara-khand	Jammu & Kashmir	Himachal Pradesh	Total
2005	1026.1	414.6	800.4	23.0	86.1	NA	**2350.2**
2006	773.1	399.0	561.2	20.0	49.9	NA	**1803.2**
2007	1761.0	805.6	1660.7	14.7	84.2	NA	**4326.2**
2008	2256.9	2568.9	1279.6	38.5	81.0	NA	**6224.9**
2009	2432.5	3164.7	986.4	58.3	79.5	NA	**6721.4**
2010	2209.0	2568.9	1279.6	50.5	79.7	10.3	**6197.97**
2011	2676.8	2695.5	2066.2	80.1	94.9	42.8	**7656.27**
2012	2261.3	2282.2	1428.5	53.9	96.1	5.7	**6127.7**
2013	2899.0	2293.0	1270.0	54.2	92.6	3.4	**6667.0**
2014	3702.0	3499.0	1261.0	66.4	240.8	2.15	**8774.0**
2015	3234.0	3541.0	1066.0	45.6	152.2	7.34	**8058.0**
2016	2797.0	2337.0	817.0	42.3	133.8	29.7	**6154.0**
2017	2535.0	2142.0	763.0	39.0	132.0	30.0	**5644.0**
2018	2226.0	2005.0	739.0	37.0	135.0	20.0	**5165.0**
2019	NA	NA	NA	NA	NA	NA	NA
2020	NA	NA	NA	NA	NA	NA	NA
2021	NA	NA	NA	NA	NA	NA	NA
Kharif-2022	NA	NA	NA	NA	NA	NA	NA
Kharif-2023	4096.6	4200.3	2745.7	61.4	146.3	3.9	**11262.9**

Source: Basmati Export Development Foundation (APEDA), New Delhi.

breadth, before and after cooking; Singh et al., 2000). Basmati is aromatic rice, but not all aromatic rice is Basmati. For a rice variety to be called Basmati rice, it has to meet all the minimum standards of Basmati rice quality traits. Export of Basmati Rice (Quality Control and Inspection) Rules, 2003, defines which genotype of aromatic rice can be called as Basmati. According to this rule, Basmati is of two types:

i. **Traditional Basmati**: These are pure line selections from the existing Basmati landraces which are six in number.

   Traditional Basmati varieties grown are Basmati-370, Basmati-386, Taraori Basmati, Type3, and Ranbir Basmati. These varieties are tall (plant height 140–148 cm), with a growing period of 145–150 days, except Ranbir Basmati which matures in 120 days. Ranbir Basmati is grown only in the Jammu region of Jammu and Kashmir state.

ii. **Evolved Basmati**: Evolved Basmati varieties are developed through hybridisation or any other breeding methods in such a way that at least one of the parents used to develop them was a traditional Basmati variety or pure line Basmati variety (Sharma et al., 2021).

Evolved Basmati is the recently bred Basmati varieties, having traditional Basmati as one of their parents. Pusa Basmati-1, Pusa Basmati-1121, CSR-30, and Pusa Basmati-1509 are the widely grown varieties in this group. These varieties are of shorter plant height and shorter crop duration. Due to their higher yield, they are being grown in a larger area now and are slowly replacing traditional Basmati varieties. The list of traditional Basmati and evolved Basmati varieties is given in Table 2.4.

TABLE 2.4
Traditional Basmati and Evolved Basmati Varieties of India

Traditional Basmati	Year of Release	Evolved Basmati	Year of Release
Basmati-217	1973	Punjab Basmati-1	1984
Basmati-370	1976	Pusa Basmati-1	1989
Type-3	1978	Kasturi	1989
Ranbir Basmati	1996	Haryana-1	1991
Taraori Basmati	1996	Mahi Sugandha	1995
Basmati-386	1997	Improved PusaBasmati-1	2007
		Pusa Basmati-1121	2008
		Pusa Basmati-6	2010
		Punjab Basmati-2	2012
		CSR-30	2012
		Pusa Basmati-1509	2013
		Basmati-564	2015
		Vallabh Basmati-24	2015
		Punjab Basmati-3	2016
		Pusa Basmati-1728	2016
		Pusa Basmati-1718	2019
		Haryana Basmati-2	2019
		Pusa Basmati-1692	2021
		Pusa Basmati- 1847	2022
		Pusa Basmati-1885	2022
		Pusa Basmati-1886	2022
		Pusa Basmati-1985	2022
		Pusa Basmati-1979	2022
		Pusa Basmati-1882	2022
		Punjab Basmati-7	2022

Source: APEDA and http://drdpat.bih.nic.in. Reference: Sharma et al., 2021 (https://apeda.gov.in/apedawebsite/Announcements/34_basmati_varieties_2021.pdf).

2.5 CURRENTLY CULTIVATED AROMATIC BASMATI RICE VARIETIES IN INDIA

In India, 45 varieties of Basmati rice so far have been notified under the Seeds Act of 1966. These are Basmati-370, Type-3 (Dehraduni Basmati) Punjab Basmati-1 (Bauni Basmati), Pusa Basmati-1, Kasturi, Haryana Basmati-1, Mahi Sugandha, Taraori Basmati (HBC 19/Karnal Local), Ranbir Basmati, Improved Pusa Basmati-1 (Pusa-1460), Pusa Basmati-1121 (after amendment), Vallabh Basmati-22, Pusa Basmati-6 (Pusa 1401), Punjab Basmati-2, Basmati CSR-30 (after amendment), Vallabh Basmati-21 (IET 19493), Pusa Basmati-1509 (IET 21960), Pusa Basmati-1847, Pusa Basmati-1637, Pusa Basmati-1728, Pusa Basmati-1718, Pusa Basmati-1885, Pusa Basmati-1886, Punjab Basmati-7, Haryana Basmati-2 and Pusa Basmati-1692.

In India, several indigenous aromatic rice varieties grown in different states are: *Kalanamak* (popularly called the "black pearl of eastern Uttar Pradesh"), *Shakarchini* and *Hansraj* of UP; *JauPhool*, *Vishnu Bhog*, *Sheetal Bhog*, and *Do Dana* from Chhattisgarh, *Kalajoha* of North East; *Ambemohar* of Maharashtra, and *Randhunipagal* of Orissa and West Bengal region (Bano et al., 2017). The Pusa Basmati-1121 and CSR-30 are major Basmati varieties grown in Punjab, Haryana, and Uttar Pradesh having higher yields as well as remunerative prices. The yields of Pusa Basmati-1121 and Pusa Basmati-1 are in the range of 47 to 52 quintals per hectare whereas traditional varieties such as HBC-19 (Taraori) yield 22 to 25 quintals per hectare. Basmati-370 and Taraori Basmati are the widely cultivated varieties of export quality for decades. Breeding efforts have resulted in the development of the world's first high-yielding, semi-dwarf Basmati variety—Pusa Basmati-1, combining the quality features of traditional varieties and yield potential of high-yielding semi-dwarf varieties. In the GI area of Basmati rice in India, Pusa Basmati rice varieties namely, Pusa Basmati-1121, Pusa Basmati-1718, Pusa Basmati-1509, Pusa Basmati-1692, Pusa Basmati-1847, and Pusa Basmati-6, occupy more than 90% of the area under Basmati rice cultivation and account for 90% of the Basmati rice exports from India.

2.6 AREA, PRODUCTION, AND PRODUCTIVITY OF AROMATIC BASMATI RICE IN INDIA

The major Basmati rice-producing states are Haryana, Punjab, Uttar Pradesh, Jammu and Kashmir, Uttarakhand, and Himachal Pradesh (Thakrar and Ahuja, 1990). The area under Basmati rice in India was 678.7 thousand hectares in 2005–06 and increased almost 4-fold to 2442.1 thousand hectares in 2023–24. The acreage and production gradually increased during the last decade due to the introduction of high-yielding and short-duration varieties like Pusa Basmati-1121, Pusa Basmati-1718 (long duration segment), and Pusa Basmati- 1509, Pusa Basmati-1692, Pusa Basmati-1847 (short duration segment) subsequently (APEDA, 2019). The recently released Basmati varieties are disease tolerant. In the year 2023, almost 48% of the area under long-duration varieties has been replaced by short-duration varieties. The farmers in IGP grow wheat immediately after the rice crop and get very little time to prepare the field. Thus, the short-duration varieties contributed not only to the higher income of the farmer by giving high yields but also reduced the cost of cultivation.

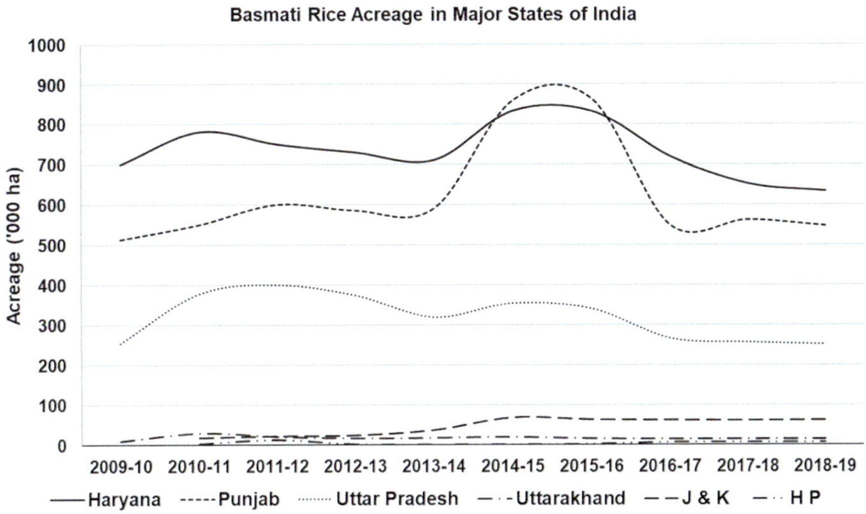

FIGURE 2.3 Aromatic Basmati rice acreage in major states of India.

The groundwater potential has become a challenge in intensively irrigated areas of Punjab, Haryana, and Western Uttar Pradesh. A lesser number of irrigations is required in the case of short-duration varieties, which mature in 120–125 days. Also, the farmer gets a little more time to prepare the field for wheat. And even in potato-growing areas in Punjab, the farmers quickly adopted the short-duration varieties (Pachauri et al., 2010) adding a three-fold value to their income. The acreages under aromatic Basmati rice in the major states of India are given in Figure 2.3. The production of aromatic Basmati rice in the major states of India during the *kharif* seasons from 2008–09 to 2018–19 is given in Figure 2.4.

2.7 AROMATIC BASMATI RICE CULTIVATION PRACTICES

Basmati rice is normally cultivated during the *kharif* season (May to November), and the optimum time for nursery sowing in the states of Punjab and Haryana is during the first and second fortnights of June (Kamboj et al., 2021). Traditional Basmati varieties are tall, prone to lodging, very low yielding, photoperiod sensitive, and temperature sensitive, and these varieties flower when a specific day length is reached. The time of transplanting is a critical factor in determining the yield and quality of the Basmati aromatic varieties; therefore, the optimum time of transplanting for some of the varieties, like Pusa Punjab Basmati-1509, Basmati-386, and Basmati-370, is the second fortnight of July. Seedlings of Basmati/aromatic rice varieties are ready for transplanting when they attain the 5- to 6-leaf stage or are 25–35 days old after sowing. Harvesting is typically carried out as soon as rice grains mature, which is the proper stage of harvesting, and in the case of early and medium duration varieties, 25 to 30 days after flowering, and 35–40 days after flowering in late varieties. In Punjab, traditional Basmati varieties (Basmati-386) are in the flowering stage in the last week

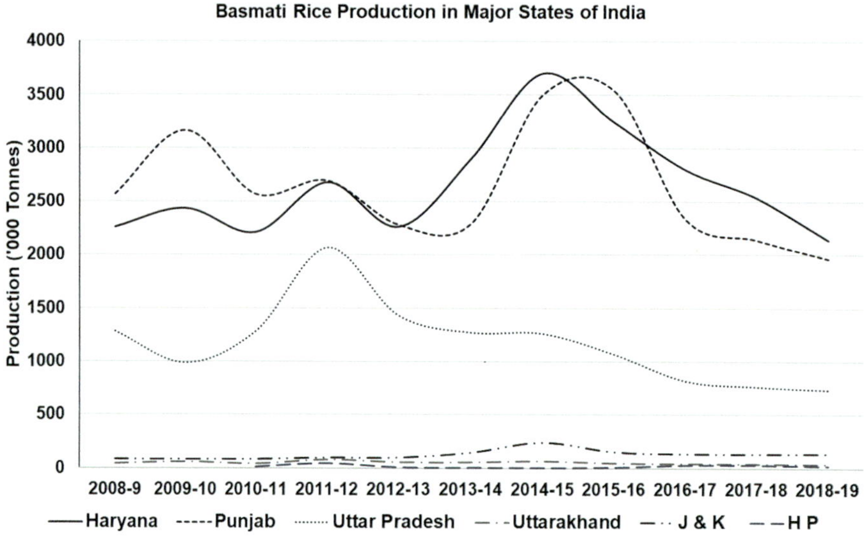

FIGURE 2.4 Aromatic Basmati rice production in the major states of India.

of October and harvested in the fourth week of November. Evolved Basmati (Pusa Basmati-1 and PB-1121) flowers in the second week of October and is harvested in the third week of November. At harvesting, the moisture content of paddy should be 17–18%, and the harvested grain is dried under shade to bring down the moisture content to 14% for storage and milling. Optimum climatic requirements for Basmati rice cultivation and its normal growth are similar to those of high-yielding rice, which include 20–35°C temperature throughout the crop duration, high humidity (70–80%) during the vegetative phase, clear sky during the day, low night temperatures for reduced maintenance respiration, and equitable distribution of rainfall. Basmati varieties with superior cooking and eating characteristics can be produced if the crop matures in relatively cooler temperatures (Rice Today, 2022).

2.8 CROP PHENOLOGY OF AROMATIC BASMATI RICE

Crop phenology defines the physiological development stages of crop growth from planting to harvest and it also indicates adaptation of a crop to its environment. The crop growth stages or crop phenology are affected by the local climate, and it varies from place to place from year to year. The crop growth stages start with seed sowing/crop emergence and end with physiological maturity and harvest. Crop seed sowing or crop planting dates normally depend on soil moisture, soil temperature, and weather conditions. After planting, emergence is the first vegetative growth stage; the harvest is the last growth stage and represents the end of the crop season. The crop growing period is defined by the crop emergence and harvest dates, and the other various crop growth stages during the crop growing season differ depending on the crop type. The generalised phenological growth stages of aromatic Basmati rice observed in the major Basmati growing districts of Punjab and Haryana states are given in Figure 2.5.

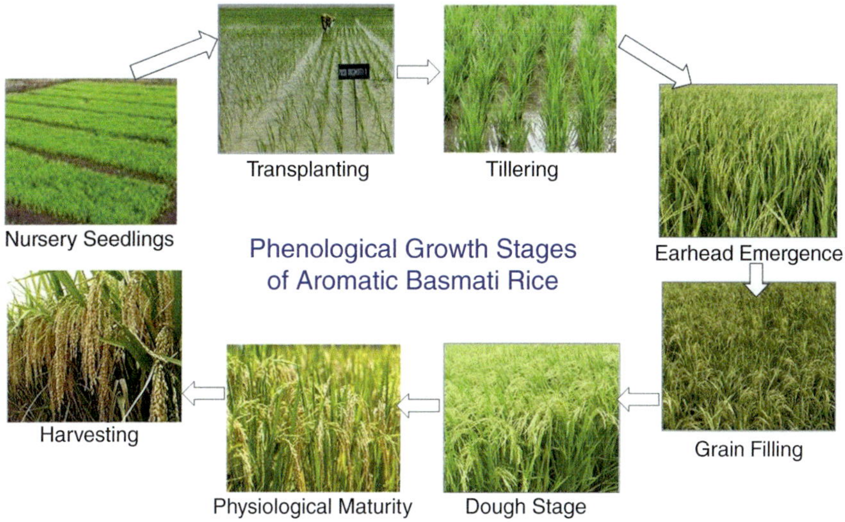

FIGURE 2.5 Phenological growth stages of aromatic Basmati rice.

2.9 AROMATIC BASMATI RICE CULTIVATION IN PAKISTAN

In Pakistan, aromatic Basmati varieties are cultivated mainly in the area between the Ravi and Chenab Rivers. The important aromatic Basmati growing districts are Gujranwala, Hafizabad, Sheikhupura, Sialkot, Narowal, Nankana Sahib, Chiniot, Gujrat, Mandi Bahaudin, and Kasur, which comprise 30% of the total rice production in Pakistan (Akhter et al., 1997; Akhter et al., 2007). The climate of this region is subhumid, subtropical with annual rainfall during the monsoon season (Jul.–Aug.) of 400–700 mm, with a hot summer and moderate cold winter which are extremely favourable for the production of the best quality aromatic rice. As per the reports of the Statistical Bureau of Pakistan (2017–2018), Basmati rice was cultivated over an area of about 3.50 million hectares, which was about 6% higher as compared to the previous year, 2016–2017. In Pakistan, Punjab province is the major producer of rice, with about 68% of the rice area and about 56% of the total rice production of Pakistan (ADB, 2018), and approximately 70% of the total rice area is under Basmati varieties. The major approved Basmati rice varieties cultivated in Pakistan are KS-282, IRRI-6, KSK-133, KSK-434, Basmati-385, Super Basmati, Basmati-2000, Shaheen Basmati, Basmati-515, PK-1121, Aromatic PK-386, Kissan Basmati, Punjab Basmati, Super Basmati-2019, Super Gold, and so on (Sabar and Sabir, 2020).

The increase in area under Basmati cultivation was due to increased demand for Pakistani Basmati rice in the international market. The production of Basmati rice increased by about 12% due to an increase in average yields from 2 t/ha to 2.2 t/ha during 2016–2017 and 2017–2018, respectively, which was mainly due to the release of improved high-yielding Basmati rice varieties in the Punjab province. During 2015–16, Pakistan exported about 4.262 million tons of rice (about 0.503 million tons of Basmati and about 3.759 million tons of non-Basmati), and the Basmati rice export during 2016–2017 was 0.446 million tons (Akhter and Haider, 2020). The Basmati rice area, yield, and production in Pakistan from 1990–91 to 2023–24 Crop

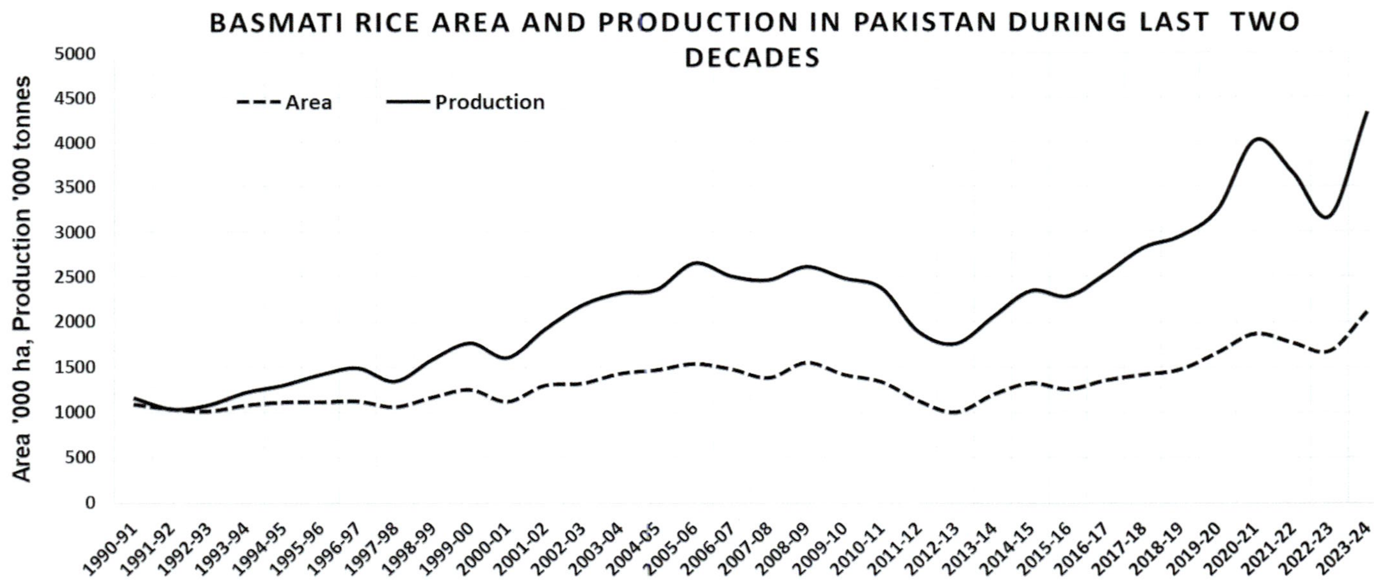

FIGURE 2.6 Basmati rice area and production from 1990–91 to 2023–24 in Pakistan.

Source: URL of Pakistan Dept of Agriculture, https://crs.agripunjab.gov.pk/.

Reporting Service is given in Figure 2.6 (URL of Pakistan Dept. of Agriculture, https://crs.agripunjab.gov.pk/).

REFERENCES

ADB. 2018. ADB Central and West Asia Working Paper Series Investment in Research and Development for Basmati Rice in Pakistan Environment, Natural Resources, and Agriculture Division, Central and West Asia Department No. 7|December 2018.

Ahuja, S. C., U. Ahuja, and S. Ahuja 2019. History and folklore of Basmati rice. *Journal of Cereal Research* 11(3): 206–214. https://doi.org/10.25174/2249-4065/2019/95207

Akhter, M. and Haider, Z.. 2020. Basmati Rice Production and Research in Pakistan. In *Sustainable Agriculture Reviews*, ed. E. Lichtfouse, 39, Cham, Switzerland: Sustainable Agriculture, Reviews 39, https://doi.org/10.1007/978-3-030-38881-2_5

Akhter, M., H.Z. Iqbal, W.S. Khan and S. Ali. 1997. Genotypic and phenotypic correlation coefficient studies of yield and economic traits in cotton. *Pakistan Journal of Scientific Research* 49: 1–2.

Akhter, W., M. Sharif, and N. Akmal. 2007. Analysis of economic efficiency and competitiveness of the rice production systems of Pakistan's Punjab. *Lahore Journal of Economics* 12(1): 141–153.

APEDA. 2019. Basmati Crop Survey Report. 2019. Basmati Export Development Foundation, APEDA, New Delhi. Vol. 2. 11.

Bano, D. A., S. P. Singh, and S. A. Waza. 2017. Generation mean analysis for yield and quality traits in aromatic genotypes of rice (Oryza sativa L.). *International Journal of Pure & Applied Bioscience 5*: 870–878. https://doi.org/10.18782/2320-7051.6055

Bhattacharjee, P., R.H. Singhal, and P. R. Kulkarni. 2002. Basmati Rice: A Review. *International Journal of Food Science and Technology* 37: 1–12.

Kamboj, P., S. Kumar, and D. Sanjay. 2021. Economic analysis of basmati cultivation in Haryana. *The Pharma Innovation Journal*; SP-10(11): 2948–2995.

Maclean, J.L., Dawe, D.C., Hardy, B. and G.P. Hettel. 2002. *Rice almanac*. 3rd Edition, International Rice Research Institute, Los Baños, 253.

Nene, Y. L. 1998. Basmati Rice: A distinct variety (cultivar) of the Indian subcontinent. *The Asian Agri-History Foundation* 2, 321–330.

Pachauri, V., M. K. Singh, A. K. Singh, S. Singh, N. A. Shakeel, and V. P. Singh. 2010. Origin and genetic diversity of aromatic rice varieties, molecular breeding, and chemical and genetic basis of rice aroma. *Journal of Plant Biochemistry and Biotechnology* 19(2): 127–143.

Rice Today. 18-Oct-2022. Basmati, the queen of perfumed rice. International Rice Research Institute (IRRI) and CGIAR's Research Program on Rice.

Sabar, M. and A. M. Sabir. 2020. *Production Technology of Rice*. Kala Shah Kaku, Punjab, Pakistan: Rice Research Institute, pp. 1–148.

Shamim, M., K.K. Singh, B. Gangwar, S. Kumar, and V. P. Mandal. 2013. Aromatic rice: An overview. www.riceportal.in/print/15969

Sharma, A., S. Sandhya, S. Akanksha, M. Snehanshu, S. M. Subhash, S. Chhavi, S. Akanksha, A. Kuma, and H. K. Jaiswal. 2021. Aromatic rice of India: Its types and breeding strategies. *Integrative Advances in Rice Research*. https://doi.org/10.5772/intechopen.99232 https://cdn.intechopen.com/pdfs/78726.pdf

Singh, R. K., U. S. Singh, G. S. Khush, R. Rohilla, J. P. Singh, G. Singh, and K. S. Shekhar. 2000. *Small and Medium Grained Aromatic Rices of India*. New Delhi: Oxford and IBH Publishing Co. Pvt. Ltd. pp. 155–178.

Singh, V. P. and A. K. Singh 2009. History of Basmati rice: Research and development in India. *Indian Farming* 59: 4–8 (April issue).

Thakrar, R. and S. C. Ahuja 1990. Potential prospects for export of Basmati rice. In *New Frontiers in Rice Research*, eds. K. Muralidharan and E. A. Siddiq. Hyderabad, India: DRR (ICAR). pp. 382–383.

3 Science of Rice Aroma and Environmental Factors Influencing Basmati Aroma and Other Grain Quality Traits

3.1 INTRODUCTION

Among aromatic rice varieties, including Basmati from India and Pakistan, Jasmine from Thailand, Sadri from Iran, and Della from the USA, aroma stands out as the most valued characteristic, followed by taste, grain elongation, and other region-specific quality traits. Aroma is perceptible at various stages, during crop growth in the field, at harvest, in storage, milling, cooking, and consumption (Gibson, 1976; Efferson, 1985). Its development is shaped by a combination of genetic factors and environmental conditions (G × E interaction). While the genetic composition of a rice variety is fundamental, environmental conditions significantly influence the expression of grain quality attributes, including aroma. Cooler temperatures during the grain maturity phase are particularly favourable for enhancing aroma in aromatic rice. The characteristic scent arises from specific volatile compounds produced in both the plant and the grain. Buttery et al. (1983a) identified the principal compound present in aromatic rice grain as 2-acetyl-1-pyrroline (2-AP), which can be detected by humans even at extremely low concentrations, as its odour threshold is around 0.1 parts per billion (ppb). In aromatic rice genotypes, 2-AP content varied from 6 to 90 ppb in milled rice, and from 100 to 200 ppb in un-milled rice (brown or rough rice). A 10 times higher 2-AP content was observed in aromatic Basmati 370 as compared to non-aromatic Texas long-grained rice, indicating that non-aromatic rices also contain 2-AP but in a much lower concentration. Moreover, 2-acetyl-1-pyrroline (2-AP) is the only volatile compound in rice for which a clear correlation between its levels and perceived aroma intensity has been scientifically demonstrated (Itani et al., 2004). This compound is present throughout the aromatic rice plant, excluding the roots, and contributes to its characteristic fragrance (Maraval et al., 2010; Hinge et al., 2016).

 DOI: 10.1201/9781032659398-3

3.1.1 Sensory Analysis of Aroma and Flavour

The perception of rice aroma begins when volatile compounds are released and travel through the nasal passage, where they are detected by the olfactory cilia lining the epithelium in the upper nasal cavity (Mielgaard et al., 2007). Aroma refers specifically to the scent of a food product, as identified by the olfactory system (Butt et al., 2008). In contrast, flavour is a broader sensory experience perceived in the mouth, integrating multiple chemical senses (Caul, 1957). It encompasses aromatics, volatile compounds released during chewing and detected retro-nasally, along with basic tastes such as sweet, sour, salty, and bitter, and chemesthetic sensations that stimulate nerve endings in the oral and nasal cavities (Mielgaard et al., 2007).

3.1.2 Measurement of Aroma and Flavour

Trained sensory panels use descriptive analysis to systematically evaluate the aroma and flavour of rice. This method is effective in tracking sensory variations throughout different stages, such as cultivation, harvest, milling, storage, and shelf life (Mielgaard et al., 2007). When combined with consumer preference testing, descriptive analysis helps in accurately identifying sensory attributes that align with market-specific quality demands (Champagne, 2008). Hence, the Agriculture Products Export Development Authority (APEDA), Government of India, has identified the qualitative sensory analysis panel test as one of the primary quality criteria for Basmati rice. This sensory evaluation is carried out on cooked milled rice by a team of panellists, who give numerical scores on a scale of 1–5 on the following characteristics: appearance of cooked rice, its texture consistency, softness to the touch, ease of chewing, flavour, scent, grain elongation, and overall consumer appeal (APEDA website).

3.2 GENETIC BASIS AND REGULATION OF AROMA IN BASMATI AND OTHER AROMATIC RICES

3.2.1 Genetic Control of Aroma/Aroma Genes

The genetic architecture underlying rice aroma is intricate, involving multiple unidentified genes that may be active at various developmental stages. Additionally, it is influenced by the presence and concentration of both volatile and semi-volatile compounds (Varatharajan et al., 2021). Aroma inheritance has been reported to be mainly due to a single recessive allele (*fgr*) located on chromosome 8 (Ahn et al., 1992; Jin et al., 2003), whereas some other studies have suggested that aroma is regulated by two to four genetic loci (Dhulappanvar, 1976; Geetha and Naidu, 1994; Pinson, 1994; Vivekanandan and Giridharan, 1994). The inheritance pattern of aroma seems to be dependent on the specific cross or genotype, as both the number of contributing genes and their mode of action can vary among genetic backgrounds (Varatharajan et al., 2021). Despite this complexity, aroma is considered a highly heritable trait. This is evident from the strong aromatic expression in certain lines developed from the cross between T 142 (aromatic) and IR20 (non-aromatic), along with several

TABLE 3.1
Inheritance Pattern of Aroma in Rice

Gene Action	References
Monogenic dominant	Kuo (2005)
Monogenic recessive	Ahn et al. (1992), Vivekanandan and Giridharan (1994), Garland et al. (2000), Sun et al. (2008), Vazirzanjani et al. (2011)
Monogenic recessive with an inhibitor	Ghose and Butany (1952), Tsuzuki and Shimokawa (1990)
Digenic or trigenic dominant	Kadam and Patankar (1938)
Monogenic or digenic recessive	Pinson (1994)
Digenic recessive	Geetha (1994), Vivekanandan and Giridharan (1994)
Three recessive genes	Siddiq (1983)
Two dominant complementary genes	Tripathi and Rao (1979)
Three dominant complementary genes	Reddy and Sathyanarayanaiah (1980)
Four dominant complementary genes	Dhulappanvar (1976)
Monogenic or digenic recessive or dominant, complementary	Nayak and Acharya (2004)
Monogenic or digenic dominant, duplicate	Sarawagi and Bisne (2006)
Digenic dominant suppression epistasis interaction	Chaut et al. (2010)
Polygenic	Lorieux et al. (1996), Amarawathi et al. (2008)

Source: Extracted from Varatharajan et al. (2021).

commercially released high-yielding aromatic rice varieties (Singh et al., 2000a). Studies conducted regarding the genetic regulation of aroma in rice are presented in Table 3.1. It can be seen that the number of genes responsible for aroma has been quoted to vary from 1 to 4. Similarly, the type of gene action varied from recessive to dominant, from complementary to inhibitory, depending on the genotypes studied. These variations may be attributed to the broad allelic diversity and complex genomic architecture inherent within the aromatic rice germplasm. However, as can be observed in the table, the majority of the studies have found a single recessive allele to be contributing to the aromatic trait in rice. Apart from these, conflicting observations have also been recorded on the location of the fragrance gene. Sun et al. (2008) have reported aroma gene location in different studies to be on six chromosomes, namely, 4, 5, 8, 9, 11, and 12.

3.2.2 Molecular Mapping of Aroma Gene

Targeted molecular studies have been conducted to identify and map genes associated with fragrance using genetic markers. Ahn et al. (1992) successfully localised the recessive fragrance gene (*fgr*) to chromosome 8 through the use of RFLP (restriction fragment length polymorphism) markers. Several other molecular mapping studies on aromatic rice carried out using different markers also support a solitary gene located on chromosome 8 (Garland et al., 2000; Chen et al., 2008). However, the first QTL mapping in aromatic rice was carried out by Lorieux et al. (1996), who observed two QTLs on chromosomes 4 and 12. Multiple studies have identified distinct quantitative trait loci (QTLs) linked to aroma expression in rice. In Basmati rice, Amaravathi et al. (2008) reported three QTLs mapped to chromosomes 3, 4, and 8 controlling aroma. Recently, Talukdar et al. (2017) identified 3 QTLs for aroma, one on chromosome 5, and two on chromosome 8. Till now, QTLs associated with aroma have been mapped to chromosomes 3, 4, 5, 8, and 12. A list of studies carried out on molecular mapping of aroma genes in rice has been presented in Table 3.2.

Bradbury et al. (2005) discovered the major gene associated with rice aroma as a single recessive gene on chromosome 8 encoding betaine aldehyde dehydrogenase 2 (BADH2). Subsequent studies (Bradbury et al., 2008; Chen et al., 2008) elucidated the role of *BADH2* in the biosynthetic pathway of 2-acetyl-1-pyrroline (2-AP), the principal aromatic compound. *BADH2* normally facilitates the transformation of γ-amino butyraldehyde (GABald) to γ-aminobutyric acid (GABA); when the gene is nonfunctional, GABald accumulates and is instead converted to 2-AP. The *BADH2* gene comprises 15 exons and 14 introns, coding for a 503-amino-acid protein. The functional mutation leading to the recessive *badh2.1* allele was initially characterised by three single-nucleotide polymorphisms (SNPs), including an 8-bp deletion in exon 7, which results in a truncated, nonfunctional enzyme. This deletion was shown to disrupt protein function, thereby enabling aroma expression in varieties such as Basmati and Jasmine. Given the concordance between this loss-of-function mutation

TABLE 3.2
Molecular Mapping of Aromatic Genes in Rice

Number of Genes	Type of Markers	Chromosome Location	References
1	RFLP	8	Ahn et al. (1992)
1	RFLP	8	Yano et al. (1992)
1	RAPD	-	Tragoonrang et al. (1996)
1 major gene and 2 QTLs	RFLP, STS	8, 4 and 12	Lorieux et al. (1996)
1	SSR	8	Garland et al. (2000)
1	SSR	8	Cordeiro et al. (2002)
1	SNP	8	Jin et al. (2003)
1	SSR	8	Bradbury et al. (2005)
1	EST, SSR	8	Wanchana et al. (2005)
1	SSR	8	Chen et al. (2006)
1	SSR	-	Hien et al. (2006)
1	SSR	8	Li et al. (2006)
3 QTLs	SSR	QTLs on 3,4, and 8	Amaravathi et al. (2008)

Source: Varatharajan et al. (2021).

and the recessive inheritance of fragrance, Bradbury et al. (2005) concluded that *BADH2* is synonymous with the previously described *fgr* gene, a view supported by Sakthivel et al. (2009).

Surveys of diverse aromatic rice germplasm (Bourgois et al., 2008; Shi et al., 2008; Fitzerald et al., 2008) also support the association of *badh2* with aroma. Chen et al. (2008) demonstrated that introducing the functional dominant *BADH2* allele into an aromatic rice variety eliminated fragrance, thereby confirming *BADH2* as the principal gene regulating aroma biosynthesis in rice. However, some other studies have reported exceptions to this mutation in *badh2*. Muralidhara Rao et al. (2006) studied five indigenous non-Basmati aromatic rice cultivars that were exceptions to the 8-bp deletion and possessed a lower level of 2-AP than those carrying the deletion. Although 2-AP levels are low in these genotypes, they still exhibit fragrance, indicating the contribution of other odour-active compounds beyond 2-AP (Sakthivel et al., 2009). Fitzgerald et al. (2008) observed that some aromatic rice genotypes lacking the characteristic 8-bp deletion in *BADH2* still exhibited elevated levels of 2-acetyl-1-pyrroline (2-AP), suggesting that additional genetic loci or alternative biosynthetic pathways independent of *BADH2* may contribute to aroma production in these lines.

3.3 CHEMISTRY AND BIOCHEMICAL BASIS OF RICE AROMA

Given the significance of aroma as a key quality attribute in aromatic rice, researchers in plant breeding and food chemistry have long focused on identifying and characterising the volatile compounds responsible for this trait. Rice aroma results from the combined presence of multiple volatile organic compounds (VOCs) that interact to produce its characteristic scent. Determination of these VOCs involves their extraction from the rice grain or the cooked rice and separation of volatile organic compounds (VOCs) using gas chromatography (GC), coupled with either flame ionisation detection (FID) or mass spectrometry (MS). The GC effluent directed towards MS for compound detection can also be diverted, with part of the flow directed to a sniffer port for sensory evaluation by human assessors (Champagne, 2008). Over 100 volatile compounds have been identified as contributors to the distinctive flavour of Basmati rice (Hussain et al., 1987), encompassing a diverse range of chemical classes, including 13 hydrocarbons, 14 acids, 13 alcohols, 16 aldehydes, 14 ketones, 8 esters, and 5 phenolic compounds. Among these, Buttery et al. (1988) identified several volatiles as major contributors to the aroma of cooked rice, such as 2-acetyl-1-pyrroline (2-AP), (E, E)-2,4-decadienal, nonanal, hexanal, (E)-2-nonenal, octanal, decanal, 4-vinylguaiacol, and 4-vinylphenol. Of these, 2-AP was determined to be the principal aroma compound.

Kim (1999) identified 16 hydrocarbons, 15 alcohols, 16 aldehydes and ketones, four acids, and 10 other compounds in aromatic rice. Paraffin was found to be the most common hydrocarbon compound, whereas n-pentanol, n-heptanol, and n-nonanal were found to be the most common alcohols. The presence of n-butanol and n-hexanol was observed in cooked rice only. Kim (1999) further reported no significant differences in hydrocarbon levels between aromatic and non-aromatic rice types; however, aromatic rice exhibited elevated levels of alcohols (notably *n*-pentanol, 1-octen-3-ol, menthol, and estragole), aldehydes, ketones (*n*-pentanal, *n*-heptanal, *n*-nonanal), and various acids and other volatiles. Notably, the concentration of 2-acetyl-1-pyrroline (2-AP) in aromatic rice was approximately 15-fold higher than in non-aromatic types. A Widjaja et al. (1996) analysis of volatile profiles across aromatic and non-aromatic cultivars revealed the presence of 70 distinct compounds. In addition to 2-AP, major volatiles included alkanals, alk-2-enals, alk-2,4-dienals, 2-pentylfuran, and 2-phenylethanol. Interestingly, non-aromatic varieties showed higher levels of *n*-hexanal, (E)-2-heptanal, 1-octen-3-ol, *n*-nonanal, (E)-2-octenal, (E, E)-2,4-decadienal, 2-pentylfuran, 4-vinylguaiacol, and 4-vinylphenol. In Basmati rice, 2-AP and 2-phenylethanol were the dominant aromatic volatiles, while in Jasmine and Goolrah varieties, indole was reported as a key compound in addition to 2-AP. Yang et al. (2008) analysed volatile aromatic compounds across six genetically distinct rice cultivars (including Basmati, Jasmine, two Korean Japonica lines, black rice, and a non-aromatic control) and identified 36 odour-active compounds (OACs) in cooked rice. Of these, 13—including 2-AP, hexanal, (E)-2-nonenal, octanal, heptanal, nonanal, 1-octen-3-ol, (E)-2-octenal, (E, E)-2,4-nonadienal, 2-heptanone, (E, E)-2,4-decadienal, decanal, and guaiacol—were determined to be key differentiators in the aroma profiles of the respective varieties.

Volatile compounds present at different crop growth phases in three rice genotypes (aromatic Basmati 370, and Ambemohar-157, and non-aromatic IR 64) were extracted and analysed by Hinge et al. (2016), who conducted a comprehensive volatile profiling of aromatic rice cultivars Basmati 370 and Ambemohar-157, alongside the non-aromatic variety IR 64, resulting in the identification of 88 volatile organic compounds (VOCs) spanning 13 chemical classes: Seven types of alkanes, six types of alkenes, 12 types of ketones, four types of aromatic hydrocarbon, 11 types of terpenes, 13 types of alcohol, 16 types of aliphatic aldehyde, three types of aromatic aldehyde, three types of N-heterocyclic compound, seven types of esters, four phenol containing compounds, one type of carboxylic acid, and one type of furan.

Among these, 14 odour-active compounds (OACs) were found exclusively in the aromatic cultivars. These included 2-acetyl-1-pyrroline (2-AP), eight aliphatic aldehydes (pentanal, hexanal, heptanal, octanal, (2E)-2-nonenal, nonanal, decanal, and (2E)-2-octenal), one aromatic aldehyde (phenylacetaldehyde), two alcohols (1-octanol and 1-octen-3-ol), one furan (2-pentylfuran), and one ketone ((E)-3-octen-2-one), highlighting the complex volatile composition playing a role in the distinctive aroma of aromatic rice cultivars. Among these 14 OACs, eight were common to both the aromatic varieties Basmati 370 and Ambemohar-157; while five were specifically detected in Ambemohar-157, and one compound, Camphene, only in Basmati 370. The 14 OACs detected only in aromatic rice varieties, along with their odor types, are shown in Table 3.3.

3.3.1 2-Acetyl-1-pyrroline (2-AP): The Main Volatile Compound Responsible for Rice Aroma

Buttery et al. (1983a) were the first to isolate and characterise 2-acetyl-1-pyrroline (2-AP) as a key volatile compound associated with the distinctive aroma of aromatic rice. They proposed that 2-AP plays a central role in imparting the characteristic

TABLE 3.3
Odour-Active Compounds Detected Specifically in Aromatic Rice Plants at Different Growth Stages

Compound	Odour Type	Aromatic Variety
2-AP	Nutty or popcorn-like	Basmati 370 & Ambemohar-157
2-acetyl-1H-pyrrole	Musty and nutty	-do-
β-ionone	Raspberry, floral	-do-
(E, Z)-2,6-Nonadienal	Green, metallic	-do-
ρ-xylene	Sweetish	-do-
Methyl 2-aminobenzoate	Sweet, grape fruity	-do-
Azulene	No odour	-do-
Acetic acid,1,7,7-trimethyl bicyclo (1, 2) hept-2-yl ester	No odour	-do-
Camphene	Woody, camphoraceous	Basmati 370
4-Methyldecane	Pungent, acrid	Ambemohar-157
4-Cyclopentylidane-2-butanone	Woody, fruity	Ambemohar-157
Toluene	Sweet, pungent	Ambemohar-157
1H-Indole	Floral, sweet, burnt	Ambemohar-157
Allylcyclohexane	Fruity, pineapple	Ambemohar-157

Source: Extracted from Hinge et al., 2016.

pandan- or popcorn-like scent observed in varieties such as Basmati 370 and Jasmine KDK 105. Subsequently, Paule and Powers (1989) confirmed the presence of 2-AP in Basmati 370 and established a correlation between its concentration and the typical aromatic profile of such rice cultivars. Hinge et al. (2016) conducted a comprehensive analysis of aromatic compounds, including 2-acetyl-1-pyrroline (2-AP) and its biochemical precursors, proline and methylglyoxal, along with other volatile compounds detected in both leaves and grains at different developmental stages of the aromatic rice cultivars Basmati-370 and Ambemohar-157. These profiles were contrasted with those of the non-aromatic variety IR-64 to elucidate differences in aroma-related metabolite accumulation.

Chemical structure of 2-AP: 2-acetyl-1-pyrroline is a nitrogen-containing heterocyclic compound characterised by a 1-pyrroline ring substituted at the second position with an acetyl group bearing a methyl ketone moiety (Figure 3.1). The presence of the pyrroline ring contributes to the compound's inherent volatility and chemical instability (Buttery et al., 1983a; Wakte et al., 2017).

Characteristics of 2-AP: 2-AP is a colourless to yellow solid and is categorised as a flavour and fragrance chemical compound (Ngadi and Yahya, 2014). It has a very low odour threshold, enabling its detection by humans even at extremely low concentrations—whether in field-grown plants, crushed leaf tissues, or rice grains at various stages, including pre-cooking, cooking, and post-cooking (Buttery et al., 1983a; Kovach et al., 2009).

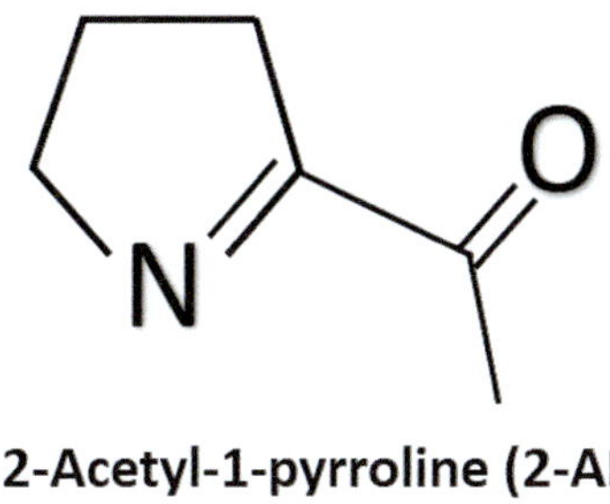

FIGURE 3.1 Chemical structure of 2-acetyl-1-pyrroline (2-AP).

3.3.2 Presence of 2-AP in Other Food Items

2-AP is also found in Pandan (*Pandanus amaryllifolius* Roxb.) leaves, commonly utilised as a seasoning in many south and southeast Asian countries when cooking non-aromatic rice in a way that introduces aroma. Buttery et al. (1983b) analysed Pandan leaves and found 2-AP to be a major volatile component in it. Pandan leaves contain significantly higher levels of 2-acetyl-1-pyrroline (2-AP), with concentrations approximately tenfold greater than those found in aromatic rice and nearly a hundredfold higher than in non-aromatic varieties. Due to their intense aroma, pandan leaves are commonly incorporated into a variety of culinary applications, including desserts, beverages, and confections. The presence of 2-AP also contributes to the distinctive flavour profiles of various foods such as popcorn, cheese, corn tortillas, green tea, and certain wines (Bradbury, 2009). 2-AP is also found in spinach, bread flowers, soybean, mung bean (*Vigna radiata*), sorghum, cucumber, and so on. The presence of 2-acetyl-1-pyrroline has also been documented in various cereal-based and plant-derived foods, including rye bread, boiled potatoes, and sweet corn confections, as well as in certain animal-origin products like heat-processed skim milk (Gaur et al., 2018).

3.3.3 Development of 2-AP in Plants and Grains/ Biosynthesis Pathway

2-AP is synthesised in all above-ground parts of the aromatic rice plant throughout its various growth stages. L-proline (an amino acid) is a precursor molecule of 2-AP and is related to the formation of 2-AP in aromatic rice (Yoshihashi et al., 2004). Hinge et al. (2016) analysed 2-AP content in aromatic rice plants at different growth stages and found it to be present at all seven stages from seedling, tillering, booting, flowering, milky grain, dough grain, and mature grains. The concentration of 2-acetyl-1-pyrroline peaked in fully developed grains, with comparatively lower levels observed during the booting stage.

3.3.4 2-AP Biosynthesis Pathway

Two pathways have been recognised for 2-AP biosynthesis in aromatic rices: (i) Enzymatic (BADH2-dependent) pathway, and (ii) Nonenzymatic BADH2-independent pathway. The biosynthesis of 2-acetyl-1-pyrroline (2-AP) predominantly follows a BADH2-dependent polyamine degradation pathway. In this route, polyamines such as arginine, ornithine, and spermidine are metabolised into gamma-aminobutyraldehyde (GAB-ald), which can spontaneously cyclise to form delta -1- pyrroline, an immediate precursor of 2-AP and a key determinant in controlling its synthesis rate (Chen et al., 2008). In aromatic rice, the inactive form of the BADH2 enzyme, encoded by the *osBADH2* gene, prevents the conversion of GABald into gamma-aminobutyric acid (GABA), thereby promoting delta -1- pyrroline accumulation and subsequent 2-AP production. In contrast, functional BADH2 in non-aromatic rice facilitates the conversion of GABald to GABA, reducing the availability of Δ^1-pyrroline and consequently limiting 2-AP formation (Bradbury et al., 2008). In non-aromatic rice, the active BADH2 enzyme, encoded by the *OsBADH2* gene, facilitates the metabolic conversion of gamma-amino butyraldehyde (GABald) into gamma-aminobutyric acid (GABA); hence, accumulation of 2-AP does not take place (Figure 3.2).

3.3.5 Alternate Pathway

In the non-enzymatic, that is, gene-independent pathway (Figure 3.2b), under stress conditions, glutamate serves as a precursor for proline biosynthesis. The accumulated proline is then transformed into delta -1- pyrroline-5-carboxylic acid (P5C) through the catalytic action of the enzyme delta -1- pyrroline-5-carboxylate synthase

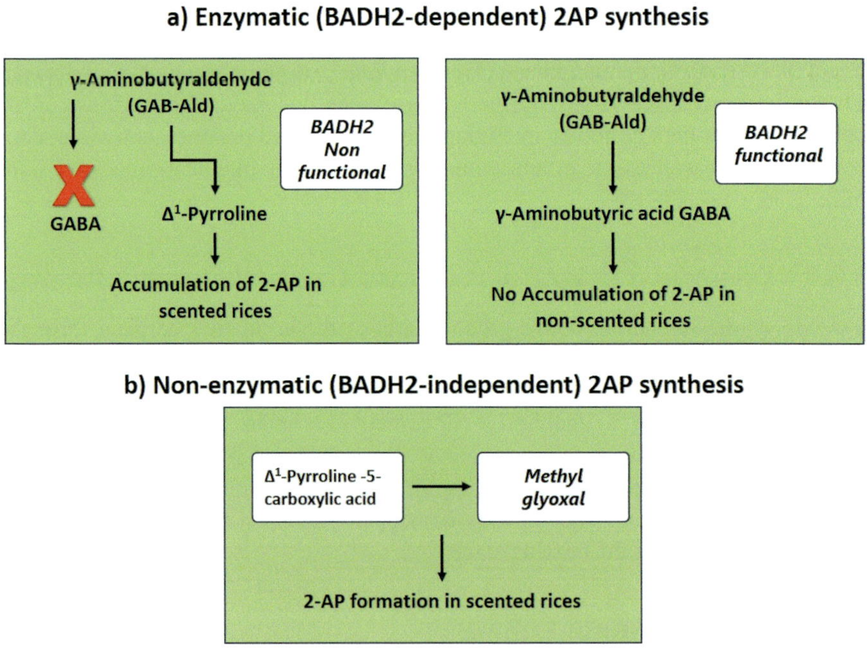

FIGURE 3.2 Rice Biosynthesis Pathway in Rice

Source: Varatharajan et al., 2021.

(P5CS). P5C can either react non-enzymatically with methylglyoxal or be converted into delta -1- pyrroline, both routes potentially contributing to elevated levels of 2-acetyl-1-pyrroline (2-AP) (Huang et al., 2008).

3.3.6 Mechanism of 2-AP Accumulation in Rice Grains

Two possible biosynthetic pathways have been proposed to explain the accumulation of 2-acetyl-1-pyrroline (2-AP) in mature grains of aromatic rice. One hypothesis suggests that 2-AP is biosynthesised in vegetative tissues such as leaves and stem sheaths and subsequently transported to the developing grains. The alternative pathway proposes that proline is mobilised from the leaves to the grains, where 2-AP is then synthesised locally (Mo et al., 2016). In support of the first model, Hinge et al. (2016) examined the temporal pattern of 2-AP accumulation in Basmati-370 and Ambemohar-157 across seven growth stages—from seedling to grain maturity—and observed peak 2-AP levels in mature grains, accompanied by reduced proline concentrations and lower expression of P5CS. These findings indicate that translocation of 2-AP from vegetative tissues may be the predominant mechanism in these cultivars.

3.3.7 Quantification of 2-AP in Basmati and Other Aromatic Rices

The growing global preference for aromatic rice cultivars has driven the advancement of analytical techniques to quantify 2-acetyl-1-pyrroline (2-AP) and differentiate aromatic from non-aromatic types (Champagne, 2008). These analytical workflows typically involve the extraction or collection of volatile organic compounds (VOCs), followed by their concentration, separation, and quantification. Conventional VOC extraction methods include static headspace (SH), purge and trap, steam distillation combined with solvent extraction (such as simultaneous distillation-extraction), and direct solvent extraction (Reineccius, 2006). Separation is usually performed using gas chromatography (GC), coupled with either flame ionisation detection (FID) or mass spectrometry (MS). In some protocols, part of the GC effluent is directed to an olfactory detection port, allowing sensory evaluation by trained personnel. Since the mid-1990s, solid-phase micro-extraction (SPME) combined with GC-MS has become a widely adopted method for accurate quantification of 2-AP (Wongpornchai et al., 2004; Hinge et al., 2016). 2-AP content in milled fragrant rice has been reported to vary depending on the method of extraction (Mathure et al., 2011). The commonly employed extraction methods by various researchers included (i) Simultaneous steam distillation-solvent extraction (Buttery et al., 1983a, Tava and Bocchi, 1999), (ii) Micro-steam distillation-solvent extraction (Tanchotikul and Hsieh, 1991), (iii) Direct solvent extraction (Bergman et al., 2000, Yoshihashi et al., 2002), (iv) Solid phase microextraction (SPME) – GC (Mathure et al., 2011), (v) Static headspace – GC (Sriseadka et al., 2006). Aromatic rice cultivars contain approximately 0.04–0.09 ppm of 2-AP, whereas non-aromatic cultivars have approximately ten times less, between 0.006 and 0.008 ppm (Buttery et al., 1983a). The concentration of 2-acetyl-1-pyrroline (2-AP) in milled Basmati rice, as measured through steam distillation techniques, has been reported to range from 0.06 (Buttery et al., 1983a) to 0.558 ppm

TABLE 3.4
Content of 2-AP in Milled Basmati Rice Grain Using Different Extraction Methods

Method of Extraction	2-AP Content (ppm)	References
Steam distillation	0.06	Buttery et al. (1983a)
"	0.588	Tava and Bocchi (1999)
"	0.061	Nadaf et al. (2006)
Solvent extraction	0.019–0.342	Bergman et al. (2000)
"	0.235	Yoshihashi et al. (2002)
Static headspace-GC	0.26–0.38	Sriseadka et al. (2006)
SPME/GC-FID	0.122–0.411	Mathure et al. (2011)

Source: Extracted from Mathure et al. (2011).

(Tava and Bocchi, 1999), with another study noting a value of 0.061 ppm (Nadaf et al., 2006). Variation in 2-AP content in milled Basmati based on different extraction methods is given in Table 3.4.

3.3.8 Quantification of 2-AP during Different Crop Growth Stages in Basmati

Hinge et al. (2016) quantified the 2-AP content in Basmati-370 and Ambemohar-157 aromatic rice varieties at seven different crop stages starting from seedling, tillering, booting, flowering, milky grains, dough grains, and mature grains and found that it continued to increase from seedling stage through tillering and booting stages and then decreased during flowering and milky grain stages. Thereafter, 2-AP content showed a steep increase during the dough grain stage and reached its maximum during the mature grain stage. The reason for the increase in 2-AP during the booting stage may be that the production of free sugar and starch is at its maximum during this stage (Yoshida, 1968). Subsequently, at the flowering and grain-filling stages, the previously accumulated starch and soluble sugars are mobilised and used for panicle and grain development, resulting in a reduction in 2-AP content (Perez et al., 1971). 2-AP, due to its volatility and lipophilic nature, binds with the lipid molecules and gets attached to starch granules and then gets translocated to grains from leaves (Yoshihashi et al., 2005). Hinge et al. (2016) also correlated the maximum 2-AP accumulation in ripened grains with a higher level of 2-AP-bound starch amount. Their results showed that the 2-AP amount in ripened grains can be enhanced by increasing its content at the booting stage of the crop. Hence, by taking suitable agronomic and physiological measures, like the addition of nitrogen at the booting stage of the crop, 2-AP levels in mature grains could be enhanced. Boontakham et al. (2019) also reported a similar trend of increase in 2-AP content in the case of Jasmine rice variety KDML 105, reaching its maximum during the booting stage, then decreasing during flowering and rising again throughout the grain filling stages. Quantification of 2-AP at different growth stages of the Basmati rice plant is given in Table 3.5.

3.4 POSSIBILITY OF AROMA IMPROVEMENT

3.4.1 Why Aroma Reduction?

The change in farming methods from traditional to modern, involving the application of nitrogenous fertilisers, the use of pesticides, and rising temperatures, has caused the loss in Basmati aroma (Singh et al., 2000b). Similar is the case reported with some other aromatic rice varieties. However, there are various methods involving genetic as well as cultural management practices that can help in improving rice aroma. Some of these are described next.

3.4.2 Gene Editing Technology

The development of new fragrant rice cultivars with higher yield, better aroma, and other characteristics using conventional or marker-assisted breeding approaches requires extended timeframes. Recently, gene editing technology has opened up new possibilities by which not only improvements in aroma and other grain qualities can be achieved through targeted mutagenesis, but aroma can also be introduced in a non-aromatic rice variety. Ashok Kumar et al. (2020) employed the CRISPR/Cas9 genome-editing system to generate novel allelic variants of *OsBADH2* with the aim of conferring aromatic traits to the high-performing, non-aromatic rice variety ASD 16. In the T1 progenies, the aroma was stably inherited, and the presence of novel aromatic compounds pyrrolidine, pyridine, pyrazine, pyridazine, and pyrazole was found in the grains, showing that the engineering of new allelic variants of *OsBADH2* has the potential to impart aromatic characteristics to otherwise non-aromatic rice

TABLE 3.5
Content of 2-AP (ppm) at Different Crop Growth Stages in Basmati and Ambemohar Aromatic Rice Varieties

Crop Growth Stage	2-AP Content Basmati 370	2-AP Content: Ambemohar 157
Seedling	0.20	0.20
Tillering	0.38	0.31
Booting	0.44	0.40
Flowering	0.16	0.16
Milky grain	0.20	0.20
Dough grain	0.26	0.58
Mature grain	0.45	0.66

Source: Extracted from Hinge et al., 2016.

varieties. CRISPR/ Cas9 gene editing technology has also been used to introduce new BADH2 alleles into the genomic composition of *japonica* and *indica* subspecies varieties for introducing aroma (Hui et al., 2022). Gene editing technologies are gaining traction for their ability to introduce highly targeted and precise genetic modifications in rice.

3.4.3 Use of Additional N Dose

Nitrogen plays an important role in supporting optimal crop development and yield, and is applied in its recommended doses, depending on the soil fertility conditions. However, applying additional N (30 and 60 kg/ha) at the booting stage has been shown to increase the proline content and the 2-AP contents in flag leaf, other leaves, and grains 17 days after flowering (Mo et al., 2018).

3.4.4 Use of Rhizobacteria for Aroma Improvement

Rhizobacterial isolates associated with rice, including strains of *Pseudomonas*, *Enterobacter*, and *Acinetobacter*, have been identified as potential producers of 2-AP. High 2-AP-producing isolates of these rhizobacteria have been shown to increase 2-AP levels in the grains of aromatic Basmati-370 (Deshmukh et al., 2016). The rhizobacteria-inoculated Basmati-370 was found to have about 1.14–1.41 times higher 2-AP levels in comparison to the non-inoculated plants. Chinachanta et al. (2021) also used plant growth-promoting rhizobacteria (PGPR) isolated from Thai Jasmine aromatic rice growing soil and observed that root inoculation with *Sinomonas* sp. ORF15-23 strain could lead to a maximum of 90% increase in 2-AP content in the leaves of KDML 105 Jasmine variety under salt stress conditions.

3.4.5 Use of Lanthanum

Soil application of Lanthanum (La), a rare metallic element, has shown varied impacts on the comprehensive growth and physiological development of rice plants. La application increased the photosynthetic rate, chlorophyll content in leaves, dry matter accumulation, and yield of aromatic rice. Its application has also shown improved 2-AP content in the leaves and grain by enhancing 2-AP biosynthesis through increased conversion of proline to 2-AP, and by reducing the conversion of GABald to GABA (Luo et al., 2021).

3.5 AREA SPECIFICITY AND INFLUENCE OF ENVIRONMENTAL FACTORS ON AROMA AND GRAIN QUALITY CHARACTERISTICS

Area specificity of aroma: Rice aroma is area specific. When Basmati is grown outside the Punjab region of Pakistan, it gives no aroma. Juliano (1972) reported that standard Basmati varieties grown in the Sindh

> province of Pakistan matured without aroma. Buu (2000) observed that when Basmati-370, introduced from India, was tested in different sites of the Mekong delta in Vietnam during the 1980s, it failed to retain its characteristic aroma and resulted in significantly reduced yield. According to Siddiq et al. (2012), the unique quality traits of Basmati rice are optimally expressed only when grown in the northwestern Himalayan foothills of the Indian subcontinent. In the case of Jasmine aromatic rice, Yoshihashi et al. (2004) reported that when Jasmine was grown in its traditional growing area, that is, the northeastern (NE) region of Thailand, it gave a high aroma, but in other regions of the country, its aroma was much reduced. Jasmine variety KDM-105, when grown in northeastern, central, and northern regions of Thailand, had the highest 2-AP content in the NE region (528–532 ppb), followed by the northern region (218 ppb). In the central region, 2-AP content was found to be the least (87 ppb). Similarly, Buu (2000) also reported that aromatic rice varieties of Vietnam, Nang Thom Cho Dao, and Tam Xaon expressed their aromatic traits exclusively in particular regions.

The area specificity of aroma in Basmati and other aromatic rices like Jasmine shows that though the aroma is likely influenced by a key gene, along with environmental factors like climate, soil characteristics, cultural practices, post-harvest storage, milling, and cooking conditions also influence the intensity of aroma and flavor. According to Tava and Bocchi (1999), variations in genotype, environmental conditions, and their interactions significantly influence the presence of 2-acetyl-1-pyrroline (2-AP), with notably high levels observed in both Basmati and certain Italian rice varieties. Prodhan and Sue (2020) hypothesised that the enzymatic processes associated with the main 2-AP biosynthesis pathway (polyamine degradation pathway) remain more active and produce more 2-AP when aromatic rice is grown at an optimum temperature range (20–30°C) and optimum environmental and growing conditions. A comprehensive analysis of the modulatory effects of environmental variables on aroma biosynthesis and grain quality parameters in aromatic rice genotypes is described below.

3.5.1 Climatic Factors

Three climatic parameters—photoperiod (sunshine hours), temperature, and humidity—have been found to affect flowering and aroma in pure Basmati rice varieties. During the vegetative stage, conditions of high relative humidity (70–80%) and temperatures between 25–35°C are considered optimal. In contrast, the flowering phase and subsequent grain development benefit from bright, clear days with temperatures ranging from 25–32°C, cooler nighttime temperatures (20–25°C), moderate humidity levels, and light wind movement—factors that collectively support effective grain filling and aroma expression in Basmati rice (Singh, 2000). Additionally, most Basmati varieties are naturally adapted to initiate flowering within a temperature window of 20–25°C (Ferrero 2004). Traditional Basmati varieties are photoperiod-sensitive, whereas newly evolved Basmati varieties are photoperiod-insensitive. However, these evolved varieties also require specific day length duration and cooler temperatures. Latif et al. (2019) demonstrated that Basmati genotypes, including advanced breeding lines, initiated flowering within a defined range of photoperiod and temperature conditions. Optimal flowering occurred when the day length was approximately 11 hours and 15 minutes, supporting the classification of Basmati as a short-day plant that responds favourably to shorter days and extended nights. The upper threshold for daily minimum temperature conducive to flowering was identified as 24°C, while the maximum day length allowing flowering was recorded at 12 hours and 18 minutes.

3.5.2 Influence of Temperature

In Pakistan, pure Basmati lines typically begin anthesis in the final week of September, coinciding with a drop in temperature below 25°C, and complete flowering by early October, before temperatures decline further below 20°C (Akhtar and Haider, 2020). Lower nighttime temperatures during the grain filling period have been associated with decreased accumulation of starch, proteins, and lipids in milled grains, while contributing to increased apparent amylose levels, greater cooked grain firmness, and elevated final and setback viscosities (Singh et al., 2014). Conversely, elevated temperatures during this stage can impair pollen viability and increase spikelet sterility, particularly in aromatic Basmati cultivars, thereby negatively impacting yield more than in non-aromatic types. Chakrabarti et al. (2010) studied the impact of temperature fluctuations on pollen viability and spikelet sterility in aromatic rice cultivars. Temperatures exceeding 33°C during anthesis were found to progressively increase pollen sterility across all rice varieties. At 35.5°C, the Basmati cultivar Pusa Sugandh 2 exhibited 17% pollen sterility along with a 26% decline in pollen germination. Reduced anther on the stigma at higher temperature, dehiscence, and less pollen deposition were reported to be the principal causes of sterility. Elevated temperatures during the grain-filling phase also led to a rise in spikelet sterility in Basmati, whereas the non-aromatic variety was found to be less affected. High temperature has been reported to shorten the grain-filling duration in Basmati rice from 32 to 26 days, leading to a yield decline of approximately 6%, along with reductions in total starch and amylose levels (Ahmed et al., 2015). The altered amylose-to-amylopectin ratio observed in the study was linked to decreased functionality of granule-bound starch synthase (GBSS), a pivotal enzyme catalysing amylose biosynthesis. Additionally, elevated temperatures during the ripening phase were associated with an increased incidence of unfilled grains in both the upper and lower sections of the panicle (Zakaria et al., 2002).

3.5.3 Soil Characteristics

Soil characteristics are believed to influence aroma and other quality characteristics, likely through their impact on nutrient dynamics and their interaction with soil-derived volatiles and aroma-related compounds. However, these effects have not been clearly characterised (Rohilla et al. 2000). Yoshihashi et al., 2004 studied the influence of soil texture and moisture conditions in Jasmine variety KDM 105 and observed that the 2-AP content varied depending on the soil and moisture conditions, and the content was very high (528–532 ppb) in sandy soils with dryland condition during the ripening phase; moderate (388 ppb) in clayey soils where soil moisture was maintained during the cropping period, and the least (218 ppb) in crop grown in flooded fields which remained submerged during the ripening phase. According to Saetung and Trelo-ges (2017), the reason for the high aroma in sandy soils is due to the higher likelihood of drought stress in such soils, which in turn activates proline synthesis to adapt and survive the moisture stress. With proline being a precursor of 2-AP synthesis, the process leads to increased 2-AP content and higher aroma.

3.5.4 Plant Nutrients

Soils rich in nitrogen are suitable for aroma development. Nitrogen has been shown to increase the amount of proline formation, which is a predecessor in 2-AP production (Yang et al., 2012), and hence increase in aroma. High potassium soils also produce higher aroma in rice, and also high whiteness in milled rice (Suwanarit et al., 1997). Zn deficiency is widespread in the soils of the rice-wheat cropping region of northern India (Prasad, 2005), where Basmati is grown. Singh et al. (2021) investigated how zinc fertilization combined with summer green manuring influences the productivity, zinc enrichment, and grain quality of Basmati rice within a rice-based cropping system. The use of EDTA–chelated zinc (containing 12% Zn) significantly enhanced grain yield and resulted in zinc-enriched grains with superior cooking characteristics. Kernel length increased by 0.27–0.86 mm in raw grains and by 0.29–1.15 mm after cooking.

3.5.5 Abiotic Stresses

Abiotic stresses like drought, salinity, submergence, and so on, generally reduce the rice crop yield and grain quality. However, in aromatic rice, drought stress at the grain formation stage has been shown to increase the 2-AP content, and so is the case with salt stress at the vegetative growth stage. The rise in 2-AP levels under stress conditions is attributed to the release of both starch-bound and free forms of the compound (Prodhan and Sue 2020). During stress, rice plants activate various defence responses involving compounds such as proline, methylglyoxal, γ-aminobutyric acid (GABA), and calmodulin, which function as key signalling molecules and stress protectants. (Tester and Devenport, 2003; Bartels and Sunkar, 2005). Mild soil salinity has been reported to increase the Basmati rice aroma. Basmati rice cultivation in *kallar* (salinity-affected) tract of Pakistan, Punjab has gained popularity due to excellent cooking quality and pleasant aroma (Latif et al., 2019). Shaheen Basmati, released by the Salinity Research Institute of Pakistan, is salinity tolerant, early maturing (95 days), with a yield potential of 60 md/acre.

3.5.6 Crop Husbandry/Cultural Practices

3.5.6.1 Transplanting Date

Transplanting rice too early or too late either prolongs or shortens the crop duration at the flowering stage and may affect grain yield and its quality. Mahajan et al. (2015) studied the optimisation of transplanting date for the newly evolved high-yielding Basmati varieties, Pusa Basmati-1121 and Pusa Basmati-1509, and compared them to the photo-period sensitive varieties Punjab Basmati-3 and Punjab Basmati-386. The research demonstrated that postponing the transplanting date to 25 July, in comparison with an earlier date of 5 July, improved water-use efficiency and enhanced grain quality attributes in both photoperiod-responsive and short-duration photoperiod-neutral Basmati cultivars. Notable improvements were observed in head rice recovery, amylose concentration, and cooked kernel length. The transplanting date also affects the nutritional, physicochemical, and milling quality of Basmati. Khalid et al. (2018) observed that delayed transplanting did not affect the flowering time, indicating that the rice lines under study were pure Basmati in nature. Maximum yield was obtained when Basmati was transplanted between 23 June and 14 July. Head rice recovery significantly increased by delaying the transplanting date. Protein and amylose content in the grain were highest when transplanted on 14 July. Mukesh et al. (2013) evaluated how different transplanting dates influenced the yield and grain quality parameters of Basmati rice varieties such as Taraori Basmati, Pusa Basmati-1, Pusa Sugandha-4, and CSR-30. Transplanting at earlier dates (25 June and 10 July) resulted in higher grain yield and improved yield-related traits compared to later transplanting on 25 July. Conversely, delayed planting enhanced quality parameters such as hulling efficiency, milling percentage, and head rice recovery. Among the varieties studied, Taraori Basmati and CSR-30 maintained consistent yields regardless of transplanting time, while Pusa Basmati-1 and Pusa Sugandha-4 experienced significant yield reductions under delayed transplanting conditions. Akhter et al. (2016) examined how transplanting dates affected the milling, cooking, and sensory qualities of both fine-grained Basmati and coarse-grained non-Basmati rice varieties in Pakistan. Their findings indicated that transplanting Basmati rice before early July minimised grain breakage during milling, with the highest head rice recovery recorded for crops transplanted on 18 June. Transplanting too early (in May) negatively impacted milling and cooking qualities in both rice types, while delaying transplanting until 10 July proved particularly detrimental to the processing and cooking traits of fine-grained Basmati varieties compared to non-Basmati types.

3.5.6.2 Shading/Solar Intensity

Shading reduces the availability of photosynthetically active radiation (PAR), potentially leading to stress in rice plants. Mo et al. (2015) investigated the effects of reduced light during the grain-filling stage on the accumulation of aroma-related compounds such as GABA, proline, and 2-acetyl-1-pyrroline (2-AP), as well as on grain yield and quality in aromatic rice varieties. Their study revealed that shading caused a notable decline in yield while simultaneously enhancing the levels of 2-AP and GABA in the grains. In the Yuxiangyouzhan aromatic rice variety, the 2-AP content in shading was found to be about two times more compared to the non-shaded environment.

3.5.7 Storage and Processing/Milling/Ageing Effects on Cooking Quality, Soaking Effect

3.5.7.1 Effect of Parboiling on Nutritional, Milling, and Cooking Quality

Parboiling has been found to transfer nutrients, particularly thiamine, from the bran layer into the endosperm, making white rice nutritionally closer to brown rice (Kyristi et al., 2011). This process is primarily undertaken to enhance milling yield, improve nutritional quality, and reduce spoilage caused by insects and molds during storage (Elbert et al., 2000). Akhter et al. (2014) evaluated milled and parboiled samples of various Basmati rice lines for nutritional, milling, and cooking quality attributes. Their findings indicated a significant improvement in mineral content, ash percentage, dry matter, and cooked grain length, along with increased levels of Vitamin B6 and better milling and head rice recovery in parboiled samples. Although reductions in crude fibre, protein, and fat were observed, these changes were not statistically significant. Moreover, parboiling led to a marked decline in quality-degrading factors such as grain breakage, bursting, and curling during cooking.

3.5.7.2 Soaking During Parboiling

Soaking is carried out during parboiling, in which the paddy is steeped in water to raise its moisture content. Soaking is essential for the gelatinisation of starch (Bhattacharya, 1985). The influence of soaking on the physical properties of Pusa Basmati-1121 was studied by Kale et al. (2017). Soaking demonstrated a significant effect on the physical characteristics of Pusa Basmati-1121. A soaking temperature of 65–70°C was found to be appropriate for soaking.

3.5.7.3 Storage of Milled Rice

Milled rice develops a "stale" flavor during storage. This stale flavor is due to the hydrolysis of surface lipids and the formation of free fatty acids, which are prone to oxidative degradation (Yamamatsu et al., 1966). Widjaja et al. (1996) reported a 40–50% decrease in 2-AP content across all processed forms of rice grain, including brown and white rice, after storage of 3 months, regardless of whether storage conditions involved atmospheric air or reduced-pressure (partial vacuum) environments. At higher storage temperatures, 2-AP content was found to decrease faster (Yoshihashi et al., 2005).

3.5.8 Influence of Atmospheric Pollution on Aroma and Other Quality Traits in Basmati

A tremendous increase has been observed in atmospheric pollution in recent years, which may also influence the aroma and other qualities of Basmati grain. Mina et al. (2018) were the first to investigate how atmospheric particulate matter (PM) deposition affects Basmati rice varieties, specifically Pusa Basmati-1509 and Pusa Sugandh-5, cultivated in the Indo-Gangetic Plains of India. Ambient particulate matter concentration at the experimental location was in the range of 162 to 660 micrograms/cubic meter during the crop-growing period from August to November. The crop was observed under three PM exposure conditions: low PM accumulation

Yano, M., E. Shimosaka, A. Saito, and M. Nakagahra. 1992. Linkage analysis of a gene for scent in the indica rice variety, Surjamkhi, using restriction fragment length polymorphism markers. *Japanese Journal of Breeding* 41: 338–339.

Yoshida, S. 1968. The accumulation process of carbohydrates in rice varieties in relation to their response to nitrogen in the tropics. *Soil Science and Plant Nutrition* 14: 153–161.

Yoshihashi, T. N., T.T. Huong, and H. Inatomi. 2002. Precursors of 2-AP, a potent flavour compound of an aromatic rice variety. *The Journal of the Science of Food and Agriculture* 50: 2001–2004.

Yoshihashi, T., T. Nguyen, and N. Kabaki. 2004. Area dependency of 2-Acetyl-1-Pyrolline content in an aromatic rice variety Khao Dak Mali 105. *The Japan Agricultural Research Quarterly (JARQ)* 38: 105–109.

Yoshihashi, T., T. T. Huong, V. Surojanametakul, P. Tungtrakul, and W. Varanyanond. 2005. Effect of storage conditions on 2-acetyl-1-proline content in aromatic rice variety Khao Dawk Mali 105. *Journal of Food Science* 70: S34–S37.

Zakaria, S., T. Matsuda, S. Tajima and Y . Nitta. 2002. Effect of high temperature at the ripening stage on the reserve accumulation in the seed in some rice cultivars. *Plant Production Science* 5: 160–168. https://doi.org/10.1626/pps.5.160

4 Remote Sensing Applications for Aromatic Basmati Rice Discrimination and Mapping

4.1 INTRODUCTION TO REMOTE SENSING

Remote sensing is the science and technology of acquiring information about objects or phenomena without direct physical contact, typically via satellite or airborne sensors. Evelyn Pruitt of the U.S. Office of Naval Research coined the term 'remote sensing' in 1960 to describe the acquisition of information from a distance. The instrument that is used in obtaining information about the object is referred to as a 'sensor'. However, how remote the instrument should be from the object is not specified, and it could vary from a few centimetres to several thousands of meters (Jensen, 2009). For example, the sensors for taking measurements could be placed on an aircraft, or satellites placed in space, or drones, or ground-based, as in a spectroradiometer. The majority of remote sensing instruments that collect data for Earth resources applications record the electromagnetic energy or radiation emanating from the Sun. However, there are other instruments/sensors (e.g., RADAR, LIDAR, etc.), which have their own energy source, and hence are not dependent on the Sun for sensing the object. Remote sensing operates on the principle that every object on the Earth emits, reflects, or backscatters electromagnetic energy differently, and by measuring those energies through a remote sensing instrument, we can identify and discriminate the object and know its properties.

4.1.1 Electromagnetic Radiation and Its Interaction with Earth's Surface Features

Electromagnetic energy from the Sun travels in the form of electromagnetic waves at a velocity of 3×10^8 m s^{-1} and reaches the Earth in about 8 minutes. The Sun emits a broad, continuous spectrum of electromagnetic radiation, extending from ultra-short, high-energy gamma rays and cosmic radiation to long-wavelength, low-energy radio waves (Jensen, 2009). The visible region of the spectrum, comprising blue, green, and red light in the 0.4–0.7 μm wavelength range, though it covers just a confined portion of the entire electromagnetic spectrum, receives 41% of the total

DOI: 10.1201/9781032659398-4

solar energy. This is the only region in which our eyes can see. The remaining 59% of the energy is received in the wavelength shorter than blue light (less than 0.4 μm), known as ultraviolet, and longer than red light (more than 0.7 μm), known as near- and far-infrared, microwave, and radio waves (Figure 4.1). The infrared range, which stretches from 0.7 to 30 μm wavelength, has been segmented into near infrared (NIR), mid- or short-wave infrared (SWIR), and thermal infrared (TIR). While the near-infrared includes energy with a wavelength of 0.7–1.3 μm, short-wave infrared covers 1.3–3 μm wavelength. The infra-red region from 0.7 to 3.0 μm is also called reflective infra-red, as the energy is reflected in this region. Thermal infrared covers 3 to 14 μm wavelength, where emission of electromagnetic energy takes place from objects on the Earth, maintaining a temperature greater than absolute zero (0 K). The microwave region, which consists of much longer wavelengths compared to the visible, near-infrared, or thermal infrared, and is generally measured in centimetres, not micrometres, covers wavelengths from 1 cm to 1 m. As the regions of the electromagnetic spectrum are identified by a beginning and ending wavelength, the wavelength interval is typically designated as a spectral band, channel, or a region in remote sensing terminology. A variety of detectors sensitive to different regions of the electromagnetic spectrum have been built for the collection of remotely sensed data. However, for studying Earth resources features, sensors in the visible, infrared (near, short-wave, and thermal), and microwave ranges are commonly used.

The Sun produces a continuous spectrum of energy ranging from gamma rays to radio waves. The wavelength increases from gamma-rays toward radio waves, whereas frequency increases in the reverse order. The visible region lies between the ultraviolet (UV) and the infrared (IR), beyond which is the microwave region. Imaging from remote sensing instruments is carried out in the visible, IR, and microwave regions of the electromagnetic spectrum.

Incident solar radiation is intercepted by the atmosphere, interacts with it, and then reaches the Earth's surface. In the atmosphere, solar energy is partially absorbed by various atmospheric gases—ozone, oxygen, carbon dioxide, and others. The presence of gas molecules, smoke, dust, and atmospheric water vapour also leads to scattering of the incident radiation. Reflection also occurs from the tops of clouds and various atmospheric constituents. The integrated effects of absorption, scattering, and reflection significantly diminish the intensity of solar radiation incident on the Earth's surface. A part of the solar energy reaching the Earth is reflected back from various surface features on Earth, whereas a part is absorbed by them and re-emitted. Modern remote sensing systems facilitate the acquisition of data on the absorbed, reflected, and emitted energy from Earth.

4.1.2 Absorption, Reflectance, Scattering, and Emission of Radiant Energy

Absorption is the mechanism through which incident radiant energy is assimilated by a material and subsequently converted into alternative energy forms (Jensen, 2009). Absorption of incoming solar radiation may occur within the atmosphere or at the

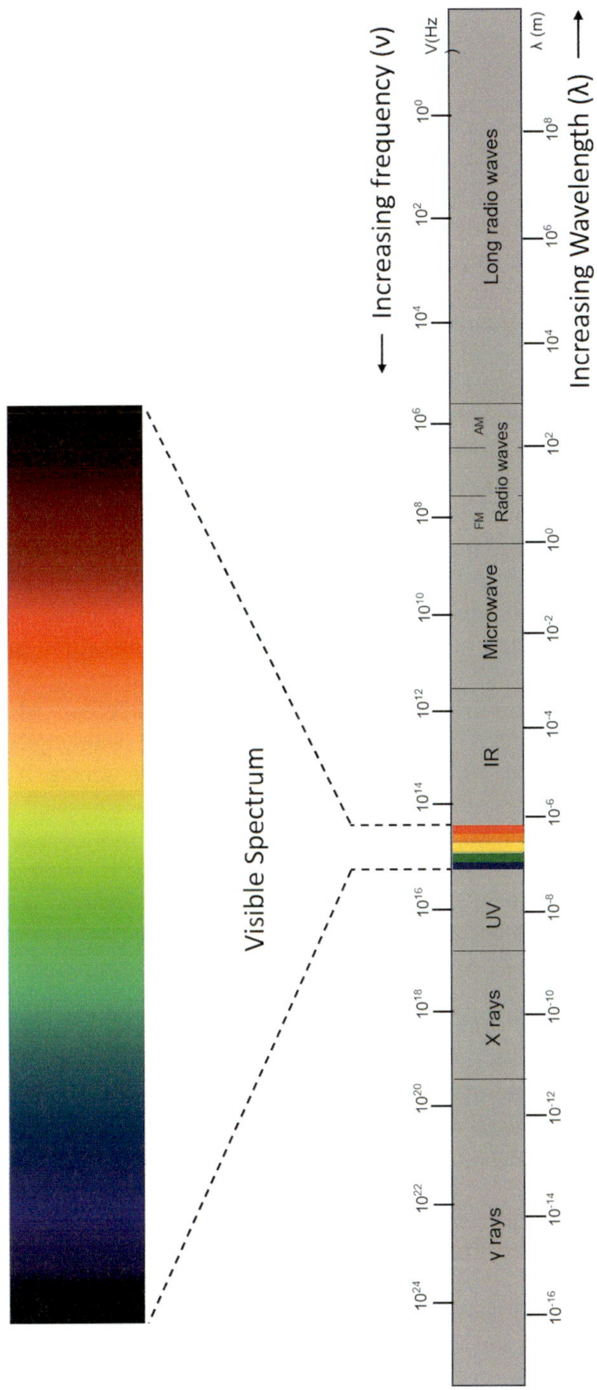

FIGURE 4.1 Electromagnetic spectrum, wavelength (m), and frequency (v).

Earth's surface. In the atmosphere, water, carbon dioxide, oxygen, ozone, and nitrous oxide are the main constituents that cause the absorption of radiant energy, resulting in closing down specific spectral regions utilised in remote sensing (for example, less than 0.3 μm, and between 14 to 30 μm), since no energy is available. Fortunately, there are some regions of the spectrum, mainly in the visible and part of the infrared region, where radiant energy is transmitted effectively to the ground. These regions are called 'atmospheric windows', where remote sensing is carried out. Reflectance is the process whereby radiation 'bounces off' an object like the top of a cloud in the atmosphere, a water body, vegetation, or bare soil at the Earth's surface (Jensen, 2009). Depending on the type of reflecting surfaces, reflection can be 'specular' as from a smooth surface like a calm water body, or 'diffused' as from a tree or crop canopy due to its stem, branches, leaves, and flowers. Figure 4.2 shows the behaviour of a perfect specular reflector, a near-perfect specular reflector, a perfect diffuse reflector, and a near-perfect diffuse reflector surface on the Earth. Reflectance in remote sensing is measured by multi-spectral sensors sensitive to the visible and reflective infrared ranges (0.4 to 3.0 μm wavelength) of the spectrum.

Scattering is the reflection of incident energy in all directions. Scattering is different from reflection in that scattering involves the redirection of energy in multiple, often unpredictable directions, whereas reflection occurs in a well-defined, predictable direction according to the laws of optics (Jensen, 2009). Scattering of solar energy occurs both within the atmosphere and at the Earth's surface. The intensity varies according to the size of the scattering particles of its interaction and the wavelength of light. Short wavelengths scatter more light compared to longer wavelengths. The scattering is of three types: Rayleigh, Mie, and non-selective. Mie scattering occurs when the size of the scattering particles exceeds the wavelength of the incident radiation, like smoke or dust particles present in the atmosphere. If the concentration of smoke and dust particles in the atmosphere is high, enhanced scattering of violet and blue light causes primarily orange and red light to be transmitted to the human eye. Rayleigh scattering occurs when the particle size is much smaller than the wavelength of the incident light, as is the case with atmospheric gases such as oxygen, nitrogen, and other trace gases. Thus, Raleigh scattering is responsible for the appearance of blue sky on a clear, cloudless day, due to the differential scattering of shorter wavelength blue light. Nonselective scattering takes place irrespective of wavelength, that is, all wavelengths experience equal scattering intensity. Cloud water droplets cause non-selective scattering across the visible spectrum, which accounts for the white coloration of clouds.

Emission of electromagnetic energy takes place by all surface features of the Earth having a temperature above absolute zero (0 K). The emission takes place in the 3–14 μm wavelength of the spectrum, also called as thermal infrared region. The thermal emission is not seen by human eyes but only felt when we are exposed to the hot Sun or are near a fire. The emission from the Earth peaks at 9.66 μm. Two primary thermal infrared windows are available, one at 3–5 μm and another at 8–14 μm. A substantial amount of thermal radiation emitted from the ground surface features is absorbed by the Earth's ozone layer due to an absorption band existing between 9.2 and 10.2 μm wavelengths; hence, satellite thermal sensors generally record emission in the 10.5–12.5 μm wavelength range. Thermal infrared sensors record the thermal characteristics (emissivity) of the landscape features.

4.2 REMOTE SENSING SYSTEMS

A remote sensing system includes a sensing instrument or 'sensor', a platform on which the sensor is deployed, collection and retrieval of the sensed data, and finally, its processing for object identification and discrimination. The remote sensing platforms could be orbital, like a satellite orbiting the Earth, or an aircraft, or an unmanned aerial vehicle (UAV), in which sensors are placed to collect and retrieve data from the Earth's surface (Figure 4.2). The advantages of the remote sensing

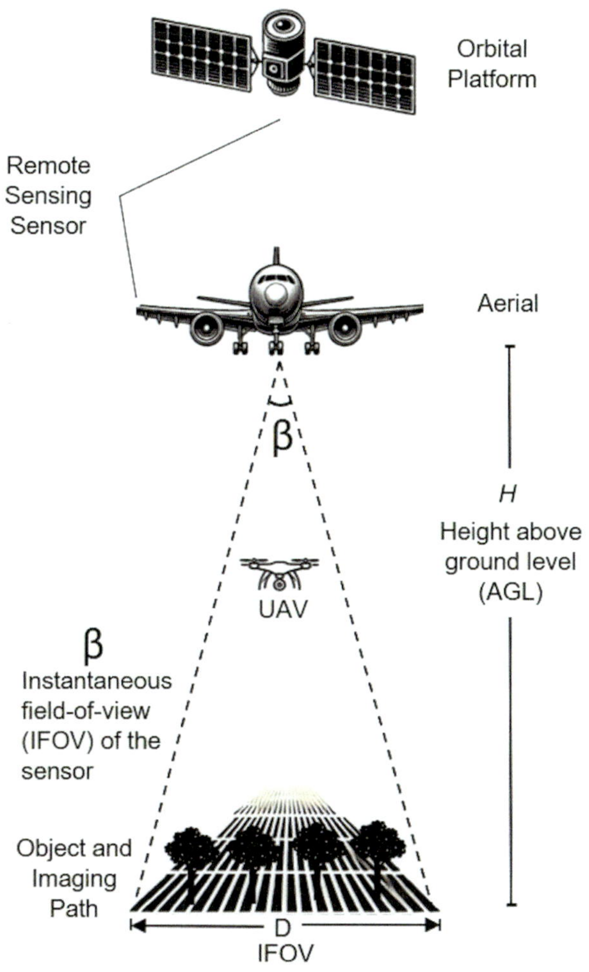

FIGURE 4.2 Remote sensing platforms and systems.

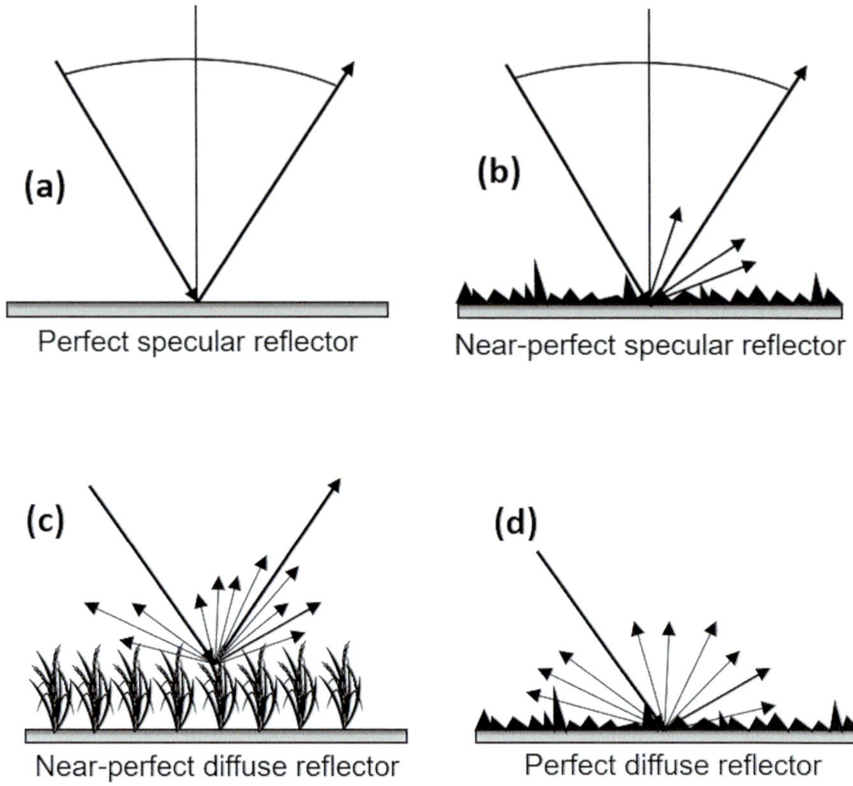

FIGURE 4.3 Specular and diffused reflectance from different types of earth surfaces. *Clockwise*: (a) Perfect specular reflection from a cool water surface, (b) near-perfect specular reflection from……, (c) Perfect diffuse reflection from a white paper or white powder, and (d) Near-perfect diffuse reflection from a crop canopy.

system have been its multi-approach capability (Colwell, 1973), for example multi-spectral sensing using different bands of the spectrum, multi-date (temporal) data collection, multi-satellite and multi-sensor data processing, multi-disciplinary (thematic) mapping of the same data by different thematic specialists like a geologist, a hydrologist, a vegetation analyst, an agronomist, a soil scientist, and others, to derive comprehensive information on the Earth's resources and its environment.

4.2.1 Remote Sensing Instruments (sensor systems)

Surface-reflected and thermally emitted electromagnetic radiation from the Earth can be recorded using various remote sensing instruments. These sensors can work either in the optical range (0.4–14 μm) or in the microwave range (1–100 cm wavelength) of the spectrum (refer Figure 4.1). However, before discussing the sensors working in the optical and microwave ranges, it is necessary to understand various terminologies related to remote sensors, as follow.

4.2.1.1 Terminologies Commonly Used in Describing Sensor Systems

Spatial resolution: Spatial resolution defines the minimum ground area that can be individually resolved by a sensor. In satellite remote sensing, it is also called a 'pixel' or picture element, which represents the minimum resolving area in terms of square meters (or in square centimeters in some recent satellite sensors). The spatial resolution of the first civilian satellite, Landsat-1, was 79 × 79 m, whereas the currently orbiting Landsat-9 has a spatial resolution of 15 m in the panchromatic band, and 30 m in multispectral bands. The spatial resolution of the Indian Remote Sensing Satellite Cartosat-2 is 1 m, which means that the sensor can resolve an object of 1 x 1 m size.

Radiometric resolution: Radiometric resolution defines a sensor's ability to discriminate between subtle differences in the intensity of reflected or emitted electromagnetic energy, emitted, or back-scattered energy from the Earth features. It is commonly termed the quantisation level. Higher radiometric resolution means an increase in the accuracy of signal recording. Radiometric resolution is often expressed as the number of bits per pixel, generally spanning from 8 to 16 bits. The radiometric resolution in some of the earlier satellites like Landsat-4 and -5 was 8-bit, which means brightness values recorded by their sensors were on the scale of 2^8 (i.e., 256), ranging from 0 to 255. Recent Landsat-9 OLI-2 has a 14-bit quantisation. IRS Resouresat-2 LISS-3 and -4 sensors have 10-bit quantisation, whereas AWiFS has 12-bit. The Sentinel-2 MSI sensor also has 12-bit quantisation, meaning that brightness values are recorded in the range of 0–4095. The recorded pixel intensity by the sensor is also known as the Digital Number (DN).

Spectral band: Spectral band refers to a specific range of wavelengths of the EM portion of the electromagnetic spectrum from which the sensor acquires information about the target object. Each band represents a particular type of radiation—visible, infrared, microwave, and so on. However, each type of radiation may further be subdivided into discrete bands, that is, visible radiation is further subdivided into blue, green, and red bands; infrared radiation into near-infrared, mid-infrared, and so on. Each band represents a specific wavelength range, which may slightly vary in different satellite sensors (refer to Table 4.3).

Spectral resolution: Spectral resolution is the size or dimension of the wavelength interval or bandwidth of a sensor. For example, the Landsat-1 MSS band 4 worked in the spectral range of 0.5–0.6 μm, so its spectral resolution is 0.1 μm, whereas band 7 worked in the range of 0.8–1.1 μm, and its spectral resolution is 0.3 μm.

Temporal resolution: Temporal resolution refers to how often the sensor records the image of a particular area. It is also called the repetivity of the imaging sensor. However, there are now several remote sensing satellites having pointable or movable sensors which can obtain images not only in vertical or nadir viewing, but also off-nadir viewing and can image the same area earlier than the repeat cycle of the satellites.

4.2.1.2 Optical Sensor Systems and Aerial Camera

Among the optical sensors, the camera has remained one of the most reliable remote sensors until now, collecting data across the visible region and a minor portion of the near-infrared spectrum. Cameras are optical systems and could have analogue (which uses a photographic film) or digital format. An aerial camera, used extensively until the mid-twentieth century, records the reflected energy received from the Earth's objects via the camera lens and projected onto the photographic film. The sensitivity of the panchromatic or colour film used in an analog aerial camera is the entire visible light (blue, green, and red colour, covering 0.4 to 0.7 μm). Later, during World War II, several developments in aerial photographic techniques took place, the prominent being the use of colour infrared film for the detection of camouflage of various military installations. It may be noted that the first false-colour infrared film was patented by the Eastman Kodak Company, USA, in 1942 (Jensen, 2009). The sensitivity of the colour-infrared film is blue, green, red, and near-infrared region (covering 0.4 to 0.9 μm wavelength) of the EM spectrum. While the traditional panchromatic / colour/colour IR films provide a single frame aerial photograph of 23 × 23 cm size covering their entire spectral sensitivity range, the multi-band camera, which arrived subsequently, started providing photographs of terrain simultaneously in different regions or bands of the visible and near IR range. For example, a Hasselblad multiband aerial camera simultaneously photographs the same area in blue (0.4–0.5 μm), green (0.5–0.6 μm), red (0.6–0.7 μm), and near-infrared (0.7–1.0 μm) bands and provides four 70 mm photographs in a single frame. Digital aerial cameras use a charge-coupled device (CCD) linear or area arrays, instead of film, to collect images of the Earth's features. Light reflected from surface features is received by a detector array, either linear or two-dimensional, and gets converted to a digital brightness value. The brightness value may range from 0–255 in 8-bit data to 0–4095 in 12-bit data. These brightness values can be stored in a matrix format (rows and columns) and read by computer systems (Jensen, 2009).

4.2.1.3 Optical Sensors in Remote Sensing Satellites

Optical sensors commonly used in remote sensing satellites can be classified into panchromatic (PAN) and multispectral imaging systems.

Panchromatic imaging systems: Panchromatic (PAN) sensors record the reflected energy from Earth features in the visible portion of the spectrum in a single band. The image obtained is in shades of grey, that is, black and white. PAN sensors are being used in currently orbiting satellites, like Landsat 9, and Cartosat-3 of the IRS.

Multispectral sensor systems: Multispectral sensors detect and record the reflected or emitted energy originating from a surface or object in multiple bands or parts of the electromagnetic spectrum. For example, the visible spectrum, in which our eyes see any object, can be split into seven distinct colour bands (VIBGYOR). Similarly, infrared spectrum can have sensors in near infrared (NIR), shortwave infrared (SWIR), and thermal infrared (TIR) bands. In the first Earth resources satellite, Landsat 1, the multispectral sensor called MSS had four bands: two in the visible green and red, and two in the near infrared region. The band designation and their bandwidths were: band 4 (0.5–0.6 μm), band 5 (0.6–0.7 μm), band 6 (0.7–0.8 μm), and band 7 (0.8–1.1 μm). With the new advancements in remote sensing technology and the user demand for better information and higher accuracy, the number of bands in recent satellites has increased significantly. For example, Landsat 9, launched in 2021, has 10 bands in the visible, NIR, SWIR, and TIR regions of the spectrum, apart from a PAN band. The European Commission's Sentinel-2 satellite, launched in 2016, has a multispectral imager (MSI) working in 13 spectral bands in the visible, NIR, and SWIR regions, covering the 0.443– 2.190 μm range.

Recording of the incoming radiation by these multispectral sensors involves utilising a scanning mirror system to continuously scan the Earth's surface either along or across the flight direction. During scanning, energy reflected or emitted from the Earth's objects is received by the detectors of these sensors and converted into an electronic signal as a brightness value or digital number. The number of detectors varies based on the spectral band configuration of the sensor. These scanning mirrors with discrete detectors have been used in Landsat MSS and TM sensors. Some other multispectral satellite sensors may use linear arrays of detectors, as in IRS Linear Imaging Self-scanning Sensors (LISS-I to LISS-IV; Joseph and Jeganathan, 2020), SPOT HRV sensors, and Sentinel-2 Multispectral Imager (MSI). The digital data collected by the sensors is stored in a matrix format. Each pixel's brightness level is located in the form of rows and columns for each band. While traditionally stored as 8-bit values (0–255), digital brightness levels are now frequently captured with higher bit depths, including 10-, 11-, and 12-bit resolutions. Details of different multispectral sensors available in Landsat, Indian Remote Sensing Satellite (IRS), Terra and Aqua, and Sentinel satellites that have been commonly used in rice mapping are summarised in Table 4.1.

4.2.1.4 Hyperspectral Systems

Unlike the multispectral remote sensing systems with generally about a dozen bands, hyperspectral systems record reflected or emitted energy in hundreds of bands simultaneously. Because of their requirement for concurrent imaging in adjacent spectral bands, these sensors use a two-dimensional area array of detector elements. Hyperion EO-1 was the first spaceborne hyperspectral mission launched by NASA in 2000 that provided publicly available data for studying crops and vegetation. Subsequently, PROBA-1 was launched by the European Space Agency in 2001. Recent hyperspectral satellite missions include PRISMA (PRecursore IperSpettrale Della Missione Applicativa), launched by Italy in 2019, which provides data in 239 spectral bands ranging between 400 to 2500 nm wavelength; EnMAP (Environmental Mapping and Analysis Program), launched by Germany in April 2022, covering 240 contiguous bands; Indian HySIS (Hyper-Spectral Imaging Satellite) launched in 2018; and HyspIRI (Hyperspectral Infrared Imager) launched by the USA in 2024. The use of hyperspectral data from the PRISMA satellite in mapping Basmati and other HYV rice has been discussed in Chapter 6 of this book.

TABLE 4.1
Remote Sensing Satellites and Sensors Commonly Used in Rice Crop Mapping and Monitoring

Satellite	Sensor	Spatial resolution (m) and spectral region	No. of bands	Temporal resolution (days)
Satellites with optical sensors				
Landsat 1 to 9	MSS (Landsat 1 to 3)	**MSS**-79 m in VNIR;	4	16
	MSS &TM (Landsat 4, 5)	**TM**-30 m in VNIR & SWIR; 120 m in TIR;	7	
	ETM+ (Landsat 7)	**ETM+** 30 m in VNIR& SWIR; 60 m in TIR;15 m in Pan;	89	
	OLI & TIRS (Landsat 8)	**OLI/ OLI2**- 30 m in VNIR, SWIR; 15 m in Pan	2	
	OLI-2 & TIRS-2 (Landsat 9)	**TIRS/ TIRS2**-100 m in TIR		
Terra/Aqua	MODIS	**MODIS**: 250 m in VNIR; 500 m in VIS, SWIR; 1000 m in VNIR, SWIR, TIR	36	1-2
SPOT	HRV (SPOT1-3)	**HRV**: 20 m in VNIR, SWIR, 10 m in Pan		261(VGT)
	HRVIR (SPOT-4, 5)	**HRVIR**: 10 m in VNIR, 20m in SWIR, 2.5 m in Pan		
	VGT (SPOT-4, 5)	**VGT**: 1150 m in VNIR, SWIR		
IRS-1A, 1B	LISS-1	**LISS 1**- 72.5 m in VNIR	4	22
	LISS-II	**LISS II**- 36.25 m in VNIR		
IRS- 1C, 1D	LISS-III	**LISS III**- 23.5 m VNIR, 70.5 m SWIR	4	24
	PAN	**PAN**- 5.8 m in VIS	1	5
	WiFS	**WiFS**-188 m in VNIR		5
IRS Resourcesat-1, 2 & 2A	LISS-III	**LISS III**- 23.5 m VNIR	4	24
	LISS-IV	**LISS IV**- 5.8 m in VNIR, 23 m in SWIR	4	5
	AwiFS	**AWiFS**- 56m in VNIR, SWIR	3	5
Sentinel- 2A, 2B	MSI	**MSI**-10 m in VNIR, 20 m in RE, SWIR, 60 m in NIR, SWIR-Cirrus	13	5
Satellites with microwave sensors				
ERS-1, 2	C-band SAR, VV	30 m		35
ENVISAT	C-band ASAR, VV, HH, VV/VH, HV/HH, VH/VV	20–500 m		35

RADARSAT-1	C-band SAR, HH	10–100 m, depending on operating modes	24, revisit 1–6 days
RADARSAT-2	C-band SAR, Quad-pol (HH, VV, VH, HV)	3–100 m, depending on operating modes	24
ALOS-2	PALSAR-2	25 m	14
RISAT-1	C-band SAR, single/dual/ circular polarisations	1–50 m, depending on operating modes	25, revisit 3–4 days
Sentinel-1A, -1B, -1C	C-band SAR, VV, HH, HH+HV, VV+ VH	5–40 m	12

Note: ***MSS***: *Multispectral Scanner;* ***TM***: *Thematic Mapper;* ***ETM+***: *Enhanced Thematic Mapper Plus;* ***OLI***: *Operational Land Imager; TIRS: Thermal Infrared Sensor;* ***HRV***: *High Resolution Visible;* ***VGT***: *VEGETATION;* ***HRG***: *High-Resolution Geometric Imaging Instrument;* ***HRS***: *High-Resolution Stereoscopic Imaging Instrument;* ***MSI***: *Multi-Spectral Instrument;* ***WiFS***: *Wide Field Sensor;* ***AWiFS***: *Advanced Wide Field Sensor;* ***LISS***: *Linear Imaging Self-Scanning Sensor.* ***VIS***: *Visible spectral range;* ***VNIR***: *Visible & Near infrared range;* ***SWIR***: *Short-wave Infrared range;* ***TIR***: *Thermal infrared range;* ***Pan***: *Panchromatic;* ***RE***: *Red Edge range.* ***SAR***: *Synthetic Aperture Radar;* ***ASAR***: *Advanced Synthetic Aperture Radar;* ***HH***: *Horizontal- Horizontal polarisation;* ***HV***: *Horizontal-Vertical polarisation;* ***VV***: *Vertical-Vertical polarisation.*

4.2.1.5 Microwave Sensors

Microwave wavelengths range from approximately 1 to 100 cm, making them significantly longer than those in the visible, near-, mid-, or thermal infrared portions of the electromagnetic spectrum. Sensors in the longer wavelength microwave region can penetrate clouds, rain, haze, and so on, and are not affected by atmospheric scattering. Hence, these sensors are suitable especially for mapping Basmati rice, which is grown during the rainy or *kharif* season (June–October) in India and Pakistan, when data collection from optical sensors is hampered severely due to the clouds and rain. The sensors in the microwave region are classified as active and passive. Passive sensors record natural microwave radiation at a wavelength of 1 cm to 15 cm (frequency 5 to 100 GHz). These sensors, which record the emitted microwave energy, are called a 'microwave radiometer'. An active microwave sensor has its own microwave radiation source. When a microwave beam or pulse is sent from an aircraft or satellite platform to the target (Earth surface features), the transmitted beam is returned back to its antenna after hitting the target. The returning signal, known as backscatter, is captured by the sensor's receiver. Active microwave sensors are categorised as imaging or non-imaging, with RADAR (Radio Detection and Ranging) being the most widely used imaging type. The non-imaging type sensors include altimeters and scatterometers. RADAR transmits long-wavelength microwaves in a 3 to 25 cm range, and is most widely used in Earth resources observation. The other active imaging remote sensing systems are LIDAR, an acronym for Light Detection and Ranging, which uses a shorter wavelength laser beam (1.4 μm), and SONAR (Sound Navigation and Ranging), using sound waves for measuring the water column. LIDAR and SONAR are used for mapping topography and bathymetry, respectively.

A RADAR remote sensing system comprises a pulse generator, transmitter, antenna, receiver, and supporting electronics for signal generation and data processing. The transmitter sends out brief microwave pulses via the antenna at regular intervals. These pulses strike the Earth's surface obliquely, perpendicular to the flight path of the platform. The time lag between the transmitted and returned signals is used to determine the target's distance and position. The motion of the aircraft or satellite and simultaneous recording of the backscatter signal generates a two-dimensional image of the terrain surface. In the RADAR remote sensing instruments, a synthetic aperture radar (SAR) is used instead of a real-aperture radar (the word 'aperture' means antenna). This is because a very large sized antenna will be required in a real aperture radar to get an improved resolving power from the radar sensor. Hence, engineers have designed a procedure to recreate a long antenna using electronic methods so that the antenna length can be kept manageable to use it in an aircraft or a satellite system. These radar instruments are, therefore, called synthetic aperture radar (SAR). The radar instruments used in remote sensing satellites are the SAR.

4.2.1.5.1 Radar Band, Wavelength, and Frequency

Radar transmits pulses of electromagnetic energy with a specific wavelength and duration through its antenna. Therefore, radar is conventionally described in terms of its wavelength. Based on their wavelength, radars used in remote sensing have been designated as K, K_a, K_u, X, C, S, L, and P; K band having the shortest (0.86 cm) to P having the longest (68 cm) wavelength. However, among the SAR sensors used in remote sensing satellites, the most commonly used band has been the C-band, followed by X and L. K-band is used in ground-based weather radars for tracking cloud cover and rainfall. Radar wavelength is also described in terms of its frequency, which is expressed in units of Gigahertz or GHz (10^9 cycles sec^{-1}). Wavelength and frequency have an inverse relationship with the speed of light, and hence, with increasing radar wavelength, the frequency decreases. SAR band designation used in different satellites/sensors and their wavelength and frequency are described in Table 4.2.

4.2.1.5.2 Radar Polarisation

Polarisation is caused due to the wave nature of the electromagnetic radiation. The sunlight is polarised weakly, but when the sunlight strikes a non-metal object on the Earth's surface, like grass, trees, water bodies, soil, and so on, it becomes depolarised, and the incident energy is scattered in different directions. Hence, only polarised light is sent and received by the radar antenna, which means the electromagnetic pulse is linearly polarised, with its electric field oscillating in a single, fixed plane perpendicular to the direction of wave propagation (Jensen, 2009). The radar antenna can emit polarised signals with the electric field oriented either in the vertical or horizontal plane, depending upon the type of filter used (vertical or horizontal). After the beam sent by the radar sensor interacts with the earth surface features and is returned back, it is called its 'backscatter'. This backscatter energy is again passed through a filter (vertical or horizontal) before being recorded at the radar antenna. Thus, there could be the following four possibilities of sending and receiving a polarised radar signal:

TABLE 4.2
SAR Bands, Wavelength, and Frequency Used in Some Remote Sensing Satellite Sensors

Satellite	Country	Launch Year	SAR Band	Polarisation	Wavelength (cm)	Frequency (GHz)
ERS-1	ESA	1991	C	VV	5.6	5.3
ERS-2	ESA	1995	C	VV	5.6	5.3
RADARSAT-1	Canada	1995	C	HH	5.6	5.3
RADARSAT-2	Canada	2007	C	HH, VV, HV, VH	5.55	5.405
Envisat (ASAR)	ESA	2002	C	HH, VV	5.6	5.3
JERS-1	Japan	1992	L	HH	23.5	1.275
COSMO-SkyMed	Italy	2007	X	HH, VV, HV, VH	3.12	9.6
TerraSAR-X	Germany	2007	X	HH	3.11	9.65
RISAT-1	India	2012	C	HH, HV	5.5	5.35
Sentinel-1A	Europe	2014	C	VV, VH	5.55	5.405
Sentinel-1B	Europe	2016	C	VV, VH	5.55	5.405

i. Vertically transmitted and vertically received polarisation (VV)
ii. Horizontally transmitted and horizontally received polarisation (HH)
iii. Vertical transmission with horizontal reception (VH)
iv. Horizontal transmission with vertical reception (HV)

Radar images generated using HH and VV configurations are referred to as co-polarised or like-polarised outputs. On the other hand, cross-polarised images are obtained from HV and VH configurations, where the transmitted and received polarisations differ. The use of different polarisations in radar imaging helps in getting additional information about the Earth's surface features.

4.2.1.5.3 Radar Backscatter Coefficient (σ^0)

The intensity of energy returned to the radar antenna from a specific area on the surface of the Earth is defined as radar backscatter coefficient (also termed sigma naught or σ^0). The backscatter coefficient is a dimensionless parameter that describes the scattering properties of all targets within a single pixel or ground resolution cell. It is influenced by factors such as surface topography, texture or roughness, moisture levels, and radar system characteristics including wavelength, incidence angle, and polarisation (Leberl, 1990). It is typically represented on a logarithmic scale in decibels (dB), with values generally ranging between −5 and +40 dB. The environmental parameters responsible for backscattering phenomena within a ground resolution cell (pixel) are surface roughness and electrical characteristics of the material (dielectric constant).

4.2.1.5.4 Effects of Terrain Structure and Surface Texture on Radar Backscatter

A rough surface will scatter more energy as compared to a smooth surface; hence, an urban area, due to its buildings of different heights, appears as a bright feature on the radar image. A smooth surface, like an unploughed dry soil, will scatter less and reflect more energy, resulting in a darker tone. However, since backscatter also is influenced by the wavelength and depression angle (angle between the radar pulse from antenna and the horizontal plane) of the radar, the same rough surface appearing bright in short wavelength (C- band) may appear dark in a long wavelength (L-band) radar image, as it will scatter less energy toward the radar antenna and reflect most of it (behaving like a near-perfect specular reflector). A criterion developed by Mikhail et al. 2001 to predict the strength of radar return (weak, intermediate, or strong) from a surface based on its relief is as follows:

- If the local topographic variation is less than one-eighth of the radar signal's wavelength, a weak return is produced because more of the incident energy will be specularly reflected away from the antenna, and very little will be returned to the antenna.
- If the local relief is more than one-half of the incident radar wavelength, a strong radar return will reach the antenna.
- If the local terrain variation lies within one-eighth to one-half of the radar's incident wavelength, it works as a diffuse reflector, and energy will be dispersed in all angles, and a moderate return will reach the antenna.

4.2.1.5.5 Influence of Electrical Characteristics (dielectric constant) of Earth Objects and Moisture Content

When a radar pulse is sent from an aircraft or spacecraft, different materials present on the Earth's surface conduct this electricity differently. The complex dielectric constant is a key parameter that describes a material's ability to transmit and store electrical energy. It reflects the electromagnetic properties of natural surfaces such as vegetation, soil, rock, water, and ice (Jensen, 2009). Dry materials like soil, rock, and vegetation typically exhibit dielectric constant values between 3 and 8, whereas water exhibits a much higher value, around 80. As a result, variations in moisture content within soil, rock surfaces, or plant tissues can greatly influence the intensity of radar backscatter (Mikhail et al., 2001). A moist soil reflects more radar energy compared to a dry soil, and hence, radar images have been used to measure the moisture content of surface soil. However, the soil moisture content also affects how deeply the radar signal penetrates the surface. A soil with high moisture content does not allow the radar to penetrate deeper than a few cm, and backscatters the energy, resulting in a brighter image. In the case of a dry soil, the depth of penetration would be almost equal to its wavelength. Hence, the depth penetration of C-band radar is much lower 'as compared to' L-band radar.

4.2.1.5.6 Response of Vegetation to Active Microwave Remote Sensing Data

In contrast to optical remote sensing, which primarily captures energy interactions, such as reflection, absorption, transmission, or scattering, from the uppermost layers of vegetation (leaves and stems), active microwave systems like SAR have the ability to penetrate the vegetation canopy. The depth of penetration depends on factors such as the radar's frequency (wavelength), polarisation, and incidence angle. When the incoming radar pulse falls on the top of a vegetation canopy, scattering takes place from the top layers (leaves, stems, etc.), which is called surface scattering. When the pulse is scattered multiple times, for example, from leaves to stems to branches to the trunk of a tree, it is called volume scattering. Similarly, surface and volume scattering can also occur from the soil beneath the vegetation canopy. Radar systems operating at longer wavelengths can penetrate deeper into the vegetation canopy. In the shorter wavelength radar (X- or C-band), surface scattering contributes more prominently. In the longer wavelength radar (L- or P-band), greater penetration leads to greater volume scattering from the canopy.

Polarisation plays a role in determining how radar signals interact with vegetation. Like-polarised radar signal (HH or VV) receives backscatter from single reflections from canopy components (surface scattering). A cross-polarised radar signal penetrates the canopy and receives backscatter after multiple reflections from canopy components (volume scattering). SAR data has been found advantageous in providing information on the following biophysical parameters of vegetation: vegetation type, canopy water content, canopy structure, leaf area index, biomass, and canopy structure (Carver, 1988).

4.2.2 Remote Sensing Platforms

Sensors for remote sensing of earth resources are commonly placed in an aircraft, spacecraft, unmanned aerial vehicle (UAV), or even at a few meters above ground (refer to Figure 4.2).

4.2.2.1 Aerial Platforms

An aircraft as a remote sensing platform has been in use for more than a century now. However, the first photograph from an aerial platform was taken by Gaspard Felix Tournachon (who called himself 'Nadar'), using a balloon as early as 1858. After the first flight of an aeroplane by the Wright brothers in 1903, the first known instance of aerial photography from an aircraft occurred in 1908, when L. P. Bonvillain, a Pathe motion-picture photographer, captured images alongside Wilbur Wright during a flight over Camp d'Auvours in France (Jensen, 2009). During World Wars I and II, the use of aerial photographs for military reconnaissance became common. During World War II, the colour-infrared (CIR) film for camouflage detection was developed by the Eastman Kodak Company in 1942, and aerial photography with CIR, camouflage detection film, gained significance. Subsequently, during the Cold War period, rivalry between USA and the Soviet Union led to the development of highflying U-2 aircraft by the USA in 1956, that could fly at an altitude of 70,000 ft (21.336 km) above msl, beyond the range of surface-to-air missiles available with the Soviet Union at that time, and collect aerial photo-reconnaissance (Jensen, 2009).

4.2.2.2 Space-based Platforms

The use of space platforms for the collection of information started in 1960 with the Corona satellite program of the USA and the Zenit-2 satellite of the Soviet Union. These satellite programmes were used for gathering military intelligence. However, the first civilian satellite for Earth resources applications, Landsat-1, was launched by the United States in 1972. Subsequently, several satellites have been launched by different countries, notable among them are SPOT from France, IRS from India, RADARSAT from Canada, ERS from the European Union, Terra and Aqua from the United States, Sentinel from Copernicus, the European Union's Earth Observation Programme. Details of some Earth observation satellites currently being used in rice mapping and monitoring have been provided below, and their important characteristics are summarised in Table 4.1.

4.2.2.2.1 Landsat Programme

Landsat-1, the first satellite for Earth resources observation launched by the United States in 1972, carried onboard a multispectral scanner (MSS), which collected data in four spectral bands, two in the visible range: green and red, and two in the near infrared (NIR) range of the spectrum. The ground viewing capability or the spatial resolution of MSS sensors was 79 × 79 m. However, due to resampling, it resulted in 56 × 79 m for each image picture element or pixel. Subsequent Landsat satellites carried improved sensors with a larger number of spectral bands and better spatial or ground resolution. Landsat-4 and -5 carried a sensor called Thematic Mapper (TM), apart from MSS. The TM had seven spectral bands: bands 1 through 3 in the blue, green, and red portions of the visible spectrum, respectively; band 4 in NIR; and bands 5 and 7 in the shortwave infrared (SWIR) range. Band 6, which operated in the thermal infrared (TIR) range, measured emitted energy from the object. The ground/spatial resolution of TM bands was 30 × 30 m in bands 1 through 5 and 7, whereas band 6, the TIR band, had a spatial resolution of 120 × 120 m.

Landsat-9, the most recent satellite in the series, launched on 27 September 2021, has 11 spectral bands in its sensor named Operational Land Imager (OLI-2); four in the visible, one in the near-infrared, three in SWIR, two in TIR range, and one PAN band. The spatial resolution of the visible, NIR, and SWIR bands is 30 m, whereas that of the TIR bands is 100 m. The PAN band has a resolution of 15 m. The data recorded has 14-bit quantisation (brightness values from 0 to 16,383). The repeat cycle is 16 days. However, with both Landsat-8 and -9 operating in orbit, the data is available every 8 days. The swath width is 185 km, which means each Landsat frame contains data from a 185 km × 185 km area on the Earth.

4.2.2.2.2 SPOT Satellite Programme

SPOT-1, the first satellite in the series, was placed by the French space agency CNES into a sun-synchronous orbit at an altitude of 822 km in February 1986. To capture imagery from various viewing angles and produce digital terrain models, the satellite used steerable mirrors that could scan both sides of its ground path across a 950 km swath, offering a markedly different capability from that of the Landsat sensors. The sensors are linear array push-broom type and are equipped with twin high-resolution imaging instruments that operate in the visible range of the electromagnetic spectrum working in panchromatic mode with 10 m spatial resolution, and in multispectral mode with 20 m resolution in three bands (green, red, and near-IR). The swath width at nadir is 60 km. Subsequent SPOT-2 and SPOT-3 had similar HRV sensors. SPOT-4 and SPOT-5 were placed into orbit in 1998 and 2002, respectively, with a 'VEGETATION' instrument, providing daily global coverage at 1 km spatial resolution and 2250 km swath width, and a high-resolution visible infrared (HRVIR) instrument with 4 bands, which included a short-wave infrared (SWIR) band in addition to the three bands of HRV. SPOT-6 and -7 were launched in 2012 and 2014, respectively, with a 1.5 m resolution sensor in PAN and a 6 m resolution in multispectral mode. The revisit cycle of SPOT-6 and SPOT-7 varied from 1 to 5 days. However, since SPOT-7 ceased its operation in March 2022, only SPOT-6 continues imaging currently.

4.2.2.2.3 IRS Programme

The Indian Remote Sensing Satellite (IRS) programme was initiated with the launch of IRS-1A in 1988. The IRS series expanded with the launches of IRS-1B in 1991, IRS-1C in 1995, and IRS-1D in 1997. IRS-1A and -1B carried two sensors: Linear Imaging Self-scanning Sensors (LISS-I and LISS-II), with a spatial resolution of 72 m and 36 m, respectively, a swath width of 146 to 148 km, and a revisit capacity of 22 days. IRS-1C and -1D carried three sensors—an improved LISS-III sensor with a spatial resolution of 23 m, a panchromatic (PAN) sensor with a spatial resolution of 5.8 m, and a wide field sensor (WiFS) with a resolution of 188 m. The swath width of the three sensors was 141 km for LISS-III, 70 km for PAN, and 692 km for WiFS. The revisit capability of LISS-III was 24 days, whereas that of WiFS was 5 days due to its wider swath. PAN sensor had a revisit of 5 days, and a stereoscopic imaging capability due to its +26° off-nadir viewing. IRS-1A and -1B were positioned in a sun-synchronous orbit approximately 904 kilometres above the Earth, whereas IRS-1C and -1D are at an altitude of 817 km.

4.2.2.2.4 IRS-P6 (Resourcesat-1)

Launched in 2003, carried three sensors: LISS-III, LISS-IV, and AWiFS (Advanced Wide Field Sensor). LISS-IV is a high-resolution (5.8 m) sensor having spectral bands similar to LISS-III and a 5-day revisit capability. LISS-III is a medium resolution (23 m) sensor, and a slightly coarser resolution (56 m) AWiFS sensor. AWiFS operates in four spectral bands similar to LISS-III, and covers a swath of 740 km. Resourcesat-2 and -2A were subsequently launched in 2011 and 2016, respectively, with similar sensors on board as in Resourcesat-1, and are currently operative. The combined presence of Resourcesat-2 and 2A in orbit has enhanced the sensors' revisit frequency, now approximately every 2 to 3 days for AWiFS, 12 to 13 days for LISS-III, and 25 to 26 days for LISS-IV. The advantage of improved revisit is extremely useful in monitoring vegetation, croplands, and water resources (NRSC website).

4.2.2.2.5 Terra/Aqua MODIS

The MODIS instrument, developed by NASA, is mounted on the Terra satellite launched in 1999 and the Aqua satellite launched in 2002 as part of the Earth observation program, which images the entire Earth's surface every one to two days. The orbits of both the satellites have been timed such that the Terra passes from north to south across the equator in the morning, while Aqua passes from south to north above the equator in the afternoon. The data from MODIS is collected in 36 spectral bands, with a swath width of 2330 km, so as to help in improving our understanding of global dynamics and processes occurring on the land, in the oceans, and in the lower atmosphere (MODIS website). MODIS operates as a whiskbroom-type radiometer, using a side-to-side scanning mirror, optical elements for signal collection, and multiple linear detector arrays with spectral filters arranged across four focal planes. (Jensen, 2009). The data collection by MODIS is done in 36 co-registered bands, of which 20 bands are in the 0.4 to 3 micrometres, and 16 bands in the 3 to 15 micrometres wavelength range, at three spatial resolutions—250 m, 500 m, and 1000 m. Bands 1 and 2 cover the visible and near-infrared regions at 250 m resolution; bands 3 through 7 extend across visible, near-infrared, and shortwave infrared wavelengths at 500 m; while bands 8 to 36 span visible, NIR, mid-infrared, and thermal infrared ranges at a coarser 1000 m resolution. Due to its high temporal resolution of 1–2 days, large swath width, and free data availability, MODIS data has been used extensively despite its low spatial resolution. Besides, due to the availability of a set of specialised data products like 8- and 16-day composited NDVI, MODIS data has been used in mapping and monitoring continental and global land use and vegetation changes, and in rice crop mapping at national and multi-country levels (Gumma et al., 2011a).

4.2.2.2.6 ERS Programme

ERS-1 was the first environmental monitoring satellite launched by the European Space Agency in 1991. The satellite was inserted into a near-circular, sun-synchronous polar orbit approximately 785 kilometres above Earth's surface, with an active microwave SAR instrument (AMI) working in C-band with VV polarisation. The spatial resolution of imaging SAR ranged between 10 and 30 meters in spatial resolution, covering a swath of 100 kilometres. The local crossing time of the satellite at the equator was 10:30 hours, and the repeat cycle was 35 days. Later, ERS-2 was launched as an enhancement of the ERS-1 mission in 1995 with a similar C-band sensor. The ERS-1 mission was concluded in the year 2000, and ERS-2 in July 2011. Hence, both these SAR missions have ended.

4.2.2.2.7 Sentinel-1 and -2

Sentinel-1 and -2 satellites are operated under the Copernicus programme, jointly implemented by the European Commission (EC) and the European Space Agency (ESA) (Sentinel website). Sentinel-1 operates with a microwave (SAR) imaging sensor, whereas Sentinel-2 operates with an optical imaging sensor. The Sentinel-1 mission includes a pair of polar-orbiting satellites in sun-synchronous paths, designed to revisit the same location every 12 days (with both satellites operating, repeat cycle of 6 days). Positioned in the same orbital plane but separated by a 180° phase difference, the satellites conduct C-band SAR imaging of Earth continuously, regardless of day or night conditions (Copernicus website). The system supports four distinct imaging modes, offering resolutions as fine as 5 metres and swath widths reaching 400 kilometres, along with dual polarisation functionality. Sentinel-1A was launched on April 3, 2014, followed by Sentinel-1B on 25 April 2016. After a successful operation for several years, Sentinel-1B experienced a system anomaly in 2024, and the mission had to be ended on 3 August 2024. In order to ensure consistent data coverage to its users globally, the Sentinel-1C was successfully launched on 5 December 2024 and is currently disseminating data.

The **Sentinel-1 satellite** is placed at an altitude of 693 km from the Earth and completes its orbit in 98.6 minutes, with a local time at ascending of 18:00 hours. Because orbital track spacing changes with latitude, revisit intervals vary: the Arctic experiences coverage in under 24 hours, regions such as Europe and Canada are revisited approximately every 2 days, and equatorial areas every 3 days. The C-band SAR instrument aboard Sentinel-1 operates at a central frequency of 5.405 GHz, equivalent to a wavelength of about 5.55 centimetres in 4 polarisations: VV, HH, HH+ HV, and VV+ VH. Sentinel-2, like Sentinel-1, is also a constellation of two satellites forming a part of the Copernicus, the European Union's Earth Observation program, orbiting the Earth, but with optical imaging sensors. Sentinel-2A was launched in June 2015, whereas Sentinel-2B was launched in July 2016. Sentinel-2C was launched on 5 September 2024 as a replacement for Sentinel-2A and is operating successfully. The satellites have been placed in a sun-synchronous orbit at an altitude of 786 km from the Earth and cross the equator at 10:30 hours local time in the descending node. The main imaging instrument on Sentinel-2A, 2B, and 2C is the Multispectral Imager (MSI). The MSI sensor captures imagery across 13 spectral bands spanning the optical range from 0.443 to 2.19 micrometres, covering a ground swath of 290 kilometres. Details of Sentinel-2 bands, their wavelength, and spatial resolution are given in Table 4.3.

TABLE 4.3
Resourcesat-2 and Sentinel-2 Satellite Sensor Characteristics

Resourcesat-2 and -2A			Sentinel-2A, -2B, -2C MSI		
Sensor/Bands	Spectral Resolution (μm)	Spatial Resolution (m)	Bands	Central Wavelength (μm)	Spatial Resolution (m)
LISS-4 Band 2 Green	0.52–0.59	5.8	1 (Coastal aerosol)	0.443	60
3 Red	0.62–0.68	5.8	2 (Blue)	0.490	10
4 NIR	0.77–0.86	5.8	3 (Green)	0.560	10
LISS-3 Band 2 Green	0.52–0.59	23.5	4 (Red)	0.665	10
3 Red	0.62–0.68	23.5	5 (RE 1)	0.705	20
4 NIR	0.77–0.86	23.5	6 (RE 2)	0.740	20
5 SWIR	1.55–1.70	23.5	7 (RE 3)	0.783	20
AWiFS Band 2 Green	0.52–0.59	56	8 (NIR)	0.833	10
3 Red	0.62–0.68	56	8A (RE)	0.865	20
4 NIR	0.77–0.86	56	9 (Water vapour)	0.945	60
5 SWIR	1.55–1.70	56	10 (SWIR-Cirrus)	1.375	60
			11 (SWIR1)	1.610	20
			12 (SWIR 2)	2.190	20
Swath Width (km)	70 (LISS-4), 141 (LISS-3), 740 (AWiFS)		290 (MSI)		
Repeat cycle (days)	24 days, AWiFS 5 days, LISS-IV with steerable camera (± 26 degrees) can achieve 5 days revisit		10 days with one satellite, 5 days with two satellites		
Launch	R-2 launched in July 2011, R-2A in December 2016		Sentinel-2B launched in July 2016, Sentinel-2C on 5 September 2024		

Source: ISRO (Indian Space Research Organisation), and Sentinel websites
Note: NIR—Near-infrared; SWIR—Short Wave Infra-red; RE—Red Edge

4.2.2.3 Unmanned Aerial Vehicles (UAVs)

The use of unmanned aerial vehicles (UAVs) or drones for gathering information has recently gained importance. The UAVs range from small, light-weight handheld units (e.g., Desert Hawk) to behemoths that weigh thousands of kilograms (e.g., Global Hawk) (Jensen, 2009). Some UAVs can stay aloft for a long time, constantly monitoring the same geographic area. In the case of an inaccessible terrain, UAVs flying at low altitude can acquire high spatial resolution images that can be used in place of ground information. Multispectral image sensors for UAVs have been developed in the visible, infrared, thermal, and other ranges of the spectrum, and are currently being used in imaging an area. Remote sensing using UAVs offers a cost-effective, adaptable, and efficient method for tracking crop development with high spatial and temporal detail, supporting informed decision-making in crop management. Several studies involving UAV–based imaging have also been utilised to derive estimates of plant height, crop density, biomass, and grain yield across different crop types (Bending et al., 2015; Watanabe et al., 2017; Jin et al., 2017; Du and Noguchi, 2017; Li et al., 2019). Cen et al. (2019) have used a lightweight UAV with a dual-image frame snapshot camera with 25 spectral bands in the 0.6–1.0 μm wavelength range for estimating rice biomass during its entire growth stage. In view of these studies, UAVs are likely to play an important role in the remote sensing of crops in the future.

4.2.2.4 Ground-based Instruments

Ground-based remote sensing instruments are used for detailed *in situ* information collection from different objects—crops, soil, rock, water, forests, and so on. These instruments generally collect information from radiometers or other sensors placed at a height of one to a few metres above the ground. Sensors may be mounted on a ladder, crane, and so on. The information thus collected may be used to know the spectral characteristics of various objects, their correlation with aircraft- and satellite-based measurements, the development of sensors for aircraft or satellite missions, or the development of spectral libraries of different objects. Commonly used handheld spectroradiometers measure spectral reflectance in the visible, NIR, and SWIR range (350–2500 nm or 0.35–2.5 μm wavelength). A near-continuous spectrum of reflectance from an object can be obtained from these radiometers at a sampling interval of 2 nm (ASD, 2006).

4.3 ADVANCES IN RICE CROP MONITORING THROUGH SATELLITE REMOTE SENSING

Satellite and airborne remote sensing have been applied to map and monitor rice cultivation across various global regions. In rice-producing areas, these data sources are used to extract crop traits and biophysical indicators such as:

- Mapping rice-growing area
- Monitoring rice growth and crop health
- Estimating crop growth parameters leaf area index (LAI), biomass, and fraction of absorbed photosynthetically active radiation (fAPAR)
- Yield estimation
- Incidence of pests and diseases, and damage assessment
- Assessment of the cropped area affected by droughts and floods
- Temporal and spatial shift in rice growth pattern
- Modelling crop growth
- Assessment of post-harvest biomass burning
- Precision agriculture (application of fertiliser and pesticides)
- Estimation of climate-change–related emission of methane fluxes

Several new application areas are also being developed, depending on the user's requirements (Verma et al., 2025). However, only mapping and related aspects are being discussed here.

4.3.1 Spectral Reflectance Properties of Vegetation and Rice in Optical Satellite Data

4.3.1.1 Leaf Cell Structure and Pigments Present in Leaves

To understand the spectral characteristics of vegetation, let us first understand the leaf internal cellular structure, pigments present in it, and leaf water content, as these are the three factors controlling the spectral characteristics of a plant. Leaf cellular structure varies significantly across plant species and is influenced by environmental conditions. The outermost layer, known as the upper epidermis, is coated with a cuticle that scatters light but reflects only a small portion. Photosynthesis in green leaves primarily occurs within two specialised cell types: palisade parenchyma and spongy mesophyll cells. Incoming solar radiation in the visible and near-infrared regions penetrates the epidermal layer to reach these photosynthetically active tissues. Palisade parenchyma is located in the upper part of the leaf, whereas spongy mesophyll is located in the lower part. In the plants with horizontal leaves (planophile), the palisade is towards the upper surface, but in vertical leaves (erectophile) like the rice plant, palisade cells may form on both sides. Palisade parenchyma cells contain chloroplasts with chlorophyll pigments. Chloroplasts are densely concentrated in the palisade cells near the upper surface of the leaf, resulting in a darker tone on the top side compared to the lighter underside. The pigments present in spongy mesophyll cells are yellow carotenes and pale-yellow xanthophyll. When a plant undergoes senescence or is under environmental stress, the chlorophyll may decrease or disappear, and yellow carotenes and other pigments become dominant and more visible to our eyes. Below the spongy mesophyll cells, the lower epidermis is located, which also contains stomata, which are tiny pores through which carbon dioxide present in the atmosphere enters the leaf.

4.3.1.2 Absorption, Transmittance, and Reflection of Radiant Energy from Leaves

Healthy green vegetation absorbs incoming sunlight and diffuse skylight that falls upon it. This radiation interacts with various components of the plant, including pigments, moisture content, and the air-filled spaces within and between leaf tissues. A part of this radiation is absorbed by the plant leaves, a part is reflected, and a part is transmitted through the leaf to the lower leaf or surface. Since plant leaves are primarily responsible for photosynthesis, the leaf pigments like chlorophyll-*a*, chlorophyll-*b*, β-carotene, phycoerythrin, and phycocyanin effectively absorb the majority of the radiation falling on them in blue, green, and red wavelengths. Absorption of radiation in the infrared wavelengths is negligible. In the 0.7–1.3 μm wavelength NIR region, scattering from spongy mesophyll cells leads to higher reflectance. In the later part of the IR, called mid- or shortwave infrared (1.3– 2.6 μm wavelength), leaf water content is the primary factor controlling leaf reflectance. Water absorption bands at 1.45, 1.94, and 2.7 μm significantly reduce the reflectance, resulting in a dip in the reflectance curve at these wavelengths (Figure 4.4).

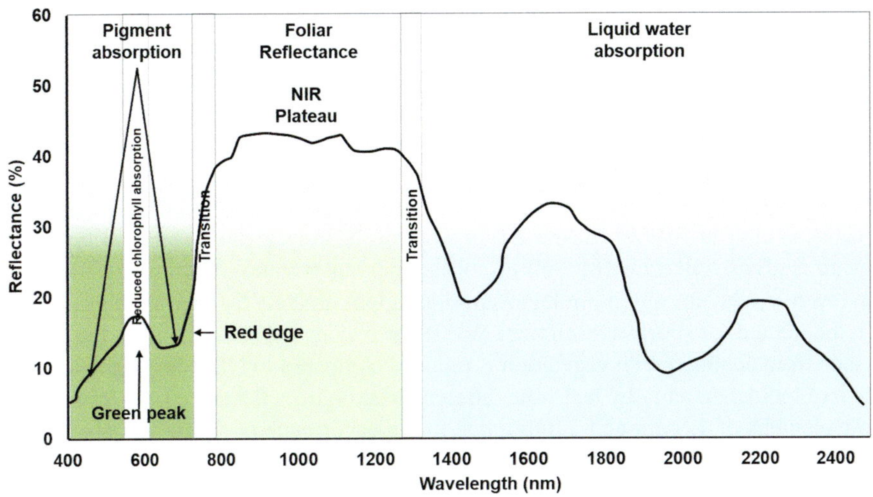

FIGURE 4.4 Spectral reflectance characteristics of healthy green vegetation in the visible (400–700 nm), near-infrared (700–1300 nm), and short-wave infrared (1300–2500 nm) wavelengths.

Transmittance is the part of radiant energy that penetrates or passes through the leaf to the other leaves or the ground beneath the plant canopy. The reflectance and transmittance curves are almost mirror images of one another throughout the visible and near-infrared portions of the spectrum. Due to strong absorption in the visible region of radiation, both reflectance and transmittance are much less compared to the near-infrared (NIR) region, where both reflectance and transmittance from the leaf suddenly increase dramatically and continue throughout in this region. Absorption in the entire NIR region is negligible (~ 5%). Reflectance is the part of radiant energy that is neither absorbed nor transmitted but reflected back from the leaf surface. Since the sensor records this reflected energy, it is of prime significance in remote sensing, hence will be discussed in detail below.

4.3.1.3 Reflectance Characteristics of Green Vegetation in Different Wavelengths

Three factors controlling reflectance of solar radiation from healthy green vegetation are leaf pigments in the visible wavelength, scattering in the leaf mesophyll cells in the NIR wavelength, and leaf water content in the SWIR wavelength. When the solar radiation is intercepted by the green vegetation, the blue and red wavelength radiation centred at 0.45 and 0.67 µm is strongly absorbed primarily by pigments within the mesophyll cells of the leaf like chlorophyll *a* and *b*, and β-carotene and used in photosynthesis, resulting in less reflectance at these wavelengths (Figure 4.4). On the other hand, the green wavelength radiation centred at 0.55 µm is reflected back from chlorophyll pigments inside chloroplasts located in palisade cells, making the leaves appear green to human vision.

The near-infrared (NIR) radiation (0.7 to 1.3 μm wavelength) penetrates the leaf as deep as the spongy mesophyll cells, which lie below the palisade parenchyma, and contain many intercellular air spaces. These spongy mesophyll cells control the amount of NIR reflectance from the green leaf. The exchange of oxygen and carbon dioxide for respiration and photosynthesis takes place in these mesophyll cells. Multiple reflections at the interfaces between cell walls and internal air spaces within the leaf result in strong diffuse reflectance in the near-infrared (NIR) region (Gausmann et al., 1969). In the NIR spectrum, healthy green vegetation typically exhibits strong reflectance (40–60%), substantial transmittance (also around 40–60%) to lower leaf layers, and minimal absorption (approximately 5–10%) (Jensen, 2009). In the middle- or shortwave-infrared (SWIR) wavelength (1.3 to 2.5 μm), the reflectance from healthy green vegetation is reduced compared to NIR, due to the absorption of incident energy by leaf water content. Maximum reflectance is observed near wavelengths of 1.6 μm and 2.2 μm, due to the occurrence of significant absorption features caused by atmospheric water vapor are centered around 1.45 μm, 1.94 μm, and 2.6 μm (Figure 4.4).

4.3.1.4 Reflectance from Stressed or Senescent Vegetation

Plant stresses are commonly caused by low moisture, inadequate nitrogen, disease, and pest infestation. In the case of stressed or senescent vegetation, the decreased chlorophyll production leads to a lesser absorption of radiation in blue and red wavelengths. Almost a similar amount of reflectance in the blue, green, and red regions gives a yellowish or brownish appearance to plant leaves to our eyes. Carter (1993) has suggested that higher reflectance is within the visible wavelength range in the 0.535–0.640 and 0.685–0.700 μm wavelength, which is the most consistent leaf reflectance response to plant stress, as it indicates decreased absorption by pigments. Increased reflectance due to various stresses near 0.700 μm represents the 'blue-shift' at the red-infrared transition, also known as the red edge region. Reflectance in the infrared region shows a noticeable change only after significant leaf dehydration has occurred due to advanced stress. Moisture stress in plant leaves generally increased reflectance throughout the visible and reflective infrared (0.40–2.5 μm) wavelengths (Carter, 1991).

4.3.1.5 Red Edge Region in Vegetation Reflectance and Its Importance

Red edge is the region of the electromagnetic spectrum where the spectral reflectance of green vegetation changes rapidly, resulting from the interplay of intense chlorophyll absorption and the scattering of light within the leaf tissues. It roughly spans 670–760 nm, placing it between the visible red and NIR bands (refer to Figure 4.4). The red edge band was first introduced by Collins (1978). Vegetation stress shows up in this region first. An increase in reflectance near 700 nm signifies a 'blue shift' of the red edge, a displacement of the red-to-infrared transition curve toward shorter wavelengths in plants under stress, when reflectance is plotted against wavelength (Cibula and Carter, 1992). Clark et al. (1995) emphasised that the long-wavelength tail of chlorophyll absorption contributes to one of the steepest spectral slopes observed in naturally occurring materials. Chlorophyll absorption is very high, resulting in < 5 per cent reflectance near 680 nm to NIR reflectance of 50%or more at about 730 nm.

When the chlorophyll absorption is reduced, the absorption band becomes narrower, causing a red-edge shift caused by water deficiency, natural senescence, or biotic and abiotic stresses. Carter et al. (1996) also reported a high sensitivity of 684 ± 3 nm reflectance to vegetation stress, and found that reflectance increases due to stress in two pine species, *Pinus taeda* L. and *Pinus elliottii* Engelm, could be detected at least 16 days before visible signs of stress appearance. Subsequent researches conducted on the vegetation red edge reflectance (Gitelson et al., 2005, Delegido et al., 2011) emphasised the utility of red edge band and its ratio with NIR band in remote evaluation of canopy-level chlorophyll and LAI (leaf area index), and subsequently led to the deployment of three red edge bands centred at 705, 740, and 783 nm in the multispectral imager (MSI) sensor onboard Sentinel-2A and -2B satellites launched by ESA in 2015 and 2016 (refer to Table 4.3).

4.3.1.6 Green Leaf Reflectance vs. Canopy Reflectance

While the green leaf reflectance is governed mainly by leaf pigments, leaf cell structure, and water content, the remote sensing instruments deployed in aerial or space platforms measure reflectance from the entire canopy, which takes into account the canopy architecture (arrangement of leaves), leaf area index (LAI), leaf angle distribution, sun angle and sensor angle geometry, and the soil background reflectance.

4.3.1.7 Rice Crop Phenology and Reflectance Characteristics

Rice plants have a growth cycle ranging from 3 to 6 months, which depends on the variety and the growing environment (GRiSP, 2013). The aromatic Basmati varieties presently grown in India and Pakistan have a duration ranging from 120 to 150 days. The duration of rice can be categorised into three distinct phases: vegetative, reproductive, and ripening. In a typical 120-day rice variety grown under tropical conditions, approximately 60 days are allocated to the vegetative stage, followed by 30 days each for the reproductive and ripening stages (GRiSP, 2013). However, in the cool temperate regions of Hokkaido in Japan or Yanco in Australia, the ripening phase may extend to 65 days. The vegetative phase begins with seed germination and includes seedling development and tillering. During this phase, there is a gradual increment in plant height, and new leaves and tillers emerge at regular intervals. The tillers that emerge the central stem are termed primary tillers, which may further generate secondary and tertiary tillers. The reproductive growth phase is subdivided into flag leaf (the last leaf) emergence, booting (panicle initiation and development), heading (panicle exertion), and the flowering of spikelets. Panicle initiation typically occurs around 25 days prior to heading (GRiSP, 2013). Flowering begins with the panicle exertion or heading and continues until most of the spikelets in the panicle have bloomed. Rice primarily undergoes self-pollination, and anthesis or spikelet opening and pollination take place on the same day, generally from 10:00 to 13:00 hours in the tropical environment. The ripening phase follows fertilisation and is subdivided into milky, dough, and maturity stages, based on the texture and colour of the growing grains.

Transplanting of rice seedlings brings a shock to the plant, and it takes about 2–3 weeks for it to recover. During the early transplanting stage, at about 3 weeks after transplanting, the vegetation cover is less than 15%, and the reflectance in the visible region is only slightly lower than the NIR region because of more contribution from the underlying soil and water complex (Chang et al., 2005). As the crop grows to the active tillering stage, the leaf cover and the number of tillers increase rapidly. During the vegetative stage at about 6–9 weeks after transplanting rice, the reflectance across the visible wavelength range decreases due to the chlorophyll absorption and increases rapidly in the NIR region due to high scattering from the leaves and stems. During the reproductive stage of booting (panicle formation) through heading or flowering, the reflectance in the visible region is generally lowest, while in the NIR region, it is the highest at about 12 weeks after transplanting (Chang et al., 2005). During the ripening phase, when the leaf yellowing and plant wilting start, the reflectance in the visible region increases due to the lowering of plant chlorophyll content and decreases in the NIR region due to wilting (Figure 4.5). Casanova et al. (1998) have reported that the red reflectance in rice foliage increases as chlorophyll declines, rising from about 2% at flowering to roughly 16–18% at maturity due to reduced green absorption by leaves and stems and the yellowing of grains. In contrast, near-infrared (NIR) reflectance, which tracks total biomass, starts near 15% during early tillering, peaks around 50% at flowering when vegetative mass is greatest, and then falls to about 33% by maturity as leaves, panicles, and grains dry and lose moisture.

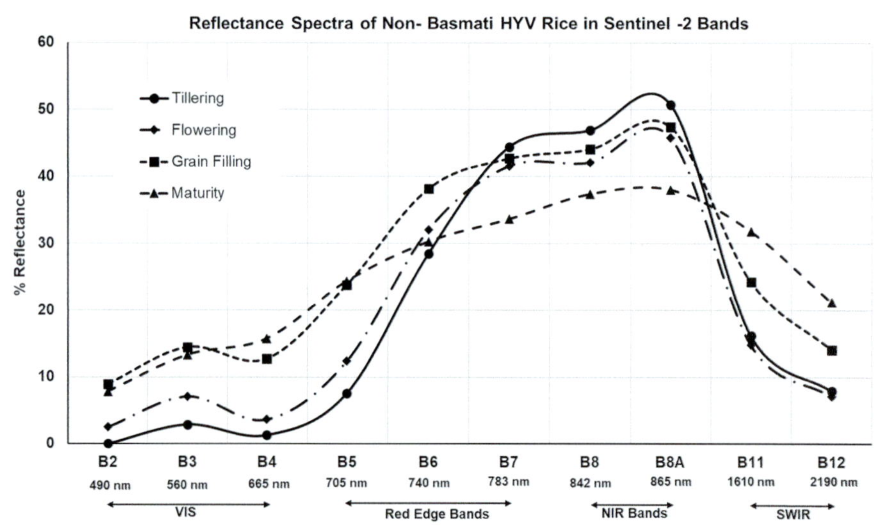

FIGURE 4.5 Spectral Reflectance spectra of HYV Rice observed at different growth stages in Sentinel-2 MSI bands in Haryana.

Note: *B2, B3, and B4 are blue, green, and red bands, respectively, whereas B5, B6, and B7 are red-edge bands. B8 and B8A correspond to NIR, and B11 and B12 to SWIR bands. During the tillering stage, the reflectance in the visible bands is lowest, in the three red-edge bands, it is increasing, in the NIR bands it reaches the highest, and then drops in the two SWIR bands. At the maturity stage, the visible band reflectance is high; in the red-edge and NIR bands, it decreases due to the crop senescence, and in the SWIR bands, higher than tillering and flowering stages due to the loss of leaf water content.*

4.3.2 Backscatter Response of Rice in Microwave Remote Sensing Data

4.3.2.1 Characterising Rice Backscatter Signatures in SAR Data

Understanding backscatter characteristics of rice crop at different growth stages is essential for rice identification and mapping. Traditionally, both irrigated and rainfed lowland systems, rice seedlings are first raised in seedbeds and later transplanted into the primary field, which is puddled and kept flooded to control weeds and pests. Therefore, in addition to volume scattering within the canopy and ground scattering that is weakened by the overlying vegetation, there is also significant multiple scattering between the canopy and water layer at the ground becomes important in rice (Bouvet et al., 2009). Before rice transplanting, the SAR signal remains low as a result of specular reflection caused by standing water and puddled soil (Le Toan et al. 1997; Panigrahy et al. 1999). At this stage, rice fields can be easily discriminated from other non-rice fields because of a significant backscatter difference between flooded areas and non-flooded other crops (Chen and McNairn, 2006; Choudhury et al., 2007). As the rice crop grows from seedling to vegetative growth through its maturity stage, there are shifts in plant stature, biomass production, water content, and morphological structure, which in turn influence the radar backscatter coefficient (Ribbes and Le Toan, 1999; Dong et al., 2006). Moreover, with the crop growth, there is a change in the amount of vegetation covered area and the water level underneath. Hence, in rice fields, the dominant source of radar backscatter is the interaction between the radar waves and the combined vegetation-soil-water system (Le Toan et al., 1997). The scattering mechanism in rice canopy and the canopy-soil-water interaction are given in Figure 4.6.

The volume backscatter originating within the rice canopy and canopy-soil-water interaction results in an increase in the backscatter coefficient after transplanting, which continues during the vegetative stage, and reaches a maximum at the flowering stage (Chen and McNairn, 2006). During the ripening stage, the reduction in water content of the plant and the leaf density results in a decrease in the backscatter coefficient, which continues till the crop is harvested (Chen and McNairn, 2006; Lim et al., 2008). Apart from the temporal dynamics of crop growth parameters, backscatter characteristics are also related to the frequency, SAR polarisation, and the radar's incidence angle. Studies conducted on the influence of polarisation, incident angle, and frequency/wavelength on rice scattering show the following:

(i) Backscattering coefficient (σ^0) from VV-polarisation is higher than HH-polarisation in the early growth stage of rice (Kim et al., 2008) due to short erect vertical leaves and stems at this stage (Lim et al., 2008). As the crop grows to tillering, booting, and flowering stages, HH-polarised backscatter becomes higher than the vertical architecture of the rice canopy attenuates the radar wave, thereby reducing VV-polarised backscatter (Kim et al., 2008; Bouvet et al., 2009).

(ii) Backscattering coefficient (σ^0) from HH- and HV-polarisations increases at the reproductive (flowering) stage and is quite stable at the ripening stage (Le Toan et al., 1997).

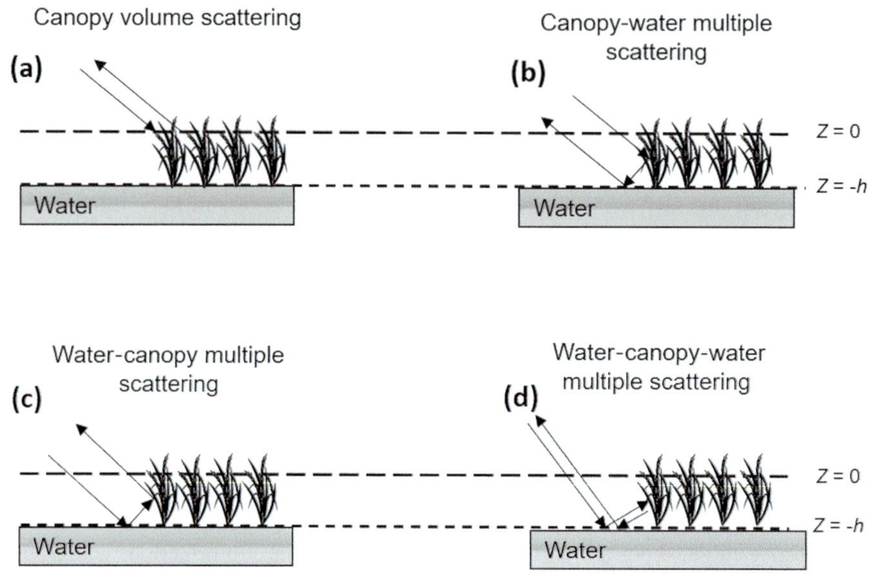

FIGURE 4.6 Scattering mechanisms in rice crop field due to radar wave-vegetation-soil-water interaction. Modified from Le Toan et al. 1997.

(iii) Temporal change in backscatter coefficient for Ka and Ku bands exhibits greater variability at smaller incidence angles, whereas for C- and L-bands, the variation is more pronounced at larger incidence angles. (Inoue et al., 2002). In the C-band, HH-polarised signal reaches saturation at the heading stage, probably due to a decrease in water in the fields and a change in the physical size of the plant at the maturity and ripening stage (Chen et al., 2007). However, in the case of ERS-1 C-band VV-polarised data, Le Toan et al. (1997) have noted that towards the end of the productive phase, the backscatter coefficient increased to between –8 and –6 dB and then stabilised through the remainder of the crop cycle. In L-band observations, HH-polarisation records low backscatter values during the early growth stage following transplanting, which increases at the tillering and vegetative growth stages, and levels off at heading and ripening stages (Inoue et al., 2002).

(iv) The backscatter in the X-band during the ripening stage showed a reverse trend with the radar incident angle; it decreased at 25°, but increased at 55° (Inoue et al., 2002).

(v) Among the biophysical variables, LAI correlates best with HH- and cross-polarisation in C-band, whereas fresh biomass correlates best with HH- and cross-polarisation of L-band. Lower wavelength bands (Ka, Ku, and X) are poorly correlated with rice LAI and fresh biomass (Inoue et al., 2002).

4.3.3 Spectral Vegetation Indices Used in Rice Mapping

As vegetation absorbs electromagnetic energy in the visible band (0.45–0.70 μm) and reflects in the NIR range (0.70–0.90 μm), researchers have developed vegetation indices (VIs) using reflectance measured in these spectral ranges. These VIs indicate the amount of healthy green vegetation, and also normalise the external effects of sun angle, sensor viewing angle, and the atmosphere in spectral data, and help in making consistent comparisons between temporal datasets. They also normalise internal canopy effects by accounting for background variability caused by terrain elevation changes, soil reflectance differences, and the presence of non-photosynthetic elements such as senesced leaves or woody stems (Running et al., 1994; Huete and Justice, 1999). Biophysical parameters like above-ground biomass, leaf area index (LAI), and absorbed photosynthetically active radiation (APAR) have been quantitatively linked to vegetation indices (VIs) calculated from satellite-derived spectral reflectance data (Jensen, 2009). Several VI have been developed, of which the commonly used indices are- normalised difference vegetation index (NDVI), normalised difference water index (NDWI), land surface water index (LSWI), soil adjusted vegetation index (SAVI), and the enhanced vegetation index (EVI). A list of vegetation indices commonly used in rice mapping and monitoring has been summarised in Table 4.4.

4.3.3.1 Normalised Difference Vegetation Index (NDVI)

Developed by Rouse et al. (1974), NDVI is the most commonly used vegetation index, and is calculated as:

$$\mathbf{NDVI} = (\rho\mathbf{NIR} - \rho\mathbf{Red}) / (\rho\mathbf{NIR} + \rho\mathbf{Red}),$$

where ρRed and ρNIR are the spectral reflectance values in red and NIR bands. NDVI can provide information on seasonal progressive changes in leaf area and plant height and help in its monitoring. It also reduces noise present in multi-band and multi-date data due to variations in solar illumination angle, the presence of cloud-induced shading, atmospheric scattering and absorption effects, and changes in terrain elevation and slope orientation. NDVI's main advantage is its high correlation with the leaf area index (LAI) of crops and other vegetation. However, the correlation often weakens during peak LAI periods, likely due to the saturation effect of NDVI, which becomes less sensitive to further increases in canopy density beyond a certain threshold (Wang et al., 2005). Disadvantages of the NDVI include its sensitivity to canopy background variations, like the colour of the background soil. Dark-coloured soil background increases NDVI values. To overcome this problem, a soil adjusted vegetation index (SAVI) has been developed by Huete (1988).

4.3.3.2 Soil Adjusted Vegetation Index (SAVI)

SAVI has incorporated a soil background and/or atmospheric adjustment factor in the NDVI (Huete, 1988), which is calculated as:

$$\mathbf{SAVI} = (\mathbf{1} + \mathbf{L})(\rho\mathbf{NIR} - \rho\mathbf{Red}) / (\rho\mathbf{NIR} + \rho\mathbf{Red} + \mathbf{L})$$

TABLE 4.4
Spectral Vegetation Indices (VIs) Commonly Used in Rice Mapping and Monitoring

Vegetation Index (VI)	Equation	References
RVI: Ratio Vegetation Index	ρNIR/ρRed	Birth and McVey, (1968)
NDVI: Normalised Difference Vegetation Index	(ρNIR − ρRed)/(ρNIR + ρRed)	Rouse et al. (1974)
EVI: Enhanced Vegetation Index	2.5 × (ρNIR − ρRed) / (ρNIR + 6 ρRed − 7.5 ρBlue + 1)	Huete et al. (1997)
SAVI: Soil-Adjusted Vegetation Index	(1 + L)(ρNIR − ρRed) / (ρNIR + ρRed + L)	Huete (1988).
LSWI: Land Surface Water Index	ρNIR − ρSWIR1 / ρNIR + ρSWIR1	Chandrasekar et al. (2010).
NDWI: Normalised Difference Water Index	ρNIR − ρSWIR1 / ρNIR + ρSWIR1	Gao (1996)
MNDWI: modified **NDWI**	ρGreen − ρSWIR / ρGreen + ρSWIR	Xu (2006)
NDFI: Normalised Difference Flood Index	ρSWIR − ρRed / ρSWIR + ρRed	Cian et al. (2019).
TVI: Triangular Vegetation Index	0.5 × [120 × (ρ*NIR* − ρ*Green*) − 200 × (ρ*Red* − ρ*Green*)]	Broge and Leblanc, (2001)
RGVI: Rice Growth Vegetation Index	1 − (ρBlue + ρRed) / (ρNIR + ρSWIR1 + ρSWIR2)	Nuarsa et al. (2011)
	Red Edge Vegetation Indices using Sentinel-2 MSI data	
Chlorophyll Index Red Edge (CIred-edge)	(ρ NIR / ρ Red Edge) − 1	Gitelson et al. (2005)
Chlorophyll Index Green (CIgreen)	(ρ NIR / ρ Green) − 1	Gitelson et al. (2005)
GCI	(NIR) / (Green) − 1	Gitelson et al. (2003)
	Sentinel-1 SAR VH and VV Indices	
VH–VV Difference Index	VH − VV polarisations	Saad El Imanni et al. (2022)
SAR Rice Index	(VV − VH) / (VV + VH)	Sun et al. (2022)
VH- VV Ratio Index	VV / VH VH / VV	Kobayashi and Ide (2022)
Radar Vegetation Index	4* VH / (VH + VV)	Nasirzadehdizaji et al. (2019)

Note: ρ is the surface reflectance values for blue (B), green, red (R), near-infrared (NIR), Shortwave infrared (SWIR1 and SWIR2) bands.

where ρRed and ρNIR are the spectral reflectance values in the red and near-infrared (NIR) bands used along with the canopy background adjustment factor, L, which compensates for differential attenuation of red and NIR radiation within the canopy (Huete, 1988). An L value of 0.5 has been shown to effectively reduce the influence of soil brightness and eliminate the need for additional calibration across varying soil types (Huete and Liu, 1994).

4.3.3.3 Normalised Difference Moisture or Water Index (NDWI/NDMI)

NDMI has been developed to get information about vegetation water content. Gao (1996) found that the ratio of near- and middle-infrared (also called short-wave infrared, SWIR) bands in Landsat TM, bands-4 and -5, respectively, provides information on canopy water content. NDMI is calculated as:

$$\mathbf{NDMI = (\rho NIR - \rho SWIR) / (\rho NIR + \rho SWIR)}$$

where ρNIR and ρSWIR are the spectral reflectance values in NIR and SWIR bands. NDMI has been found to have a high correlation with canopy water content and is widely used for the detection of water in plant leaves.

4.3.3.3.1 Chlorophyll Index (CI)

Two chlorophyll indices, namely, Chlorophyll Index-Green (CI-green), and Chlorophyll Index-Red Edge (CI-red edge), have been used to assess the amount of chlorophyll present in different crops using spectral reflectance in green, red-edge, and NIR bands (Gitelson et al. 2003; Gitelson et al. 2005). The differences are mainly because of the differences in the chlorophyll levels in the leaves, so that the spectral signature of agricultural crops presents a unique phenological profile. These indices are calculated as:

$$\mathbf{CI - green = (\rho NIR / \rho Green) - 1}$$

$$\mathbf{CI - red\,edge = (\rho NIR / \rho Red - edge) - 1}$$

where ρNIR, ρGreen, and ρRed – edge are the spectral reflectance values in NIR, green, and red-edge bands, respectively.

In recent years, vegetation indices derived from remote sensing data have been increasingly utilised to map and quantify rice cultivation areas at regional and national scales (Thi et al. 2012; Boschetti et al. 2014; Boschetti et al. 2017). These indices are often used with low spatial resolution multispectral data, like SPOT vegetation (VGT), and Terra/ Aqua MODIS (Gumma et al. 2011b; Clauss et al. 2016). Some research studies have also used time-series remotely sensed vegetation on index products from Landsat, IRS 'Sentinel-1 and 2', and other satellites with medium spatial resolution (Nuarsa et al. 2011, Panigrahy et al. 2009, Verma et al., 2025).

4.3.3.4 SAR Data-based Vegetation Indices

Similar to the VIs developed using reflectance values in optical sensor data, indices have been developed using backscatter coefficient values in different polarisations in microwave SAR data.

4.3.3.4.1 Radar Vegetation Index (RVI)

Radar Vegetation Index (RVI) is a measure used to assess the health and condition of vegetation using radar data. It has been used for predicting crop growth over time (Nasirzadehdizaji et al., 2019). RVI is calculated as:

$$\textbf{Radar Vegetation Index}\left(\textbf{RVI}\right) = \textbf{4} \times \textbf{VH} / \left(\textbf{VH} + \textbf{VV}\right)$$

where VH and VV are the backscatter coefficient values in VH and VV polarisation, respectively. RVI values range from 0 to 1; values close to zero indicate bare soil or young plants, whereas values close to 1 indicate fully developed vegetation. The RVI is influenced by the water content within the vegetation canopy and exhibits high sensitivity to biomass (Kim and van Zyl, 2004). RVI quantifies the contribution of volume scattering within the vegetation canopy, as reflected by the strength of the cross-polarised (e.g., HV or VH) backscatter signal (Nasirzadehdizaji et al., 2019).

4.3.3.4.2 VH–VV Difference Index

VH–VV polarisation difference index has been used for predicting the phenological stages of 'rice (Verma et al., 2025) and wheat' by (Saad El Imanni et al., 2022), who reported that the gap between SAR backscatter coefficients $\left(\sigma^0\right)$ in VH and VV polarisations for wheat widened from the germination phase to the heading stage, but narrowed again as the crop progressed through maturity and harvest.

4.3.3.4.3 VH–VV Ratio Index

The ratio of VV and VH has been used by Kobayashi and Ide 2022, who observed that the VV / VH ratio observed at a low incidence angle serves as a more effective indicator for diagnosing the ripening stage and can help in determining the optimum time of harvest.

4.3.4 Methods for Rice Mapping and Monitoring Using Optical Remote Sensing Data

Satellite remote sensing has been widely used in mapping rice fields, monitoring crop health, and estimating biomass and grain yield globally. Various methods have been employed in mapping rice area and yield estimation in different rice-growing countries. Extensive reviews are available on rice mapping methods using satellite remote sensing (Kuenzer and Knauer, 2013; Mosleh et al., 2015; Dong and Xiao, 2016; Zhao et al., 2021). Dong and Xiao (2016) reviewing the literature published during1980s to 2015, have grouped rice mapping methods into four groups: (1) based on reflectance values and image statistics, (2) utilising vegetation indices (VIs) and image statistics, (3) temporal analyses using VIs or SAR backscatter data, and

(4) phenology-driven approaches aligned with critical rice growth stages. (e.g., transplanting, flowering, and others). In the phenology-based mapping methods, remote sensing imagery captured throughout the full growing season or during key developmental stages, such as transplanting, tillering, booting/heading, and flowering, has been utilised (Mosleh et al. 2015).

The spectral reflectance data and image statistic-based method have been used since the beginning, in which the remotely sensed images and sample training data from the field are collected, and decision rules are determined to classify rice from other land cover classes. These methods include both supervised and unsupervised classification techniques, including methods like ISODATA, decision trees (DT), Random Forest (RF), and support vector machines (SVM) (Zhao et al. 2021). In the supervised classification, training sites of known classes are provided to the computer, statistics of the training sites is generated, and each image pixel is classified using the training statistics based on suitable algorithms, of which maximum likelihood has been the most commonly used classifier in mapping rice with medium-resolution remotely sensed data (McCloy et al. 1987; Panigrahy et al. 1991; Turner and Congalton 1998; Choudhury and Chakraborty, 2004; Panigrahy et al. 2009). Unsupervised classification is an image analysis technique that does not rely on predefined class labels. Instead, it derives feature patterns through automated clustering and statistical analysis to generate classification rules (Zhao et al. 2021). Some studies have used both supervised and unsupervised classification to achieve higher accuracy (Turner and Congalton 1998). Advancements in classification techniques have led to the introduction of newer algorithms, such as neural networks (NNs), SVMs, RF, and deep learning, for rice mapping applications (Dong and Xiao, 2016; Mosleh et al., 2015). However, these approaches often depend on localised prior information to achieve accurate rice mapping at finer spatial resolutions.

Vegetation indices (VIs), like NDVI, enhanced vegetation index (EVI), and moisture stress index (MSI), have been used in mapping rice. NDVI is sensitive to chlorophyll content and primarily reflects changes in the visible spectrum, whereas EVI is more responsive to the near-infrared region and performs better in areas with dense vegetation or high biomass (Cheng 2006). Oguro et al. 2001 used NDVI and EVI for monitoring rice and concluded that the difference between NDVI and EVI is best suited to differentiate rice from forests, yielding a relative error of under 2%. Panigrahy et al. (2009) evaluated multiple indices, including NDVI, MSI, and a custom-designed Three-Band Index (TBI) incorporating red, NIR, and SWIR wavelengths, for crop classification, and found that integrating the SWIR band, which reflects vegetation water stress, into the vegetation index enhanced overall classification accuracy.

The phenology-based approaches rely on the knowledge of the seasonal growing pattern and the development of different growth stages of the rice crop during its life cycle. The phenological characteristics of rice, like flooding and transplanting, tillering, flag leaf, flowering, grain filling, and maturity, are unique. During the vegetative growth stages, the rice field is covered with flooded water, growing rice plants, and some exposed wet soil. During the reproductive stage, the leaves start turning yellow, leading to a decline in vegetation coverage due to the maturity of the rice plants. Time-series remote sensing data and their various spectral indices are utilised in rice discrimination using phenological methods. The time series data of EVI and LSWI can effectively monitor the changes in rice during the entire growth process from transplanting to harvesting. Qiu et al. (2015) proposed an index called the Ratio of Change Amplitude of LSWI to EVI2 (RCLE), derived from EVI2 and LSWI values recorded at specified tillering and heading dates, effectively capturing the spectral response dynamics between these critical phenological stages. This index is calculated using the following equation:

$$\mathbf{RCLE} = \left(\mathbf{LSWI}_{\max} - \mathbf{LSWI}_{\min}\right) / \left(\mathbf{EVI2}_{\mathbf{heading}} - \mathbf{EVI2}_{\mathbf{tillering}}\right)$$

Here, $LSWI_{max}$ and $LSWI_{min}$ represent the maximum and minimum LSWI values observed between the tillering and heading stages, while $EVI2_{heading}$ and $EVI2_{tillering}$ denote the EVI2 values corresponding to the heading and tillering dates, respectively. The heading date of the rice field was identified based on the maximum value of EVI2 achieved during the growing period, and the tillering date was 40 days before the heading. The minimum LSWI value ($LSWI_{min}$) observed between the tillering and heading stages is used to represent the lower bound that should be maintained for rice. For effective discrimination of rice fields, the RCLE must fall below a defined threshold, while $LSWI_{min}$ should exceed a separate, predetermined threshold (Qiu et al. 2015).

Remotely sensed data of medium spatial resolution from satellites like Landsat (MSS, TM, ETM+, OLI), SPOT (XS), IRS (LISS-II, LISS-III, AWiFS), and others have been used for rice area mapping in different countries and have employed various data analysis techniques. Recently, Sentinel-1 and -2 data with higher spatial (10 m) and temporal resolution (5 days) have become freely available, and these are being increasingly used in rice mapping. The free access to Landsat data from 1972 onwards has offered unique opportunities for monitoring long-term changes in rice areas (Dong et al., 2016).

Apart from these satellites with medium resolution sensors, low spatial resolution data from AVHRR, MODIS, and SPOT VEGETATION have also been used in mapping rice; however, their coarse spatial resolution (250 m to 1 km) makes it difficult to map smallholder rice fields in Asia, which are typically smaller than 1 hectare. On the other hand, these sensors have the advantage of high temporal resolution due to their higher swath width (2200 km for SPOT VEGETATION, 2330 km for MODIS, 2500 km for AVHRR), and have been successfully used to map rice at the country, multi-country, or continental scale (Gumma et al. 2011a, Salam and Rahman 2014, Clauss et al. 2016). Time-series vegetation indices have been utilised in change detection studies. Gumma et al. (2011b) used time-series MODIS NDVI data of 10 years (2000–2009) for tracking changes in rice cultivation areas in Nepal. Singh et al. (2006) used AVHRR NDVI to monitor changes in rice growth patterns in Punjab (India) between 1981 and 2000 and noted a shift in the rice growing pattern, advancing by approximately 3 to 4 weeks.

Examples of studies on rice mapping conducted in different countries using optical remote sensing data have been described along with their methodology, classification accuracy, and other details in Table 4.5.

TABLE 4.5
Studies Using Optical Remote Sensing Data in Mapping Rice in Different Countries

Sensor	Method	Outcome
Landsat MSS (G, R, two NIR bands)	Evaluated two methods of data analysis; the first was the use of a maximum likelihood classifier for generating rice maps, and the second was the use of a vector classifier using water and green canopy reflectance.	The vector classifier showed higher agreement during calibration (about 94%) in 1983/84 and about 95% in 1984/85 rice season in New South Wales, Australia (McCloy et al. 1987).
Landsat MSS/TM/ IRS-1A LISS-I (VNIR bands)	A stratified random sampling approach comprising a 10% sample fraction was used to reduce the cost and time required for analysis. Selected sample segments (10 × 10 km each) were classified using supervised maximum likelihood classification to estimate *kharif* rice acreage for three years (1986–87 using Landsat MSS,1988–89 using IRS-1A LISS-I, and 1989–90 using IRS-1A LISS-I supplemented by Landsat MSS/TM), for the entire state (geographical area 15.54 M ha, *kharif* rice area 4.22 M ha)	Accuracy was evaluated against the BES (Bureau of Economics and Statistics) estimate. 90/90 accuracy was obtained at the state level, but relative deviation (RD) and CV were found to be large at the district level. State-level RD was high in the first year, 1986–87 (26%), but reduced during 1988–89 to 4.31%, and during 1989–90 to 6.64%. Cloud-free data availability during the critical month (October) remained a major problem, resulting in no estimate possible during 1987–88. Orissa, India (Panigrahy et al. 1991).
Landsat ETM+	Used temporal images from transplanting to harvesting period; everal VIs including NDVI, SAVI, TVI, and rice growth vegetation index (RGVI) were used. The concept of RGVI was introduced using blue, red, NIR, SWIR 1 and SWIR 2 bands of ETM+, which was highly correlated with the rice age. RGVI was then used to develop a model to rice mapping. Two images in sequential acquisition dates were used to calculate rice plant age difference and to prepare rice map.	High correlation ($r^2 = 0.90$) between RGVI and rice age observed during calibration phase. Rice area was compared with available district statistics, which showed an r^2 of 0.97 in the study area in Bali, Indonesia (Nuarsa et al. 2011).
Landsat-8 OLI	Employed phenology to classify rice under the object-based image analysis framework. The first step included image segmentation, which was performed using multi-resolution segmentation algorithm in e-Cognition software. It was then classified based on neural network classification method. An object-based RF classifier, which was tested to produce accurate rice maps, showed the contribution of each object-based feature so that the users can understand the feature's importance.	Classification result with overall accuracy of 91%, and kappa coefficient of 0.82 could be achieved using only a bi-temporal dataset, which is important in view of the difficulties faced in monsoon season data acquisition from optical sensors in China (Su, 2016).

(Continued)

TABLE 4.5 (Continued)

Sensor	Method	Outcome
IRS P-6 (Resourcesat-1A) AWiFS (VNIR, SWIR bands)	Four dates AWiFS data were collected and analysed using multivariate statistics and classification to map *rabi* season crops, including rice. Principal component analysis (PCA) and separability measures were used for best band selection for crop mapping. Ratio indices like NDVI, MSI, and the self-developed TBI were also used in classification.	Incorporation of SWIR band in MXL classification improved overall accuracy. A three-band ratio index (TBI) was developed with Red, NIR, and SWIR bands, which showed improved discrimination in multi-date data classification compared to other available SWIR band indices. (Panigrahy et al. 2009).
SPOT-XS (VNIR bands)	Images were acquired during pre-flooding, and first half of the flood period (early stages of rice). Two classification approaches were followed. The first approach used the integration of both supervised and unsupervised classification of a database formed by the principal component reduction. The second method included unsupervised classification, followed by stratification, and then supervised classification.	The second approach showed higher classification accuracy (~70%) in Niger Delta, Mali (Turner and Congalton, 1998).
Terra-Aqua MODIS	Time series of 8-day, 500-m spatial resolution composite images from the MODIS sensor were used to prepare a rice map for 2000–2001. A suite of methods, including spectral matching techniques, decision trees, and an ideal temporal profile data bank, was used to map rice area over a large spatial extent, covering six South Asian countries, with 12 rice classes. Ancillary data sets included spatial data (elevation, precipitation), national statistics, maps, and a large number of field-plot data.	Rice maps and statistics were compared against independent field-plot points, and available sub-national statistics on main season rice (*kharif* / rainy season). A fuzzy classification accuracy testing showed that 12 rice classes were mapped with absolute accuracy ranging from 67 to 100% for individual classes and an overall accuracy of 80% for all 12 classes. The intermixing was only within rice classes. When all the 12 classes were pooled into a single rice class, overall accuracy was ~ 95%. MODIS–derived area statistics was also correlated with sub-national statistics of these countries, which showed an R^2 value of 97% at district level, and 99% at the state level for the six countries of South Asia (Bangladesh, Bhutan, India, Nepal, Pakistan, and Sri Lanka) (Gumma et al. 2011a).
Terra-Aqua MODIS	Time-series MODIS products MOD13Q1 and MYD13Q1, and a one-class SVM to detect and classify rice in continental China. Rice maps were prepared for 2002, 2005, 2010, and 2014 at 250 m resolution.	Overall classification accuracy of 0.90 and a kappa coefficient of 0.77 were observed compared to the reference dataset of 2014. The maps correlate highly with the national statistics (R^2 of 0.92 in 2010, 0.90 in 2005, and 0.90 in 2002) of China (Clauss et al. 2016).
NOAA AVHRR	Used NDVI images, one image in 1989 acquired during the peak greenness stage, and four images in 1999 acquired during the vegetative to peak greenness stage, to map *boro* (summer) rice. The density slicing approach was adopted for area estimation.	Over-estimation of 23.3% and 27.5% in rice area was observed during 1989 and 1999, respectively, in Bangladesh (Salam and Rahman, 2014).

4.3.5 Microwave Remote Sensing Techniques Applied to Rice Area Mapping

Microwave remote sensing has two major advantages: its all-weather capability (not being affected by cloud cover, rain, snow, and solar irradiance) and the specific physical interaction between the radar waves and the rice canopy over inundated or moist soil surfaces. In such situations, most studies have demonstrated the usefulness of SAR data at C, L, and X bands from satellites ERS-1 and -2, JERS-1, RADARSAT-1 and -2, ENVISAT ASAR, ALOS PALSAR, RISAT-1, COSMO-SKYMED, Terra SAR-X, Sentinel-1, and so on, for delineating rice cultivation areas and extracting biophysical parameters of rice, with potential applications in yield prediction and production forecasting (Bouvet and Le Toan, 2011; Shao et al., 2001; Ribbes and Le Toan, 1999). The growing accessibility of free Sentinel-1 data has significantly facilitated the use of SAR imagery for rice detection and classification. The use of ERS-1 data, the first operational microwave satellite launched in 1991, in mapping rice showed encouraging results (Patel et al. 1995, Ribbes and Le Toan, 1996, Panigrahy et al. 1997). These ground-breaking studies were constrained to small spatial scales due to limitations such as insufficient high-quality ground-truth data, reliance on single polarisation, low temporal resolution, or challenges associated with handling large data volumes (Zhao et al. 2021). Later studies shifted their focus toward the utilisation of SAR data from RADARSAT-1, JERS-1, RISAT-1, Sentinel-1, and other satellites with sensors having different frequencies and polarisations, to enhance rice mapping accuracy across extensive regions using different algorithms (Ramadhani et al., 2020).

Rice mapping methods using microwave SAR data can be broadly divided into two groups: machine learning and empirical model–based methods. In recent years, machine learning techniques have been widely adopted, leveraging extracted features from backscatter coefficients as inputs to classification models for rice mapping (Zhao et al. 2021). Both traditional machine learning models (decision tree, support vector machine, random forest) and deep learning models (CNN, RNN) are used, which are similar to those used in optical remote sensing data analysis. The only difference lies in the input data for classification, which is the backscatter coefficient values in the case of microwave SAR analysis. The incorporation of phenological data and texture features is also used to improve the classification accuracy.

In the empirical model method, a mathematical model is formulated to represent the variation in backscatter coefficients throughout the rice growth cycle, determining the threshold, coefficient, and other parameters, and extracting rice area based on these parameters. However, the development of an empirical model requires existing high-resolution land cover data to derive the equation and define classification thresholds. In the absence of ground reference data, threshold values from prior studies are adopted, which may introduce classification errors (Zhao et al. 2021). Examples of studies conducted using data from microwave satellite sensors in different rice-growing countries, along with their methodology and outcome, have been summarised in Table 4.6.

TABLE 4.6
Studies Using Microwave Remote Sensing Data in Mapping Rice in Different Countries

ERS-1 (C-band, VV polarisation)	A maximum likelihood classifier was used on multi-date SAR data acquired within 30 days of transplanting and at the flowering stage (~ 60 days)	Accuracy of 90 and 91.5% in the case of Howrah and Hughly districts, respectively, in West Bengal, India (Panigrahy et al. 1997)
RADARSAT-1 (C-band, HH polarisation)	Three multi-temporal images were used for each of the standard and fine beam modes acquired during the transplanting and reproductive stages. Backscatter coefficients of rice showed a significant temporal variation; however, the dynamic range in Radarsat was found to be lower than that of ERS data, due to higher backscatter at HH polarisation than the VV polarisation during the early crop stage.	When compared with the available land cover map, an accuracy of 87% was found in Java, Indonesia (Ribbes and Le Toan, 1999).
RADARSAT-1	Temporal images were acquired during wet and dry seasons for rice mapping. Four classification approaches were adopted: neural network classification, maximum likelihood classification, change detection, and integration of neural network and change detection.	Integration of change detection and neural network performed best with an accuracy of 96% in Munoz and Santo Domingo, Philippines (Chen and McNairn 2006).
ENVISAT ASAR (C-band with HH and VV polarisation)	Multi-temporal HH and VV polarised images were used to develop an empirical relationship between the backscattering coefficient, plant height, and rice biomass.	An accuracy of 81% was obtained in Fuzhou, China (Wang et al. 2010).
RISAT-1 (C-band SAR with HH, HV polarisations)	Multitemporal SAR data was analysed with a sample segment approach and a hierarchal decision rule-based classifier (HDRC) as an operational project.	State- and national-level statistics were generated during 2013, whereas during 2014–2016, district-level statistics were also generated for *kharif* rice in India (Jain et al. 2019).
COSMO-SKYMED and TerraSAR-X (X-band, HH polarisation)	Used a new rule-based algorithm on multi-temporal SAR to demonstrate operational mapping of rice in diverse growing environments across south- and south-east Asia (wet and dry seasons, irrigated and rainfed, direct-seeded and transplanted, short- and long-duration).	Under the RIICE (Remote sensing–based Information and Insurance for Crops in Emerging economies) project, 1.6 M ha rice area mapped with classification accuracy ranging from 85– 95% in 13 sites across 6 countries—Cambodia, India, Indonesia, Philippines, Thailand, and Vietnam (Nelson et al. 2014).

(Continued)

TABLE 4.6 (Continued)

ALOS-2 (L-band, HH and HV polarisation) and Sentinel-1(C-band, VV and VH polarisation)	Two regional initiatives, Asia-RiCE and Space Applications for Environment (SAFE) of the Asia-Pacific Space Agency Forum, carried out rice mapping and growth monitoring in Cambodia, Indonesia, Thailand, and Vietnam. A machine learning algorithm was applied to time-series ALOS-2 ScanSAR data (100 m resolution), and Sentinel-1 SAR (10 m resolution). Training data sets were selected, and SAR data were analysed into rice and non-rice classes.	Results of Phase I and II of Asia-RiCE showed that work in some countries (e.g., Indonesia) has reached operational level in SAR–based rice mapping. In Cambodia, only ALOS-2 was used, with classification accuracy exceeding 90%. Data validation was done with the field staff. In Vietnam, Sentinel-1 was used to map all three-season rice (WS—Winter-Spring, SA—Spring-Autumn, AW—Autumn-Winter) in the Mekong Delta, and ALOS-2 for only WS season rice; the accuracy was 87% In Indonesia, both Sentinel-1 and ALOS-2 data were used; the highest accuracy was 90.3%, with a kappa coefficient of 0.86 (Oyoshi et al. 2022).

4.3.6 Integrated Approaches Using Optical and Microwave Remote Sensing Data for Rice Mapping

Optical and microwave remote sensing datasets possess distinct characteristics and individual strengths in rice mapping. Studies conducted using the integration of data from both sensors have shown higher classification accuracy compared to the single data source. Two approaches have been used in these studies: the complementary approach and the comparison class approach. The complementary approach mainly includes the following three methods of integration (Zhao et al. 2021):

i. Phenological information is derived from optical remote sensing data, and the SAR data is analysed based on the phenological information for rice mapping (Verma et al., 2025). Torbick et al. (2017) followed this approach in Myanmar, where Landsat-8 and other data were used to generate a land cover map, and then SAR backscatter data were used to generate a rice map (refer to Table 4.7).
ii. Optical features and radar features of rice are input to the classifier together for mapping rice. Fiorillo et al. (2020) used Sentinel-2 time series NDVI and SAR time series data to input into an object-based RF classification together. The combined dense time series provided reliable predictors for RF classification in Senegal (refer to Table 4.7).
iii. Classification was independently performed using each data source, and their intersection was taken as the final output. Ramadhani et al. (2020) applied SVM classification to Sentinel-1, Sentinel-2, and MODIS datasets, combining the individual results to produce a rice distribution map for Java, Indonesia.

The comparative classification approach employs various data fusion strategies and classification algorithms to identify the most effective combination, utilising both optical and microwave remote sensing datasets. Onojeghuo et al. (2018) used Landsat-8 NDVI and dual-polarisation Sentinel-1 data, and RF and SVM were applied for rice mapping in China. RF, combined with multitemporal VH and NDVI, outperformed SVM in classification accuracy. (refer to Table 4.7). Examples of studies conducted using the integration of optical and microwave sensor data in different rice-growing countries, along with their methodology and outcome, have been summarised in Table 4.7.

TABLE 4.7
Studies Using Integration of Optical and Microwave Imagery in Rice Area Delineation in Different Countries

Sensor	Method	Outcome
Landsat TM (visible and SWIR bands) and JERS-1 SAR (L-band, HH polarisation)	Used unsupervised classification of TM data to map arable land during the dry season and SAR data to delineate rice area during the rainy season.	The estimated rice area was 12–14% less in Indramayu, Indonesia (Okamoto and Kawashima, 1999).
IRS-1D (G, R, and NIR bands) and RADARSAT-1 SAR (C-band, HH polarisation)	LISS-III data were used to map dry/summer season rice using supervised MXL classification, while temporal SAR data were used to map wet/rainy season rice. Both outputs were combined to produce a year-round rice map.	Rice classification accuracy improved from 81.6% using optical data to 96.6% using combined SAR–optical datasets in West Bengal, India (Choudhury and Chakraborty, 2004).
Sentinel-1 SAR (C-band, VH, VV polarisation), ALOS-2 satellite PALSAR-2 (L-band, HH, and HV polarisation), and Landsat-8 OLI (VNIR, SWIR)	Sentinel-1 SAR, Landsat-8 optical, and PALSAR data were integrated and classified using an RF algorithm to prepare an updated land use/land cover map. Time series phenological analysis of SAR was then executed to map rice by extrapolating unique indicators tied to rice (inundation and growth stages).	SAR data-based rice classification outperformed optical data, and optical and SAR combination outperformed single sensor data-based classification in Myanmar (Torbick et al. 2017).
Sentinel-1 SAR (C-band, VH, VV polarisation), and Landsat-8 OLI	A co-registered set of 10 dual polarisation (VH, VV) SAR data and Landsat-derived NDVI images depicting crop phenological stages was used as input to SVM and RF machine learning classification algorithms to map rice.	A significant increase in overall classification accuracy was obtained (overall classification accuracy 95.2%, and for rice 96.7%) when the NDVI time-series data were integrated with combinations of multi-temporal polarisation channels in Sanjiang plain, North-East China (Onojeghuo et al. 2018).

(Continued)

TABLE 4.7 (Continued)

Sensor	Method	Outcome
Sentinel-1 SAR (C-band, VH, VV polarisation), and Landsat-8 OLI	A rule-based algorithm was applied to synthetic statistical metrics (TSDs— Temporal Spectra Descriptors) computed from temporal datasets of optical spectral indices and SAR backscatter coefficient.	Overall rice mapping accuracy was 87.8% with commission and omission errors of 3.5% and 27.4%, respectively. Discrimination between dry-seeded and flooded rice was also achieved with 70% accuracy in Northern Italy (Stroppiana et al. 2019).
Sentinel-1 SAR (C-band, VH, VV polarisation), and Sentinel-2 MSI (G, R, Red-edge, NIR, SWIR bands)	Sentinel-2 time series NDVI and SAR time series data were input for an object-based RFclassification to map rice at high resolution in a complex environment of fluvial lowland valleys. Guided regularised random forest (GRRF) was used to identify the most important predictors, on which RF classifier was then applied.	Overall classification accuracy improved from 77.4% using SAR data to 79.9% using MSI data, and 87.4% using a combination of both datasets. User's accuracy also increased from 80.0% to 86.3%, and 89.4% in only SAR, only multispectral, and a combination of both datasets, respectively, in Senegal (Fiorillo et al. 2020).
Sentinel-1 SAR (C-band, VH, VV polarisation), and Sentinel-2 MSI (G, R, Red-edge, NIR, SWIR bands)	SAR time series data were used in a decision tree to produce a provisional rice map. An additional mask using optical data-based water and vegetation index was overlaid to remove non-rice fields.	Improvement in mapping accuracy was observed compared to only SAR data-based classification. A rice map for Japan was developed with a 30 m spatial resolution (Inoue et al. 2020).

4.4 REMOTE SENSING IN NATIONAL LEVEL RICE AREA MAPPING AND PRODUCTION ESTIMATION IN INDIA

India, the country with the largest rice-cropped area, started using space-based remote sensing technology for rice mapping during the 1980s. Under the Indian Remote Sensing Utilisation Program (IRS-UP) initiated by the Department of Space (DOS), systematic plot-level experiments on rice and wheat were carried out during 1984 to understand their spectral characteristics and to develop an empirical yield model using spectral data (Sahai, 1985). Concepts developed during these experiments were later utilised in satellite remote sensing-based crop area mapping and yield estimation studies. In 1988, the Crop Acreage and Production Estimation (CAPE) project was launched with broadened objectives and an expanded scope (Dadhwal et al., 2002). The Ministry of Agriculture sponsored CAPE, which was collaboratively implemented by the Department of Space, various state remote sensing centres, state agriculture Departments, and agricultural universities. The project undertook a focused effort to design a methodology suitable for application across extensive regions (Sahai and Dadhwal, 1990). Crops included wheat, rice, sorghum, groundnut, rapeseed/mustard, and cotton. Remotely sensed data of moderate spatial resolution (Landsat MSS & TM, IRS LISS-I, -II, and -III) was used. Data analysis was done using maximum likelihood supervised classification with limited ground truth to generate a training signature. A two-step stratification based on agrophysical and crop distribution was found to improve the efficiency of stratification (Parihar and Dadhwal, 2002). A 5 × 5 km sampling frame for the state was made, and 20% sample segments were analysed. Samples were randomly selected in proportion to the size of each stratum, and crop area estimates were derived based on the proportion of crop observed within the samples (Parihar and Manjunath 2013). During phase I of the CAPE project (1988–1991), district and state-level rice mapping was taken up in two states. The results indicated that the rice acreage estimates could meet the 90/90 accuracy goal at the state level (Panigrahy et al., 1991). During phase II of the CAPE project, which started in 1992, 11 states were taken up for rice mapping, which together contribute more than 80% of the national rice production (Parihar and Dadhwal, 2002).

From the 1998–1999 cropping season, multiple forecasts of *kharif* rice at the national level were initiated using SAR data in 13 states of India, representing over 93% of the national rice cultivation area and output (Parihar and Oza 2006). Radarsat ScanSAR Narrow Beam (SCNB) data were used due to its large area coverage (300 km swath width). Data acquired on three dates were used in the analysis. Around 90 scenes of multi-temporal data (covering transplanting, vegetative, and reproductive phases) were analysed every season. The CAPE initiative was subsequently expanded into a nationwide program known as Forecasting Agriculture using Space, Agrometeorology, and Land-based Observations (FASAL), which has been operational since 2007. The FASAL project aims at providing multiple pre-harvest production forecasts of crops at the national/state/district level (Parihar and Oza, 2006). Multi-date SAR data is used for both *kharif* and *rabi* rice estimation. During 2007–2012, Pre-harvest assessments of crop area and yield were carried out using data from RADARSAT-2 ScanSAR data. The FASAL methodology incorporates econometric modelling to project crop area and production before the commencement of sowing activities. FASAL is a collaborative programme involving multiple institutions such as IMD, the Institute of Economic Growth, state agriculture departments, state remote sensing centres, and SAC-ISRO. Since the 2013 rice season, it's been implemented by the Mahalanobis National Crop Forecast Centre, which operates under the Ministry of Agriculture (Ray et al. 2014).

In the year 2013, use of data from RISAT-1, an Indian microwave satellite with C-band SAR and HH polarisation began for rice acreage estimation at the state and national level, and 2014 onwards at the district, state, and national level in 14 major rice growing states of India (Ray and Neetu, 2017, Jain et al., 2019). In 2017, due to disruption in RISAT-1 data availability, the use of Sentinel-1 C-band SAR was initiated for rice acreage and production estimation, which is still continuing. Hierarchical decision rule–based classification (HDRC) approach is being used to delineate rice fields within the sample segments, and to compute district-level pre-harvest rice acreage for all major *kharif* rice growing states of India. SAR C-band VV polarisation data from Sentinel-1A and -1B have shown great potential for mapping the rice fields (Jain et al., 2019). *Kharif* rice forecast is made three times, first at the end of August, second at the end of September, and the third at the end of the third week of January

TABLE 4.8
Chronology of RS Data-based Methods Followed for Rice Mapping in CAPE and FASAL Projects in India

Year	Satellite & Sensor	Methodology	States Covered/ References
1988 (CAPE Phase I)	Landsat MSS and IRS-1 LISS- I	Single-date data analysis using MXL supervised classification; 10 × 10 km sampling frame;	Two major rice-growing states (Parihar and Dadhwal, 2002)
1992 (CAPE Phase II)	IRS LISS-I/ LISS-II	5 × 5 km sample segment, 20% segment analysis	Eleven major rice-growing states (Parihar and Dadhwal, 2002)
1996	ERS-1 SAR	District & state level, decision-rule based multi-date SAR data analysis; 5 × 5 km. sample segment	Three states (Assam, Orissa, West Bengal) (Parihar and Dadhwal, 2002)
1997–2001	RADARSAT-1 ScanSAR, narrow beam	District & state level, multi-date SAR data; sample segment-based analysis	One state (1997), 4 states (1998), 10 states (2000), 7 states (2000) (Parihar and Dadhwal, 2002)
2006 (FASAL Pilot study)	RADARSAT ScanSAR	State & national level *kharif* rice, multitemporal SAR, 5 × 5 km sample segment.	One state (Orissa) as a pilot study (Parihar and Oza, 2006)
2007 (FASAL implementation)	RADARSAT ScanSAR	Multiple time forecast, 5 × 5 km, 15% sample segment	13 major rice-growing states
2012 (FASAL Taken over by MNCFC, Dept. Agric. & FW)	RADARSAT-2 ScanSAR	State & National level, multitemporal, 5 × 5 km sample segment, HDRC–based analysis	14 major rice-growing states (Jain et al., 2019)
2013	RISAT-1 ScanSAR HH	State & National level, multitemporal, 5 × 5 km sample segment, HDRC–based analysis	14 major rice-growing states (Jain et al., 2019)
2014–2016	RISAT-1 ScanSAR HH	District, State & National level, multitemporal, 5 × 5 km sample segment, HDRC–based analysis	14 major rice-growing states (Jain et al., 2019)
2017 onwards	Sentinel-1, VV	District, State & National level, multitemporal, 5 × 5 km sample segment, HDRC–based analysis	14 major rice-growing states (Jain et al., 2019)

Note: HDRC–hierarchical decision rule–based classifier.

while *rabi* rice forecast is made in the first week of April. A chronology of CAPE and FASAL projects has been summarised in Table 4.8.

4.4.1 Discrimination of Rice Varieties

Despite several studies conducted in different rice-growing countries globally, using a variety of satellite data sources and analysis methods for rice mapping, publications on the discrimination of rice varieties are rare. This is due to the intricate interplay of biophysical and biochemical properties that shape the reflectance characteristics of a rice canopy (Yang et al., 2007). If the rice varieties have similar plant characteristics and crop growth conditions, water and nutrient supply, they show similar temporal spectral response (Patel et al., 1985), and hence are not possible to discriminate spectrally. Attempts have therefore been made to use crop duration and phenology of different varieties for their spectral separation. Phan et al. 2018 have used X-band SAR data from the COSMO-SkyMed satellite to differentiate long-duration Jasmine rice and another long-duration variety, OM4218, from a short-duration rice variety IR 50404 in two districts of An Giang province in Vietnam. The accuracy level obtained was 96% for the short- and long-duration rice varieties.

4.5 MAPPING AROMATIC BASMATI RICE AND VARIETY DISCRIMINATION IN INDIA AND PAKISTAN

Since aromatic Basmati rice is grown only in India and Pakistan, which is restricted to specific agro-climatic regions, a limited number of publications are available on Basmati rice mapping using remote sensing data. In both countries, it is grown only during the *kharif* cropping season extending from June to November, because of its specific cooler temperature requirement during the flowering and maturity stages of the crop. Due to the low yield potential of Basmati varieties, farmers often grow other high-yielding rice varieties (HYVs) on their farms side by side. The crop calendar of HYVs and aromatic Basmati rice varieties is not uniform during the growing season. These differences result in phenological stage differences between the different rice cultivars grown. Therefore, knowledge of the timeframe of various field operations, such as sowing in the seedbed, transplanting, irrigation, fertilisation, and harvesting, is necessary for mapping and monitoring rice varieties using satellite data.

4.5.1 Mapping Aromatic Basmati Rice Using Optical Remote Sensing Data

One of the initial efforts to distinguish Basmati rice from other HYVs and estimate its area was conducted using IRS-1A LISS-II satellite data in Haryana's Karnal district, India (Kalubarme et al., 1989). LISS-II data from two years (1988 and 1989), having a spatial resolution of 36 m, were analysed using the Maximum Likelihood (MXL) supervised classification technique. Variations in growth stages among rice varieties at the time of satellite data acquisition led to distinct spectral responses, which served as a key factor in differentiating Basmati rice from HYVs. The study further indicated that the satellite data collected during the second to third week of October provided the best separability between Basmati rice, HYV rice, and sugarcane in the area, as the Basmati varieties were at flag leaf to ear head emergence (early flowering) stages, whereas HYV rices were at the soft dough to physiological maturity stage, and their ear head and some leaves had started turning yellow. These phenological differences resulted in higher spectral reflectance of Basmati rice compared to HYV rices in the infrared band of LISS-II data of 22 October 1988 and 9 October 1989. The positive findings from this study stimulated further research efforts in India aimed at the spectral discrimination and spatial quantification of Basmati rice cultivation, employing satellite sensors with improved spatial and spectral resolutions. Later, Singh and Singh (2003) used multi-date IRS LISS II images to separate Basmati rice from other non-Basmati HYVs in a part of Bijnore district, Uttar Pradesh. Basmati rice could be discriminated from HYV rice in the October 2001 data due to the difference in their crop growth stages. Traditional Basmati rice, being long duration, is harvested late (last week of November) in this area, hence Basmati fields are generally left fallow during the next rabi season. Village-level crop inventory was also attempted to map rice and other crops in Gujarat state (Singh et al., 2013).

During the early years of Basmati mapping in India, non-availability of cloud-free data during June to first fortnight of September due to the rainy season in the Basmati-growing north- and north-western part of the country, and long revisit cycle of the available satellites (Landsat and IRS) led to mostly single-date satellite data analysis. After the year 2000, when data from new satellites with optical as well as microwave sensors (ERS-1, RADARSAT-1, Resourcesat-1) became available, the use of remote sensing technology moved from experimentation towards operationalisation.

The Basmati Export Development Foundation (BEDF), an organisation under the Agricultural and Processed Food Products Export Development Authority (APEDA) of the Ministry of Commerce and Industry, Government of India, took the initiative to use the remote sensing technique at operational-level Basmati rice area and production estimation in the entire Basmati-growing area of India. In 2005, BEDF commissioned RMSI Pvt. Ltd. to provide a remote sensing data–based acreage and production estimation for Punjab, Haryana, Uttar Pradesh, and Uttarakhand (Pateriya et al., 2009). During 2007, BEDF commissioned the study to AgriNet Solutions, New Delhi, to generate early-season estimates of cultivated area and expected yield for Basmati rice varieties in all growing districts spread across seven states of India: Punjab, Haryana, Western part of Uttar Pradesh, Delhi, and parts of Uttarakhand, Himachal Pradesh, and Jammu & Kashmir (AgriNet Solutions, 2007). Basmati varieties for which information was sought included traditional Basmati varieties (Basmati-370, Basmati-386, Type-3 (Dehraduni), Taraori Basmati, and Ranbir Basmati), evolved Basmati varieties (Pusa Basmati-1, Super Basmati), and non-notified varieties (Sharbati, Pusa-1121, and Permal). The methodology involved the analysis of IRS Resourcesat-1 AWiFS, LISS-III, and LISS-IV data along with GPS–based ground information, and ancillary data for the acreage assessment of two Basmati variety groups (traditional and evolved), and other non-notified varieties. The yield assessment was done based on the crop cutting experiments (CCE) and meteorological yield models. The successful results obtained during 2007 led to the continuation of the study by AgriNet Solutions until 2012 (Verma et al., 2022). A comprehensive account of the mapping methodology, results, and accuracy obtained in the study has been presented as Case Study I in the latter part of this chapter.

Rauf et al. (2022) developed a pixel-level classification framework using a deep convolutional neural network (CNN) architecture, integrating both spectral and temporal features from Sentinel-2 data, to differentiate Basmati and high-yielding rice varieties in Pakistan's Multan district. Multispectral reflectance (MSI) data in 12 spectral bands were collected at 16 different time frames to cover the entire rice cropping season. Radiometric indices, NDVI, NDWI, and Normalised Difference Moisture Index (NDMI) were used to enhance classification performance. Each input image represented the spectral reflectance profile of an individual pixel across 12 distinct bands, three spectral indices (NDVI, NDWI, and NDMI), and 16 temporal data points covering the rice cropping season. Basmati rice could be classified with achieving a superior accuracy of 99.7%, in contrast to the 95.2% observed for HYVs. An overall accuracy of 98.6% could be achieved following this methodology.

4.5.2 Mapping Aromatic Basmati Rice and Variety Discrimination Using SAR Data

Only a few studies are available on the use of SAR data in the discrimination of Basmati from non-Basmati rice. In 2011, a feasibility assessment was carried out in a region of Punjab, India, utilising RADARSAT-2 C-band ScanSAR narrow beam imagery acquired in both HH and HV polarisations. Temporal variations in HH backscatter, combined with changes in the cross-polarisation ratio (HH/HV) across different phenological stages from July to September, contributed to distinguishing Basmati rice from non-Basmati high-yielding varieties. The temporal shift in transplanting of Basmati rice, its longer duration, and difference in canopy volume backscatter formed the basis for its differentiation from non-Basmati HYVs. A classification accuracy of 91% was achieved in distinguishing aromatic Basmati rice from HYV rice; however, differentiation among individual Basmati cultivars was not pursued in the study (Jain et al., 2014). Differentiation of crop growth stages and LAI of Basmati and non-Basmati rice using C-band dual polarised RISAT-1 SAR was attempted by Mamta Kumari et al. (2015) in Karnal district, Haryana. Four data sets from July to October 2012 covering different crop growth stages of rice were analysed. The backscatter coefficient in HV polarisation was found to be most sensitive to crop growth and LAI of both rice types. However, the prediction of LAI could be possible only up to 3.8, and proved difficult at later stages due to the complex behaviour of the backscattering coefficient.

In another study conducted in a part of Karnal district, Haryana, the polarimetric target decomposition technique was attempted on temporal SAR data to distinguish Basmati and non-Basmati rice. To investigate scattering behaviour across different phenological stages of Basmati and non-Basmati rice, fine quad-polarised single-look complex (SLC) data from RADARSAT-2 were utilised. A decision tree classifier was employed to differentiate the two rice types using SAR observations acquired between July and September (Vineet Kumar et al., 2016). The findings demonstrated that various rice types and dominant land cover categories could be effectively distinguished using rule-based thresholds applied to σ values derived from the decomposition of scattering mechanisms based on Freeman's three-component model. A rule-based decision tree approach was developed to differentiate the two rice categories by leveraging temporal dynamics in scattering components. The classification yielded user accuracies of 85.2% for non-Basmati and 82.9% for Basmati rice. Delayed transplanting of Basmati leading to the difference in crop growth stages, differences in cultural practice, and longer duration of Basmati helped in discrimination of Basmati and non-Basmati rice with a high level of accuracy.

Two case studies on the use of satellite remote sensing data in Basmati rice mapping have been presented, one using medium resolution optical sensor data from IRS Resourcesat-1, and the other using combined optical and microwave sensor data from Sentinel-1 and -2 (Verma et al., 2025).

4.6 CASE STUDIES ON AROMATIC BASMATI RICE VARIETY DISCRIMINATION AND MAPPING

4.6.1 Case Study I: Basmati Rice Mapping and Varietal Discrimination in the Northern and North-Western States of India (using optical remote sensing data)

This case study documents an operational study using IRS Resourcesat-1 multispectral, multi-temporal, multi-resolution sensor data in mapping the entire Basmati rice area of India, spread in the seven northern and north-western states of the country. Different Basmati variety groups have also been separated based on their spectral data. The study was conducted at the behest of the BEDF to provide acreage and a pre-harvest production estimate of Basmati varieties (Verma et al., 2022). The major objectives were: (i) Discrimination and mapping of Basmati varieties and their separation into traditional and evolved varietal groups, (ii) Mapping non-notified rice variety Sharbati, and other HYV of rice, and (iii) District- and state-level acreage estimation of Basmati groups and other rices. The study was carried out during 2007–2012.

4.6.1.1 Study Area

The study area covers all Basmati growing districts in the seven states of India: Punjab, Haryana, Uttar Pradesh, Delhi, Uttarakhand, Jammu & Kashmir, and Himachal Pradesh. The districts in Jammu & Kashmir, Himachal Pradesh, and Uttarakhand form a part of the Himalayan submontane zone, and the remaining districts a part of the Indo-Gangetic Plains region (Figure 4.7).

The climate in the study area is subtropical monsoon type, with cold winters and hot summers, whereas in the south-western part of Punjab and Haryana, it is classified as tropical steppe, semi-arid type. The mean annual rainfall in western Uttar Pradesh is 839 mm, while in the northern part of Punjab, it is 853 mm. On the other hand, the south-western part of Punjab, lying in zone III of the state, receives only 484 mm of mean annual rain (Muruganandham and Singh, 2024). Similarly, the western part of Haryana receives much less rain due to its proximity to arid Rajasthan. The area has a well-developed irrigation system of canals and tubewells. The soil is fertile, alluvial, and affected by salinity/sodicity in places. Aromatic Basmati rice varieties grown in the study districts are of two types: Traditional Basmati and evolved Basmati. Traditional Basmati varieties, which include Basmati-370, Basmati-386, Type-3 (Dehraduni Basmati), Taraori Basmati, and Ranbir Basmati, have been grown by farmers in the area for several decades and are preferred by consumers for their pleasant aroma, good grain quality, good taste, and softness of the cooked rice. These varieties are tall, photo-period sensitive, long duration, and provide low grain yield. Evolved Basmati varieties include Pusa Basmati-1,

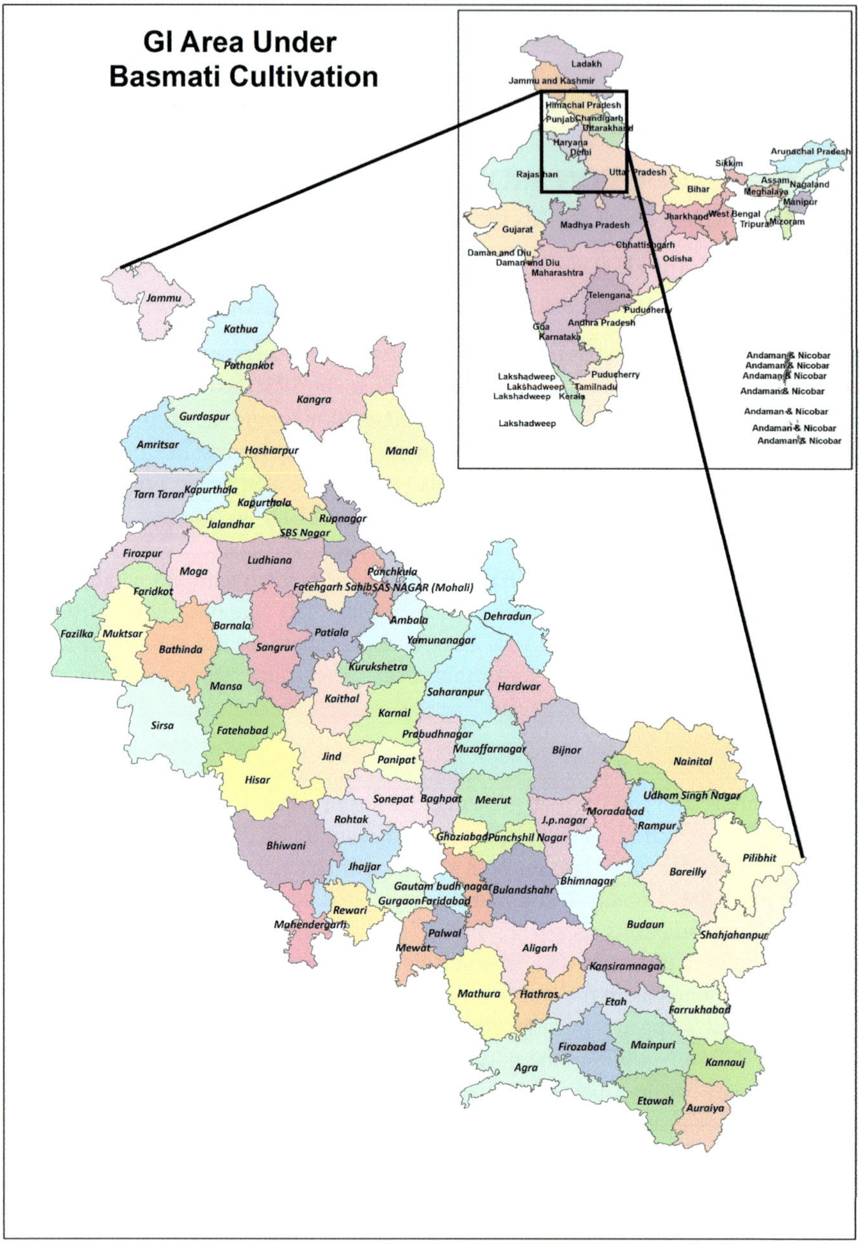

FIGURE 4.7 Location map of Basmati growing districts in seven Northern and North-Western States of India.

Note: Some of the districts have been bifurcated since these studies were carried out; hence, the district boundaries may differ.

Pusa-1121, CSR-30, and Super Basmati, and have been recently bred to overcome the shortcomings of the traditional varieties. These varieties are generally short-statured, photo-period insensitive, short duration, high grain yielders, and have one of the traditional Basmati varieties as their parent. Due to their high yield and shorter duration, evolved Basmati varieties are being preferred by farmers, hence their acreage has increased rapidly in recent years.

4.6.1.2 Remote Sensing Data

Multi-temporal cloud-free data from Resourcesat-1 Advanced Wide Field Sensor (AWiFS), LISS-III, and LISS-IV sensors during the *kharif* rice-growing season were acquired for this study during 2007–2012. Due to almost continuous cloud cover during the months of June, July, and August, generally no cloud-free data can be obtained during these months for such a large study area. During the second fortnight of September onwards, cloud-free data generally becomes available for analysis, as the monsoon starts receding from the area. Therefore, all cloud-free satellite acquisitions between September and November were utilised. AWiFS has the advantage of a large swath width of 740 km, and images the Earth once every 5 days at a spatial resolution of 56 m. Hence, the chances of getting cloud-free AWiFS data are much higher as compared to LISS-III. LISS-III covers a swath of 141 km at a spatial resolution of 23 m and has a 24-day revisit capability. During the *kharif* rice season 2007–2009, however, the availability of IRS-1D LISS-III data allowed revisit of 12 days (IRS-1D ceased operation in January 2010). The LISS-IV sensor has a higher spatial resolution of 5.8 m, but the sensor's narrow swath width of 23 km restricts the frequency of usable data acquisitions. Therefore, LISS-IV data were collected only for a few selected areas and used in the ground truth collection and validation of classification results and maps. Both AWiFS and LISS-III sensors collect data in four bands of the spectrum: green, red, NIR, and SWIR, whereas LISS-IV collects only in three bands (green, red, and NIR). The bandwidths of Resourcesat-1 are similar to Resourcesat-2 and -2A (refer to Table 4.4).

4.6.1.3 Ground Reference and Ancillary Data

The field data on Basmati varieties, HYVs, other rices, as well as other land use/land cover in the study area, were collected from the fourth week of August to the first week of October every year, coinciding with the rice flowering to grain formation/maturity stages. Data on rice sowing/transplanting dates, crop stage, crop health, and vigour were collected for all rice varieties. The GPS location of field plots as large as 300 × 300 m was collected for different HYVs and Basmati varieties spread in different states. A sample map showing the GPS location of field data points collected in Kaithal and Karnal districts of Haryana, and ground photographs of Basmati and other rice varieties are shown in Figure 4.8. The administrative boundaries of all the districts, their *tehsils* (sub-districts), and development blocks were collected, geo-referenced, and superimposed on the satellite images. Meteorological data like rainfall, temperature, occurrence of extreme meteorological events (floods, droughts, etc.), and their influences on rice varieties were also collected for different states under study.

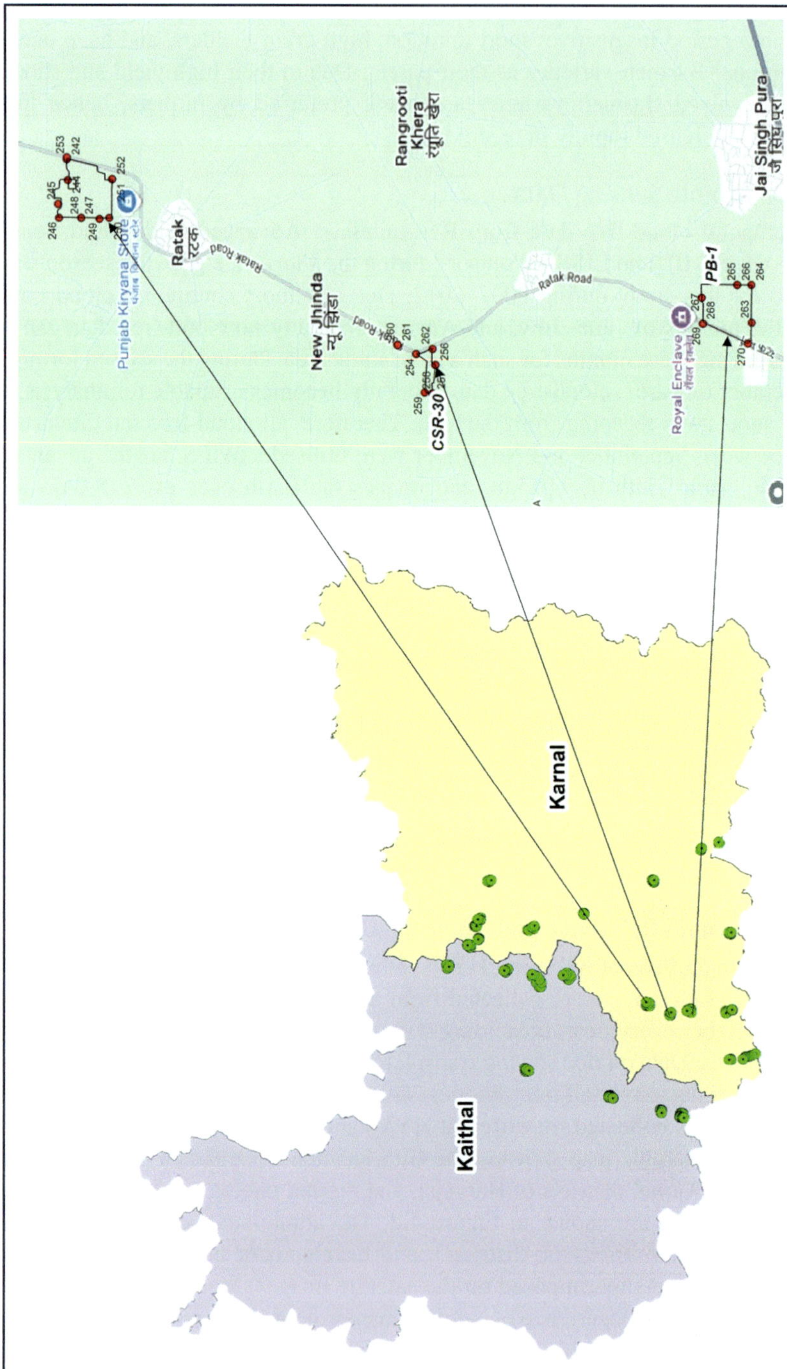

FIGURE 4.8 GPS locations and field photographs of Basmati and HYV Rice varieties in Kaithal and Karnal districts of Haryana, India.

4.6.1.4 Satellite Data Analysis

Two-stage satellite data analysis was conducted for the discrimination and mapping of Basmati and other rice varieties. Multi-date Advanced Wide Field Sensor (AWiFS) medium resolution (56 m) data was used for discriminating and mapping rice from other crops, and other land use classes. The higher resolution (23 m) LISS-III data were analysed to discriminate and map aromatic Basmati rice varieties, HYV rice, and non-notified rice variety Sharbati. A supervised classification approach was followed using the Maximum Likelihood (MXL) algorithm. The GPS locations of different Basmati rice varieties, HYV rice, Sharbati rice variety, and other land cover classes collected from the field were superimposed onto the LISS-III digital data; some of those locations were used as training sites during classification. In the initial years of study, Resourcesat-1 high-resolution LISS-IV images were also used during ground data collection in the selection of large patches of a rice variety or a land cover class to be used as a 'training site'. An example where a large Pusa Basmati-1121 rice growing area was identified in the generally cotton growing region of Bhiwani district, Haryana, on the LISS-IV image is shown in Figure 4.9. This also shows how Basmati rice is replacing cotton in some areas. Another example from Kaithal district of Haryana, where a large area growing traditional Basmati and evolved Basmati varieties (Pusa Basmati-1121 and Pusa Basmati-1) was identified on LISS-IV image for use as 'training sites', is shown in Figure 4.10. Spectral signatures of different classes were generated from the training sites to study their separability before satellite data classification.

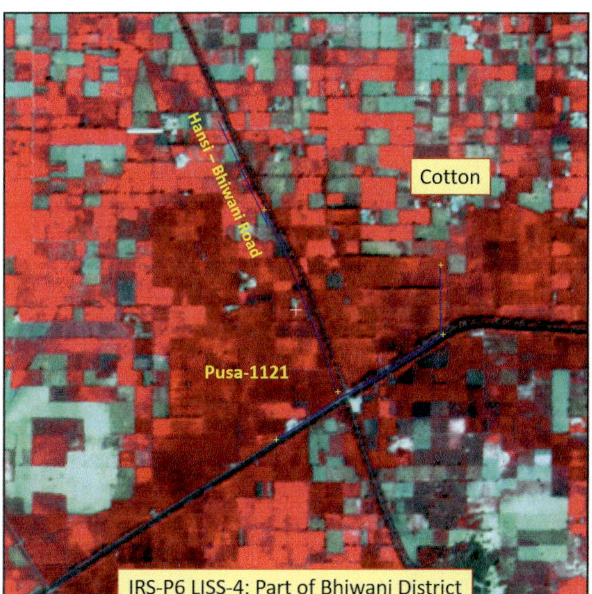

FIGURE 4.9 Resourcesat-1 LISS-IV image showing Pusa Basmati-1121 rice in the cotton growing region of Bhiwani district, Haryana.

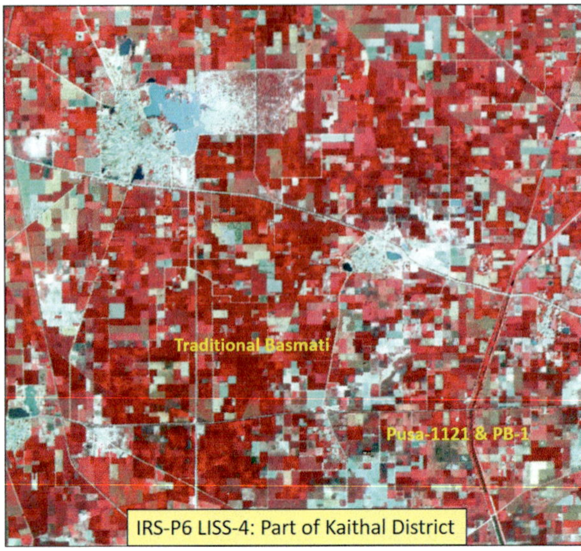

FIGURE 4.10 Resourcesat-1 LISS-IV image showing a large area under traditional Basmati and evolved Basmati varieties (Pusa Basmati-1121 and Pusa Basmati-1) in Kaithal district, Haryana.

In the case of multi-date satellite data availability for analysis, it becomes important to select a suitable data acquisition date, considering the specific objectives of the study. Figure 4.11 shows two LISS-III images collected on 31 August 2008 and 24 September 2008 of the same area in Karnal district. It is observed that the differentiation of Basmati and HYV rice is more clearly visible on the 24 September image because the two rice varieties are at different phenological stages on this date; HYV rice is at the flowering stage, whereas Basmati rice is at the booting stage. On 31 August, both rice varieties were at different stages of vegetative growth, hence lesser differentiation on satellite images, which may result in lesser classification accuracy.

4.6.1.5 Spectral Signature Generation and Separability Analysis

The variation in sowing and transplanting dates of different rice varieties results in differences in their phenological growth stage, leaf area index, and biomass content, resulting in variations in their spectral responses within multispectral data acquired on a given date. These spectral variations prove useful in the discrimination of Basmati and other rice varieties. Spectral signatures of various rice cultivars and land cover types across the green, red, NIR, and SWIR bands of LISS-III imagery have been illustrated in Figure 4.12 to study their spectral separability. Out of the four classified rice types, traditional (Basmati 1) and evolved (Basmati 2) varieties exhibited notably different reflectance patterns in the red, green, and NIR spectral regions when compared to HYVs and short-duration Sharbati rice. These differences are primarily attributed to contrasting phenological stages and canopy architecture. While HYVs and Sharbati were observed in the dough grain stage, Basmati types

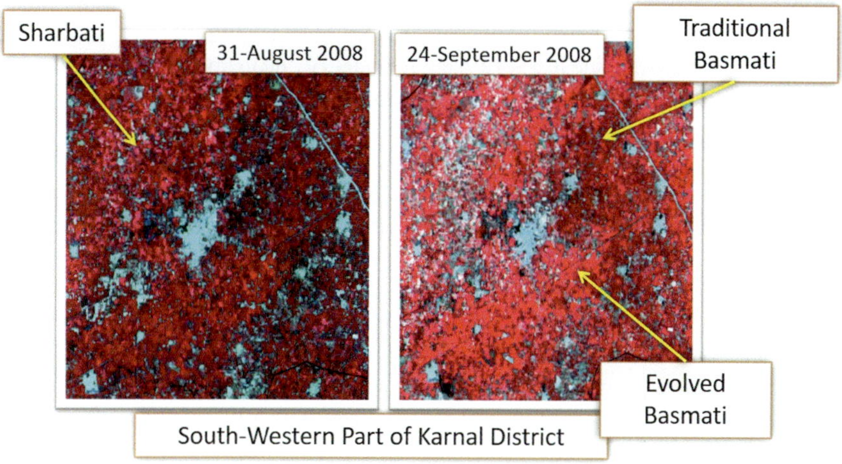

FIGURE 4.11 Resourcesat-1 LISS-III images collected on 31 August 2008 and 24 September 2008 of the same area in Karnal district, Haryana.

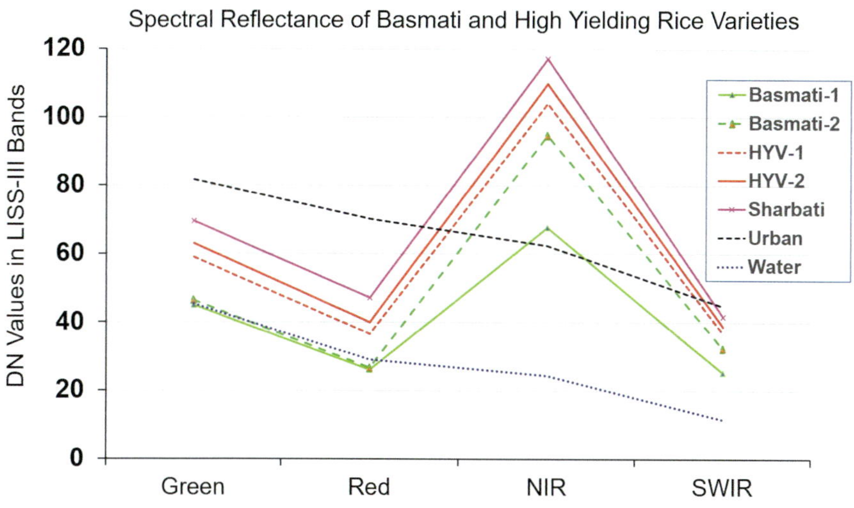

FIGURE 4.12 Spectral separability of two groups of Basmati (traditional and evolved), HYV rice, Sharbati rice, and other land features (urban, water, and fallow land).

were in earlier phases of development, ranging from flag leaf emergence to panicle initiation. Between Sharbati and HYVs, the former, being early maturing, shows higher reflectance compared to HYVs. Traditional and evolved Basmati groups show similar responses in the green and red bands but are separated in the NIR band due to differences in their plant structure and phenological growth stages. Traditional Basmati being at the flag leaf stage has a higher leaf area compared to Basmati 2 (evolved Basmati varieties) at the panicle emergence stage and thus shows higher NIR reflectance. All four rice types show a sharp increase in NIR band reflectance, followed by a sharp decline in SWIR band reflectance due to absorption by leaf water content, which is quite different from other land uses—urban, fallow, waterlogged, and water body.

Before using the spectral signatures of training sites in the supervised classification, the class separability was also studied by using the Bhattacharyya Distance. Table 4.9 shows the separability of different rice varieties and other land use classes.

4.6.1.5.1 MXL Supervised Classification

Apart from the LISS-III four-spectral-band reflectance data, the Normalised Difference Vegetation Index (NDVI) was also used in the classification. NDVI is sensitive to vegetative growth and LAI. The NDVI (Rouse et al., 1974) was computed using reflectance values (ρ) in the red and NIR bands as follows:

$$\mathbf{NDVI} = (\rho\mathbf{NIR} - \rho\mathbf{Red}) / (\rho\mathbf{NIR} + \rho\mathbf{Red})$$

Supervised classification was performed using LISS-III data comprising four spectral bands, supplemented by the NDVI. Spectral training signatures for various rice types and land-use categories were employed within a Maximum Likelihood (MXL) classification approach. The district-level rice acreage estimates obtained from supervised classification were aggregated to the state level. The Basmati and other rice variety distribution maps were generated for different districts and states. The classified Basmati rice map was then validated based on the available GPS–based field data that were not used as training sites. The maps were then finalised for the seven states under study.

4.7 RESULTS

4.7.1 Basmati Rice Map and Acreage Estimation

District-level Basmati rice maps have been prepared using Resourcesat -1/2 LISS III data. A sample map showing Resourcesat-1A LISS-III image of Gurdaspur, Amritsar, and Tarn Taran districts of Punjab acquired on 3 October 2010, and a supervised classification map showing Basmati, HYVs, non-Basmati long-grain Sharbati rice, and other land use classes in these districts have been presented in Figure 4.13 (a) and (b). Another sample map showing LISS-III image of Karnal and Kaithal districts of Haryana collected on 28 September 2010, and the classification map showing traditional and evolved Basmati varieties, and HYV rice in these two districts is shown as an example in Figure 4.14. Kaithal district shows a larger area under traditional Basmati compared to Karnal district, where evolved Basmati varieties are more prominent.

For the state-level maps, AWiFS FCC images have been shown for ease of presentation, and the images collected on 3 October 2010 for Punjab, Haryana, western Uttar Pradesh, and Jammu are given in Figures 4.15a–4.18a. Maps showing Basmati

TABLE 4.9

Spectral Separability Analysis of Rice Varieties and Major Land Cover Types

Separability Measure: Bhattacharyya Distance

Average Separability = 1.79434

Minimum Separability = 1.79434 0.76642

Average Separability = 2.00000

Signature pair with Minimum Separability: (Basmati-Evolved, High Yielding: HYV)

CLASS	Urban	Fallow	Waterlogged	Water	Basmati-Traditional	Basmati-Evolved	High Yielding Variety	Sharbati
Fallow	1.89975							
Waterlogged	2.00000	2.00000						
Water	2.00000	2.00000	1.50635					
Basmati-Traditional	2.00000	1.99999	1.99151	2.00000				
Basmati-Evolved	1.99999	1.99962	1.98733	2.00000	0.84118			
High Yielding Variety	2.00000	1.99984	1.99985	2.00000	1.12194	0.76643		
Sharbati	1.99999	1.99996	1.99997	2.00000	1.82969	1.32746	0.97988	

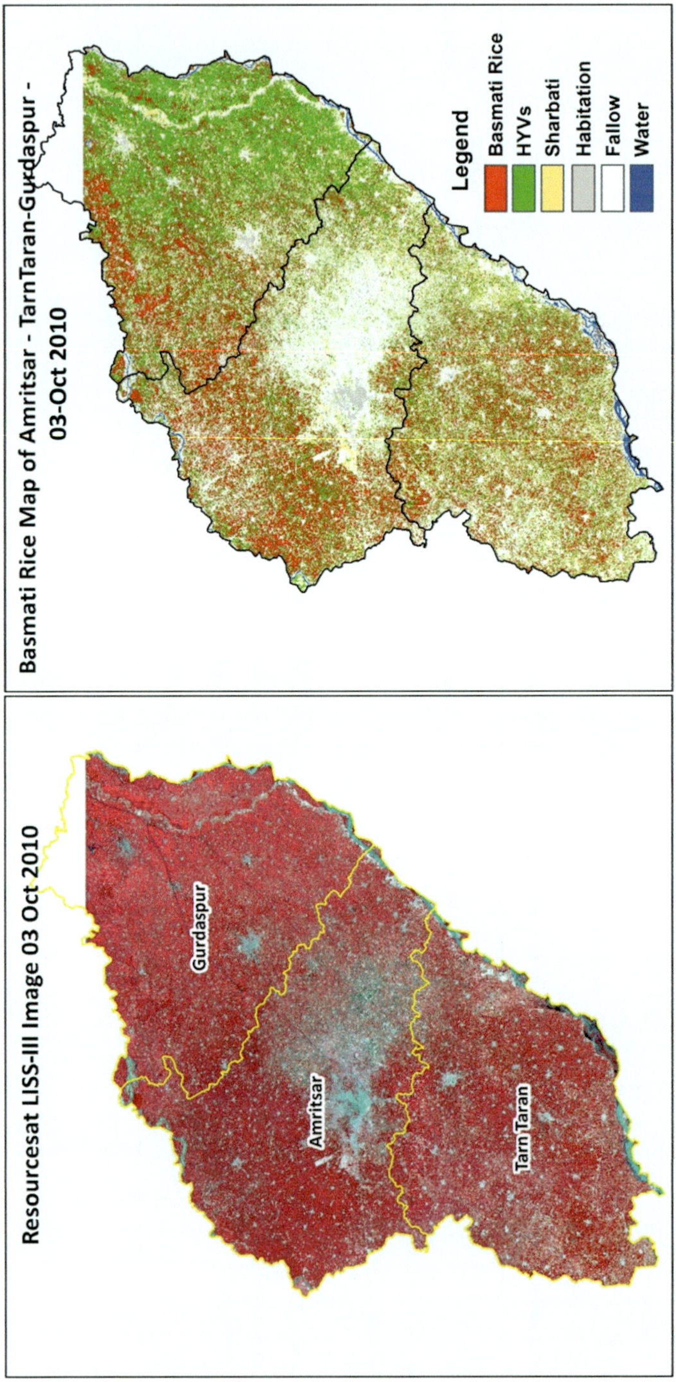

FIGURE 4.13 Resourcesat-1A LISS-III image of Gurdaspur, Amritsar, and Tran Taran districts of Punjab, India, 2010 (a), and classified map of the area (b).

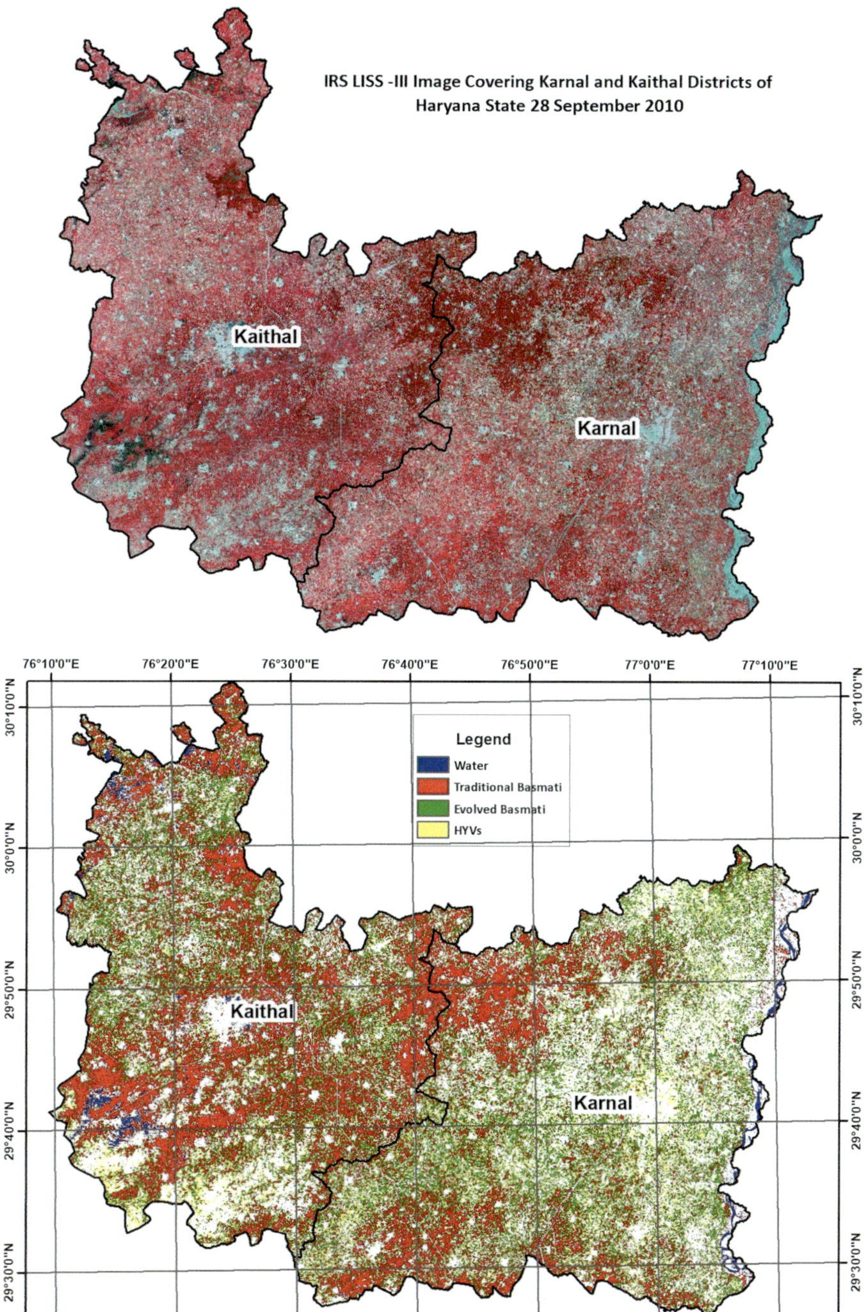

FIGURE 4.14 Resourcesat-1A LISS-III image of Karnal and Kaithal districts of Haryana on 28 September 2010 (a), and a classified map of the area (b).

varieties generated for Punjab, Haryana, western Uttar Pradesh, and Jammu have been presented as Figures 4.15b–4.18b, respectively.

The total rice acreage and the distribution of different rice varieties across states during the 2010 *kharif* season are presented in Table 4.10. The spatial variation in the area cultivated with Basmati and other rice varieties at the district level in Punjab, India, during the 2010 season is illustrated in Table 4.11.

4.7.1.1 Classification Accuracy Assessment

Classification accuracy was computed based on the confusion matrix, showing percentage of pixels correctly classified for each class. The classification accuracy assessment is shown in Table 4.12. While the overall accuracy is 88.45%, the user's accuracy for Traditional Basmati and Evolved Basmati is 88.91% and 79.76%, respectively. For HYVs and Sharbati rice, it is 78.77% and 79.09%, respectively.

4.7.1.2 Basmati Rice Scenario During 2007–2012

The results of the study indicate that it is possible to discriminate aromatic Basmati rice varieties using multi-date optical sensor data and a supervised classification technique. Based on the analysis, two groups of Basmati varieties (traditional Basmati and evolved Basmati), apart from a non-Basmati long-grained Sharbati and other HYV rice varieties, phenological stage differences enabled spectral differentiation between rice types. In the majority of the study area, HYVs are generally sown and transplanted approximately four weeks earlier than Basmati varieties, resulting in distinct temporal reflectance characteristics. The results of Basmati area estimation during 2007–2012 showed a continuous decline in traditional Basmati varieties, and a steady increase in evolved Basmati variety area (Table 4.13). During 2012, the traditional Basmati area showed a drastic decrease. This decrease is due to the longer duration of traditional varieties, which in turn delays the sowing of the next wheat crop, and also due to the lower price of Basmati rice during 2011. The steady increase in the evolved Basmati variety area is due to its higher grain yield and shorter maturity duration. Another interesting feature of the study is about fluctuation in total Basmati rice acreage. Total Basmati rice area generally shows an increasing trend from 2007 onwards, but it showed a significant decline from 1.91 million ha in 2012 to 1.71 million ha in 2011. The total Basmati area was reduced by about 16% due to the lower market price of Basmati received by farmers during the previous year.

4.7.2 Case Study II: Aromatic Basmati Rice Varietal Mapping, and Phenology Monitoring Using Sentinel 2 Optical MSI and Sentinel 1 Microwave Data

Apart from Basmati rice acreage mapping using optical remote sensing data explained in the earlier section, only a few studies involving microwave SAR data have been reported for Basmati rice mapping (Jain et al., 2014; Kumar et al., 2016). Given the complementary strengths of optical and SAR datasets in rice classification, several studies across major rice-producing regions have adopted a data fusion

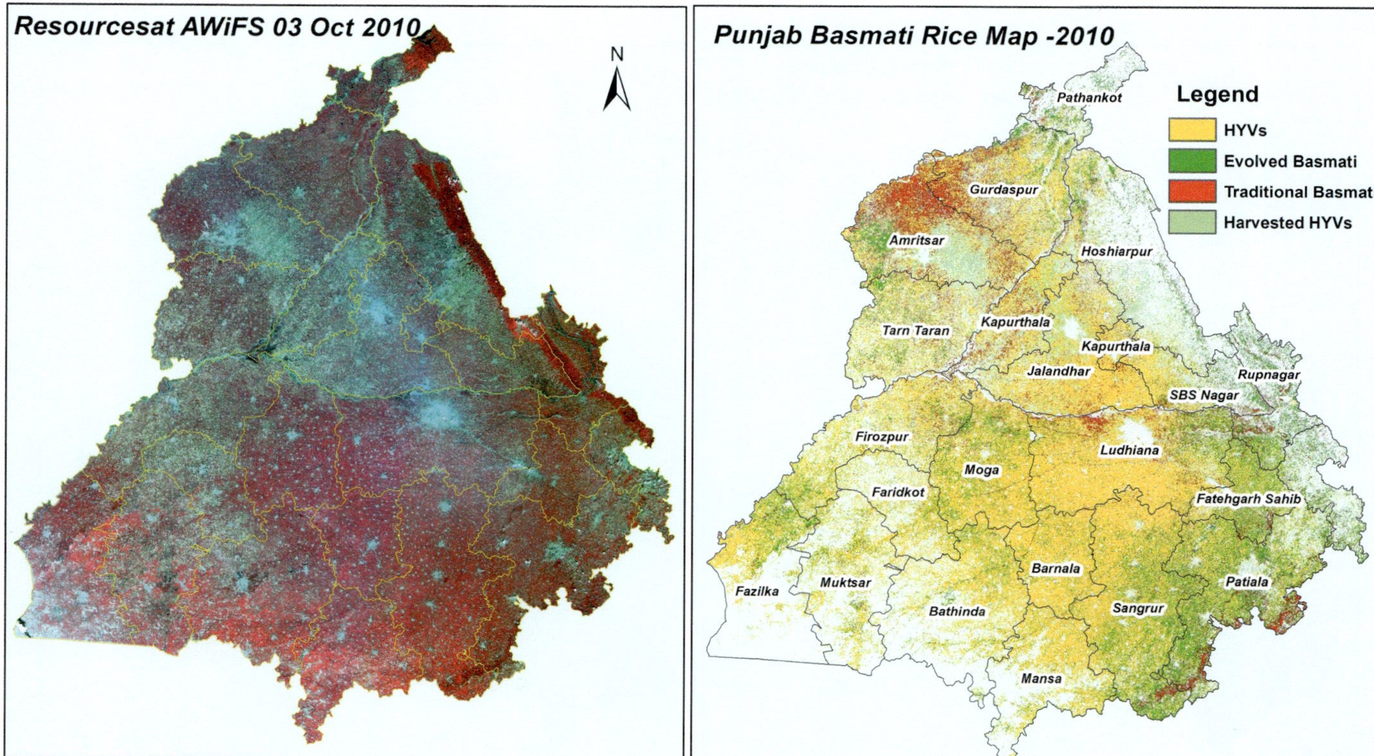

FIGURE 4.15 (a) AWiFS image collected on 3 October 2010 for Punjab, and (b) classified Basmati rice map.

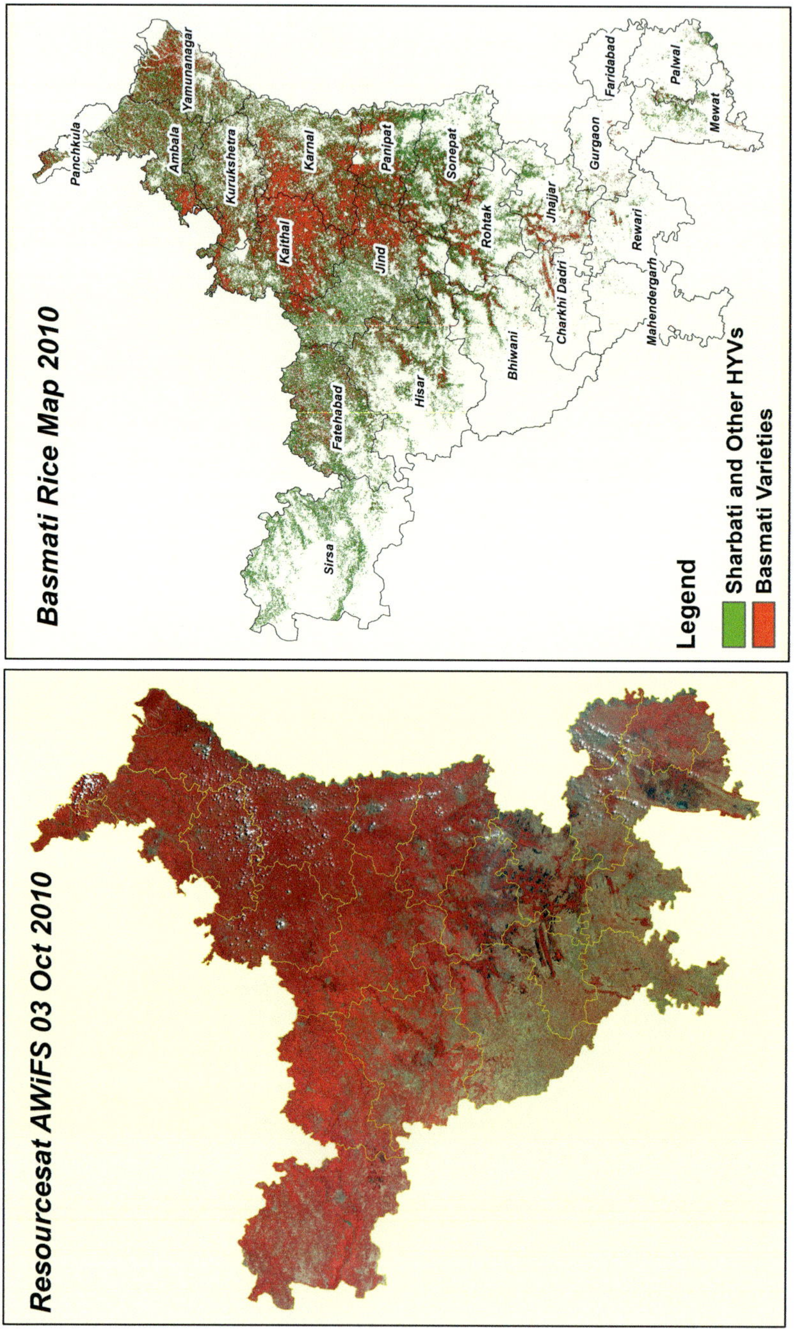

FIGURE 4.16 AWiFS image collected on 3 October 2010 for Haryana (a), and classified Basmati rice map (b).

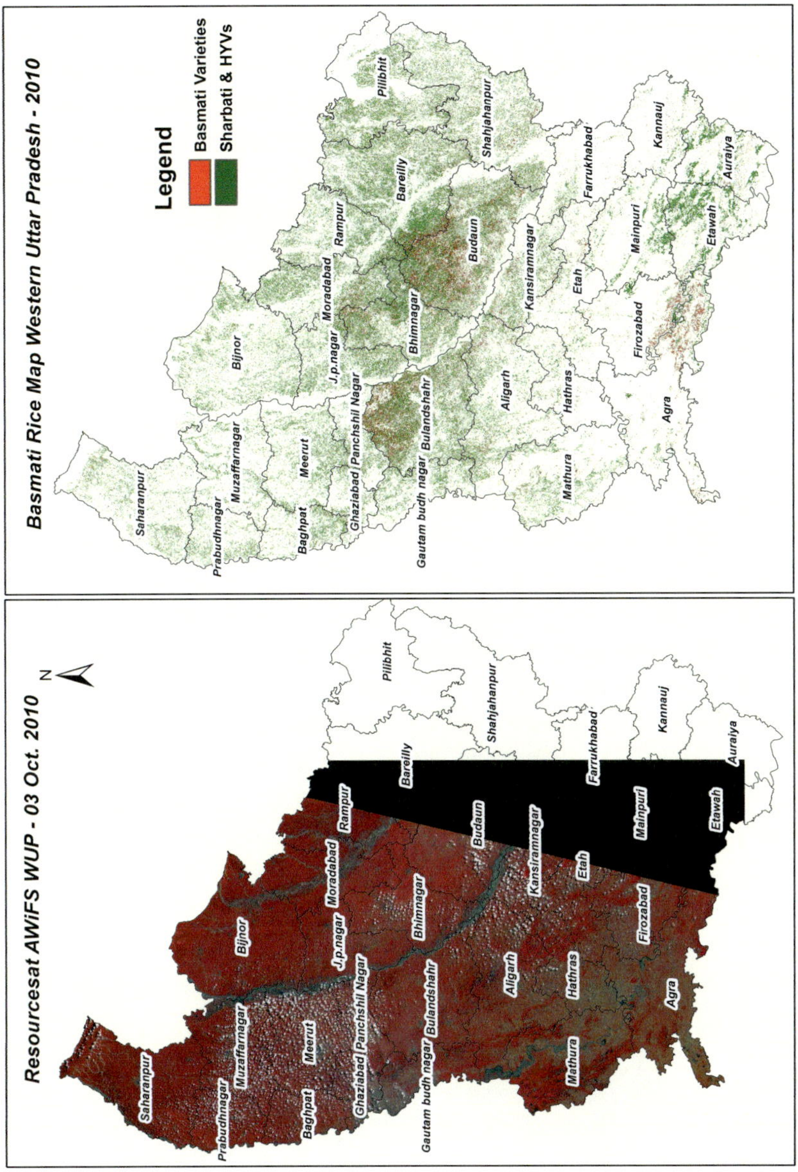

FIGURE 4.17 (a) AWiFS image collected on 3 October 2010 for western Uttar Pradesh, (b) and classified Basmati rice map.

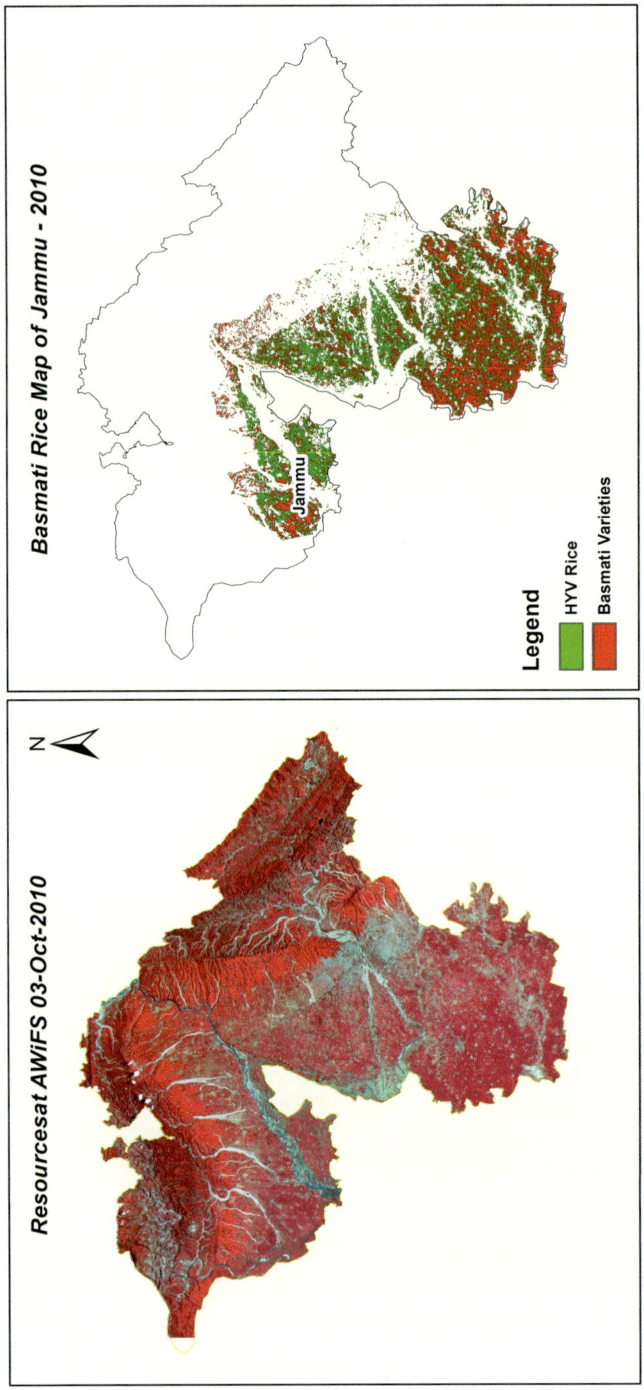

FIGURE 4.18 (a) AWiFS image collected on 3 October 2010 for Jammu, and (b) classified Basmati rice map

TABLE 4.10
Basmati Area ('000 ha) in Seven Basmati Growing States of India During 2010

S. No.	State	Basmati Area
1.	Punjab	738.92
2.	Haryana	801.40
3.	Western Uttar Pradesh	377.79
4.	Uttarakhand	29.88
5.	Jammu & Kashmir	33.00
6.	Himachal Pradesh	5.95
7.	Delhi	1.45

TABLE 4.11
District-wise Area ('000 ha) Under Basmati Varieties in Punjab, India During 2010

S. No.	District	Traditional Basmati	Pusa Basmati-1121	Pusa Basmati-1	Super Basmati
1	Amritsar	21.96	64.75		0.83
2	Barnala	5.41	9.20	5.10	
3	Bhatinda	0.58	2.64		
4	Faridkot	5.33	17.24		
5	Fatehgarh Sahib	17.00	21.25	0.85	0.85
6	Firozepur + Fazilka	15.60	40.00	7.8	
7	Gurdaspur +	14.06	90.45	0.50	3.01
8	Pathankot				
9	Hoshiarpur	4.80	15.00	1.20	0.60
10	Jalandhar	9.25	22.8	0.70	
11	Kapurthala	11.40	19.80	0.57	
12	Ludhiana	10.16	22.59	0.86	
13	Mansa	2.13	4.08	2.84	
14	Moga	1.90	6.80		
15	Mohali	0.54	1.35	0.54	
16	Muktsar	1.70	12.75	0.96	
17	Nawanshahar	2.64	17.00	0.52	
18	Patiala	18.96	35.70	7.74	1.67
19	Roopnagar	2.28	6.84		
20	Sangrur	10.68	39.75	8.01	
21	Tarn Taran	16.90	70.50		
	Total	**173.28**	**520.49**	**38.19**	**6.96**

TABLE 4.12
Classification Accuracy Assessment on Classified Image Classes

Confusion Matrix			Percent Pixels Classified by Code								
Class	**Code**	**Pixels**	**0**	**5**	**10**	**15**	**20**	**25**	**30**	**35**	**40**
Urban	5	684	0.58	97.51	1.90	0.00	0.00	0.00	0.00	0.00	0.00
Fallow	10	761	1.58	4.60	93.17	0.00	0.00	0.00	0.00	0.00	0.66
Waterlogged	15	1029	2.24	0.00	0.00	89.89	7.68	0.00	0.19	0.00	0.00
Water	20	2624	3.16	0.00	0.00	1.75	95.08	0.00	0.00	0.00	0.00
Basmati-Traditional	25	2326	0.39	0.00	0.00	0.00	0.00	88.91	4.28	6.42	0.00
Basmati-Evolved	30	1398	0.36	0.00	0.36	0.00	1.14	2.15	79.76	12.73	3.51
High Yielding (non-Basmati)	35	1545	0.00	0.00	0.00	0.00	0.00	6.73	6.34	78.77	8.16
Sharbati	40	287	0.35	0.00	0.00	0.00	0.00	0.00	4.18	16.38	79.09

Average Accuracy = 87.77
Overall Accuracy = 88.45
Kappa COEFFICIENT = 0.862
Standard Deviation = 0.00367

TABLE 4.13
State-wise Area ('000 ha) Under Basmati Rice Varieties During 2007–2012

State	2007		2008		2010		2011		2012	
	Traditional	Evolved	Traditional	Evolved	Traditional	Evolved	Traditional	Evolved	Traditional	Evolved
Punjab	119.10	118.99	108.98	539.67	173.28	565.64	80.57	620.09	4.52	704.04
Haryana	104.42	364.44	140.50	517.99	318.05	483.35	135.60	569.48	0.00	594.13
Uttar Pradesh	33.30	197.75	37.80	245.50	18.5	369.29	17.70	410.60	18.05	348.85
Uttarakhand	2.25	3.6	7.40	5.90	12.10	17.78	13.20	10.30	8.20	8.7
Jammu & Kashmir	31.10	0.00	33.40	0.00	33.00	0	38.00	0.00	34.00	8.5
Himachal Pradesh	-	-	-	-	3.39	2.56	5.60	8.60	0.00	2.85
Delhi	-	-		-		1.45	-	-	-	-
Total	**290.17**	**684.78**	**328.08**	**1309.06**	**558.32**	**1430.07**	**290.67**	**1618.98**	**64.77**	**1667.07**
Total Basmati	**974.95**		**1637.14**		**1988.39**		**1911.24**		**1731.84**	

approach, integrating multispectral optical imagery with microwave SAR observations to enhance mapping accuracy (Torbick et al. 2017, Orynbaikyzy et al. 2020). Their results showed SAR data–based rice classification outperforming optical data, and optical and SAR combination outperforming single-sensor data-based classification. Fiorillo et al. (2020) used Sentinel-1 SAR and Sentinel-2 multispectral data to map rice in a complex environment of fluvial lowland valleys in Senegal using RF algorithm, and observed improvement in overall classification accuracy from 77.4% using SAR data to 79.9% using multispectral data, and 87.4% using a combination of both datasets. Similarly, the user's accuracy also increased from 80.0% to 86.3% and 89.4% in only SAR, only multispectral, and a combination of both SAR and multispectral data, respectively. Onojeghuo et al. (2018) used RF and SVM algorithms to map rice from multi-temporal Sentinel-1A SAR data and Landsat-derived NDVI data. Integration of NDVI time-series with multi-temporal polarisation datasets (VH, VV, and their ratio VH/VV) led to a marked improvement in overall classification accuracy, as evidenced by the results. Saadat et al. (2022) developed an operational rice mapping framework for Iran by analysing time-series dual-polarised Sentinel-1 SAR data alongside multispectral Sentinel-2 imagery. Their findings indicated that the fusion of NDVI time-series with SAR data enhanced the classification of rice types. Moreover, classification using SAR time-series alone outperformed the approach based solely on NDVI temporal profiles.

However, no study has been reported on the integration of optical and microwave remote sensing data to map aromatic Basmati rice. Because of the significant accuracy improvements using the integrated approach observed in the rice classification studies reported above, the present study was taken up with the following objectives:

(i) To develop a methodology for aromatic Basmati rice mapping and varietal differentiation using multi-temporal Sentinel-1 SAR and Sentinel-2 MSI data, RF classifier, and Google Earth Engine cloud computing platform.
(ii) To use the microwave SAR data in mapping the area affected by the flood that occurred in a part of Haryana during the early rice season of 2023.
(iii) To monitor rice transplanting using microwave SAR in a part of the study area. Flooding led to complete crop loss in certain regions, necessitating the re-transplantation of rice in the affected areas. The study was conducted in the two largest Basmati rice growing districts of Haryana State, India.

4.7.3 Study Area

Karnal and Kaithal districts (Latitude 29°25'42" to 30°11'13" N, Longitude 76°08'45" to 77°14'58" E) of Haryana state were selected for this study as they have the largest area under aromatic Basmati rice. The geographical area of the two districts is 4837 sq km. The area is a part of the Indo-Gangetic Plains and falls under the Eastern Plain Zone of the state, receiving 766 mm of average rainfall during the Basmati crop growing months of June to October. The soil type is deep alluvium, medium to heavy textured, suitable for rice cultivation. The districts have a well-spread network of irrigation canals. The rice area in Haryana during 2023 was reported to be 1.52

M ha, and production was 5.51 million tonnes (MT), out of which the Basmati area was 0.812 M ha (53%), which shows the importance of Basmati rice in the state's agriculture. The location map of the study districts is given in Figure 4.19.

Haryana state is classified according to Köppen climate classification as humid subtropical, dry winter, semi-arid to sub-humid. The state of Haryana is covered by three distinct climatic zones: hot semi-arid, hot desert, and hot sub-humid, characterised by pronounced seasonal variability, including high summer temperatures and markedly cold winters. During the monsoon season 2023 (1 June–30 September), the state received 80.7 mm of rain in June against the mean of 54.7 mm, during July 236.9 mm against the normal of 149.1 mm (59% excess). Specifically, five districts showed excess rainfall in the state during this month, causing floods in some areas. However, in August, it received only 58.5 mm of rainfall against a normal of 146.1 mm (60% less) (IMD, 2023).

4.7.4 Rice Varieties and Cultivation Practices

Basmati rice varieties grown in the study area are short-duration (115–125 days maturity) Pusa Basmati-1509, Pusa Basmati-1692, and Pusa Basmati-1847; and medium to long duration (135–145 days maturity) Pusa Basmati-1121, Pusa Basmati-1718, CSR-30, Pusa Basmati-1, Pusa Basmati-1401, Pusa Basmati-1885, and Pusa Basmati-1886 (AIREA, 2023). Among these, Pusa Basmati-1509 and -1692 have the shortest duration (115 days). Besides Basmati, several non-Basmati HYVs are also grown (PR 114, PR 117, PR 118, Hybrid, etc.), which are generally of short duration. Rice transplanting begins in the second fortnight of June and continues until the end of July. The harvesting starts from October and continues until the first fortnight of November, depending upon the maturity duration of the varieties. Non-Basmati HYVs are harvested during the first fortnight of October, almost 25 to 30 days before the harvest of medium and long-duration Basmati varieties. The short-duration Basmati varieties are also harvested along with HYVs. The crop calendar and phenological stages of rice varieties are summarised in Figure 4.20.

4.8 RESULTS

4.8.1 Methodology

4.8.1.1 Remote Sensing Satellite Data

Time series optical sensor data from the Sentinel-2A/ B Multispectral Instrument (MSI), and microwave data (C-band SAR) from Sentinel-1A /B, covering the study area, were acquired from the first fortnight of August to the first fortnight of November 2023. The data were acquired based on the phenology of HYV and Basmati long and short-duration varieties grown in the area. The data were accessed in Google Earth Engine (GEE). Sentinel-2 optical data during the second fortnight of September had more than 20% cloud cover, hence not acquired. The details of data acquisition have been summarised in Table 4.14. The dates of Sentinel-1 and Sentinel-2 data acquisition are shown in Figure 4.21. A set of Sentinel-2 MSI and Sentinel-1 SAR images covering the study area is shown in Figure 4.21.

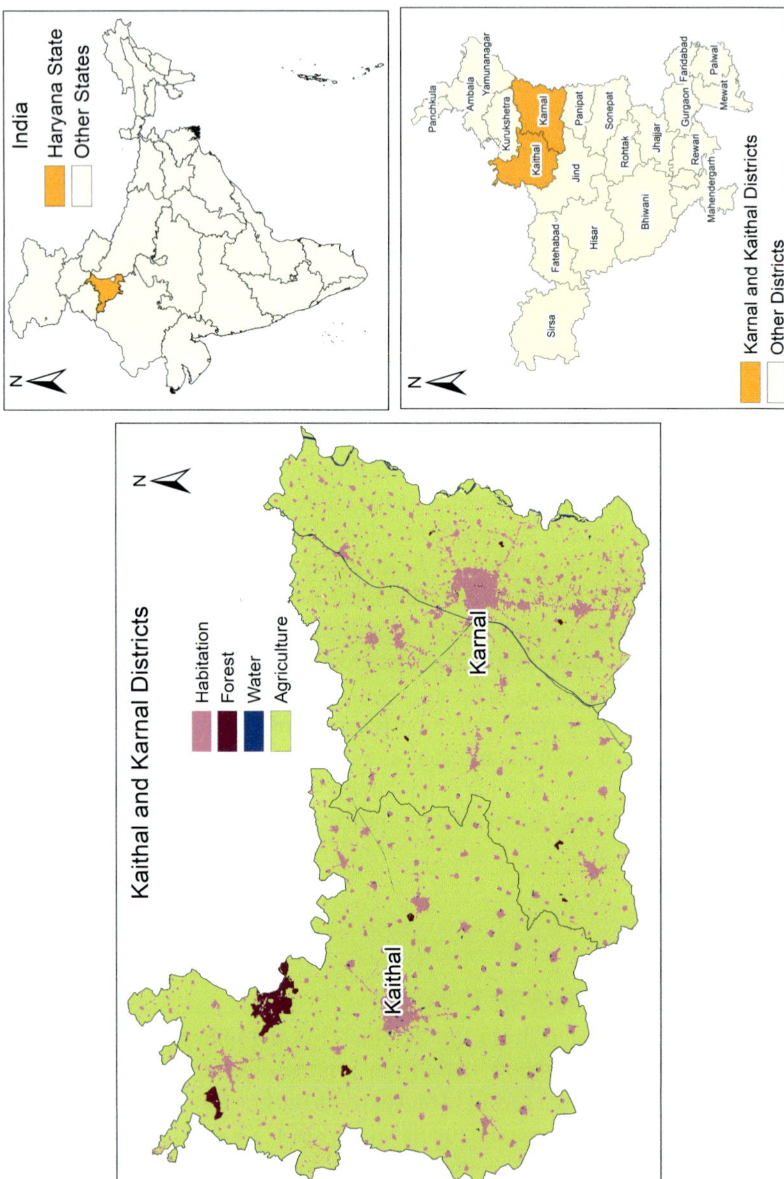

FIGURE 4.19 Location map of Karnal and Kaithal Districts in Haryana State.

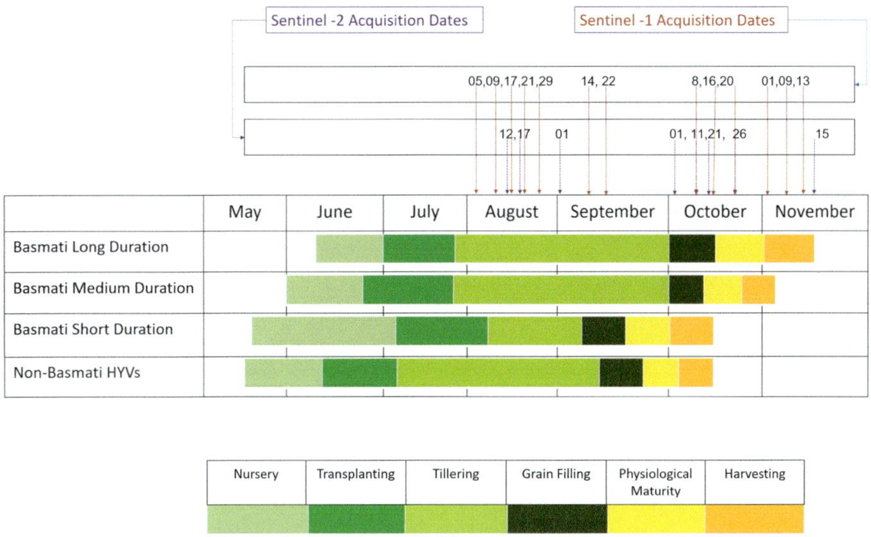

FIGURE 4.20 Rice crop calendar, phenology stages of rice varieties in the study area, and data acquisition dates of Sentinel-1 and Sentinel-2 during 2023.

4.8.1.2 Field Data Collection

Coinciding with critical crop growth stages of rice, the field data were collected during August, September, and October 2023. Large fields of different HYV rice, and of short and long-duration aromatic Basmati varieties were identified, their GPS locations recorded, and boundaries digitised. These were used as ground truth (GT) sites for repeat observations and also during validation after the satellite data classification. The GPS locations, along with field photographs, were superimposed on Google Earth images. Sample field photographs of different rice varieties are shown in Figure 4.22. Information on rice varieties, date of sowing, planting methods (direct sowing/line sowing/transplanting), crop health and condition, expected date of harvesting, irrigation availability, soil fertility status, other crops grown, and so on, was also recorded.

4.8.1.3 Pre-Processing of Remotely Sensed Data

Sentinel-1 SAR Level-1 GRD products in VV and VH polarisations were projected to ground range using the Earth ellipsoid model WGS84 with a spatial resolution of 10 meters. SAR data processing was carried out using the ESA's Sentinel Application Platform (SNAP). The pre-processing workflow consists of the following seven steps: (i) application of precise orbit files, (ii) removal of thermal noise, (iii) elimination of border noise, (iv) radiometric calibration to sigma naught $\left(\sigma^0\right)$, (v) speckle suppression using a spatial filter, (vi) terrain correction via Range-Doppler orthorectification using a high-resolution DEM, and (vii) logarithmic transformation to decibel (dB) scale. Following calibration, dual-polarisation (VV and VH) backscatter intensities were used to compute the radar backscattering coefficient $\left(\sigma^0\right)$ in dB. Polarimetric features, including the VH/VV ratio and the VV–VH difference, were subsequently derived. To

TABLE 4.14
Characteristics of Sentinel-1 SAR and Sentinel-2 MSI Data Used in the Study

Sensor	Band	Wavelength	Resolution	Product Level/ Type	Time Period
Sentinel-1A/B SAR	C-band (VV and VH polarisations)	5.6 cm	10 m	Level-1 Single Look Complex (SLC), Ground Range Detected (GRD), Interferometric Wide (IW) swath mode	August to First fortnight of November 2023
Sentinel-2A/B MSI (Multi-Spectral Instrument)	B3 (Green)	490 nm	10 m	Less than 20% cloud cover	August to First fortnight of November 2023
	B4 (Red)	665 nm	10 m		
	B5 (Red edge -1)	705 nm	10 m		
	B6 (Red edge- 2)	740 nm	10 m		
	B7 (Red edge-3)	783 nm	10 m		
	B8 (NIR)	842 nm	10 m		
	B11 (SWIR 1)	1610 nm	20 m		

reduce environmental variability and enhance temporal consistency, the backscatter time series was smoothed using the multi-temporal speckle filtering method proposed by van Tricht et al. (2018). Sentinel-2A/B MSI data with less than 20% cloud cover, accessed during August to November 2023, were used to generate bi-weekly composites using GEE. MSI 10- and 20-m resolution bands, including green, red, red-edge-1, red-edge-2, red-edge-3, NIR, and SWIR1, were used. Red-edge spectral bands are sensitive to variations in chlorophyll concentration, while the SWIR region reflects differences in vegetation moisture status. These spectral regions have demonstrated effectiveness in discriminating between crop types (Clevers and Gitelson, 2013).

4.8.1.4 Generation of Temporal Profiles of MSI Vegetation Indices (VIs) and SAR Difference Index

Vegetation indices based on the red-edge region effectively capture chlorophyll absorption, particularly under moderate to high chlorophyll concentrations (Gitelson and Merzlyak, 1996). To assess rice growth stages, this study utilised the red-edge Chlorophyll Index ($CI_{rededge}$) and the green Chlorophyll Index (CI_{green}), in addition to NDVI, as these indices are responsive to variations in chlorophyll concentration due to their sensitivity to red-edge reflectance dynamics. The details of VIs and Sentinel-2 MSI bands used are shown in Table 4.15.

Time series observations from Sentinel-1 and Sentinel-2 were analysed to generate temporal profiles as well as optical/SAR indices, and have been evaluated for monitoring the phenological stages (Saad El Imanni et al., 2022) during the wheat growing period using the GEE geospatial platform. They generated Sentinel-2

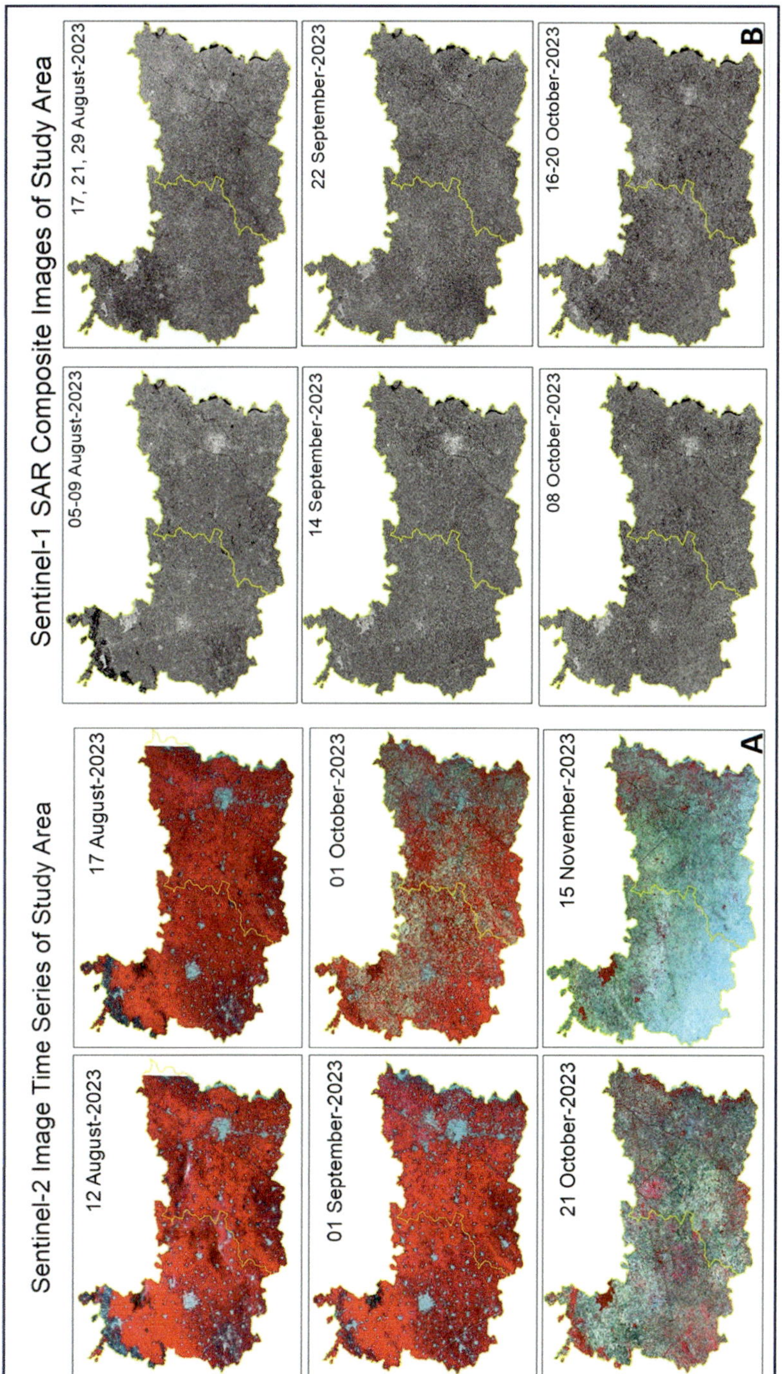

FIGURE 4.21 Sentinel-2 MSI (A) and Sentinel-1 SAR (B) images covering Karnal and Kaithal districts of Haryana.

FIGURE 4.22 Ground photographs of some rice varieties and their field boundaries demarcated on Google Earth images.

indices, namely the Weight Difference Vegetation Index (WDVI) and Sentinel-1 difference VH–VV polarisation index demonstrated the potential of remote sensing in tracking wheat phenology, achieving R^2 values of 0.70 for early stages (germination to jointing/booting) and 0.84 for later stages (heading to maturity). In this study, red-edge-based spectral vegetation indices $CI_{rededge}$ and CI_{green} were computed from Sentinel-2 time series data to monitor crop development. Additionally, a backscatter difference index (VH–VV) was derived from Sentinel-1 imagery to capture structural changes in the canopy. Saad El Imanni et al. (2022) observed that the SAR backscatter coefficient $\left(\sigma^0\right)$ in VH–VV polarisations exhibited an increasing trend

TABLE 4.15
Sentinel-2 Spectral Vegetation Indices Used in the Study

Index	Formulation (Sentinel-2 Bands)	References
NDVI	$(\text{Band8} - \text{Band4}) / (\text{Band8} + \text{Band4})$	Rouse et al. (1974)
Chlorophyll Index Red-edge (CI–red edge)	$(\text{Band7}) / (\text{Band5}) - 1$	Gitelson et al. (2005)
Chlorophyll Index Green (CI–green)	$(\text{Band7}) / (\text{Band3}) - 1$	Gitelson et al. (2005)

from the germination phase through to the heading stage, followed by a decline during the crop's maturity and harvest periods.

4.8.1.5 Composite Image Generation

A 15-day composite image using the median value was employed, taking into account data continuity, computational efficiency, and its effectiveness in enhancing classification performance. Median compositing is a robust statistical approach frequently utilised in land cover mapping to reduce noise and outlier influence in time-series datasets (Tian et al., 2019). The code for composite image generation was compiled on the GEE platform (Orynbaikyzy et al., 2020), and fortnightly composite images of MSI reflectance bands and CI–red edge and CI–green indices for each fortnight from August to November were generated. A similar process was adopted for generating fortnightly composite SAR images in VV and VH polarisations from June to November.

4.8.1.5.1 Data Analysis

A flowchart showing the methodology adopted in Sentinel-1 SAR and Sentinel-2 MSI data analysis is shown in Figure 4.23. The methodology consisted of the following steps:

i. After pre-processing of time series SAR data, the dual-polarisation backscatter signals were processed into backscattering coefficient sigma naught (σ^0) in decibel units (dB). The difference index of VH–VV polarisations was generated, which showed an increasing trend until the crop heading stage, and a decreasing trend during the maturity and harvesting stages.
ii. Temporal profiles of MSI green, red, and NIR bands, and spectral Vegetation Indices (VIs) namely, red-edge chlorophyll index ($CI_{red\text{-}edge}$) and green chlorophyll index (CI_{green}) were generated using time series MSI data.
iii. Fortnightly composite images of Sentinel-2 MSI reflectance bands, CI–red-edge, and CI–green, as well as Sentinel-1 SAR composite using VV and VH polarisations from August to November, were generated.
iv. A rice map was generated based on a fortnightly composited SAR temporal backscatter profile. This map was later used as a rice mask in further varietal discrimination of Basmati and other varieties.
v. The temporal profiles of $CI_{red\text{-}edge}$, $CI\text{-}_{green}$, and VV, VH, and VH–VV backscatter difference images were analysed for discrimination of HYV, short-duration, and long-duration Basmati varieties categorised according to differences in their phenological development stages.
vi. RF classifier, a machine learning algorithm proposed by Breiman (2001), was adopted for the classification of fortnightly composite images of Sentinel-1 and Sentinel-2 for classification and acreage estimation under different HYV rice and Basmati varieties.

4.8.1.5.2 Characteristics of Basmati and HYV Rice Varieties Observed in Sentinel-1 SAR and Sentinel-2 MSI Data

Sentinel-2 MSI band reflectance spectra of aromatic Basmati rice at different growth stages during the 2023 cropping season have been studied (Figure 4.24). It is observed that at the tillering and flowering stages, the reflectance in the visible blue, green, and red bands is the lowest, which increases in the red-edge bands, plateaus in NIR bands (7, 8, and 8A), and declines in SWIR bands 11 and 12. At the grain filling stage, the reflectance pattern is similar, but NIR reflectance is decreased slightly. However, at the crop maturity stage, the reflectance in visible bands increases, while in NIR bands it decreases, and in SWIR bands it increases again, due to the crop senescence.

Temporal FCC images of Basmati and HYV rice fields from August to November 2023, at different crop growth stages, have been shown in Figure 4.25. The change in colour on the FCC image from vegetative to harvesting stages can be seen for both Basmati and HYV rices, which is essentially because of their difference in transplanting and maturity duration. On 12 August, HYV rice was at the maximum vegetative to booting stage, while Basmati rice was at the tillering stage. On 11 September

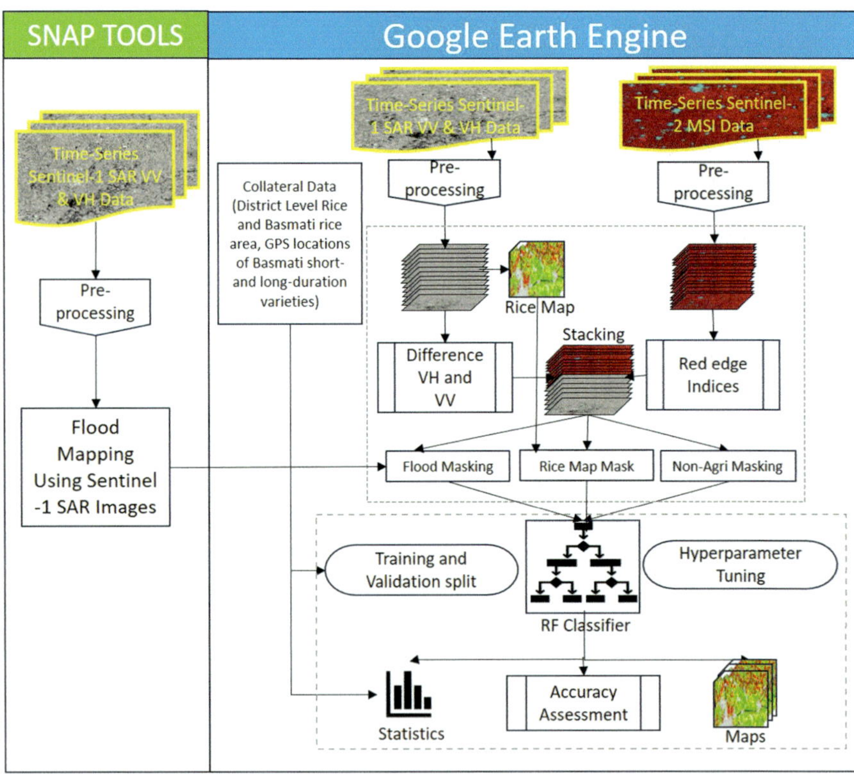

FIGURE 4.23 Methodology flow chart adopted for Sentinel-1 SAR and Sentinel-2 MSI data analysis.

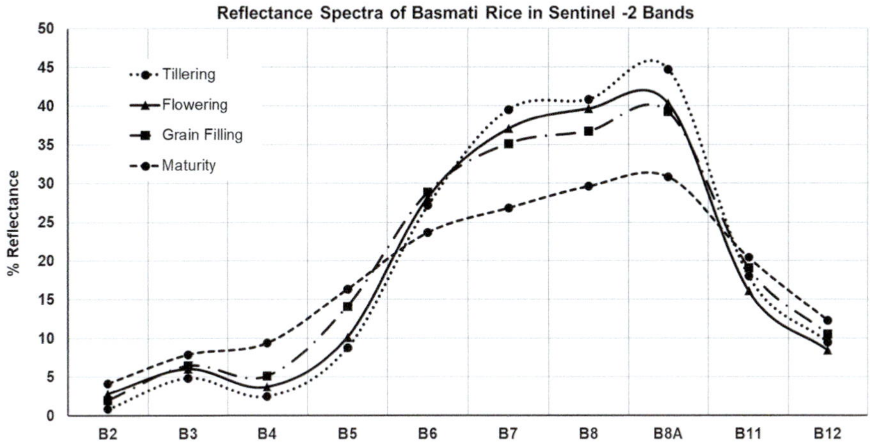

FIGURE 4.24 Reflectance spectra of aromatic Basmati rice field at different growth stages in Sentinel-2 MSI bands.

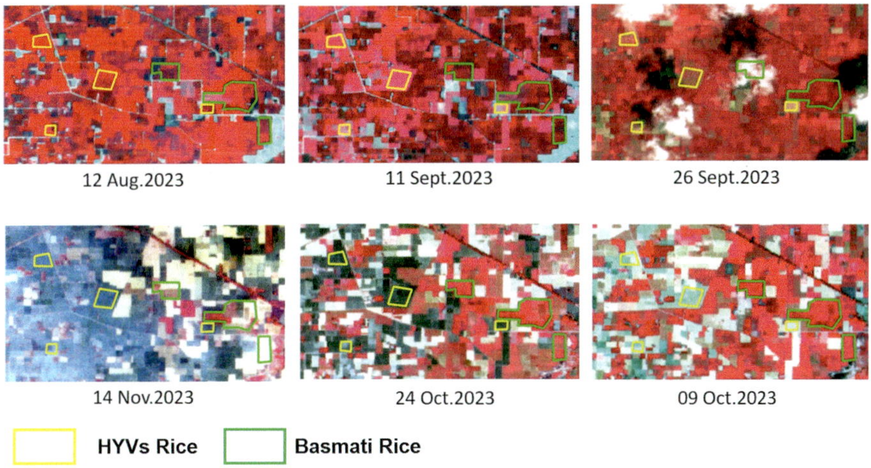

FIGURE 4.25 Sentinel-2A/B MSI temporal images showing Basmati and HYV rice fields at different growth stages during the 2023 cropping season.

image, however, the difference between the two rices on the FCC image can be clearly observed as the HYVs were at the flowering stage, whereas Basmati was at the booting stage. On the 9 October image, HYVs are being harvested, whereas Basmati is at the grain filling stage. A clear spectral colour difference in the image on 11 September and 9 October shows that the two varieties can be best separated on these dates.

Sentinel-1 C-band SAR temporal images from August to November 2023, showing Basmati and HYV rice fields observed across distinct growth phases, are shown in Figure 4.26. HYV rice fields (enclosed in yellow on the image), transplanted during the second fortnight of June, are at the maximum vegetative to booting stage on

FIGURE 4.26 Sentinel-1 C-band SAR temporal images showing Basmati and HYV rice fields at various growth stages during the 2023 cropping season.

9 August, at the flowering stage on 11 September, and at the grain filling stage on 26 September. The backscatter on these dates is high. At the maturity/ripening stage on 8 October, the backscatter is low (darker shade on the image) due to a decrease in plant water content. On 20 October, some of these HYV rice fields show dark colour on the image, possibly due to rice biomass burning after harvest; and on 14 November, the next season crop has been sown in these fields (dark on the image due to high soil moisture resulting in low backscatter). Basmati rice fields (enclosed green on the image), which are transplanted late, and are at tillering stage on 9 August, at booting and flowering stages on 11 September and 26 September, respectively; and at grain filling stage on 8 and 20 October. The backscatter from the rice crop is high during these growth stages. At the maturity/ harvesting stage on 14 November, the backscatter is low, hence the dark shade on the image is seen.

The temporal change in VV and VH backscatter coefficients of Basmati and HYVs from the second fortnight of June to September shows that backscatter in VV polarisation is always higher than that in VH (Figure 4.27). An increasing trend in backscatter values from transplanting to flowering stages of HYV rice is also seen. In the case of Basmati, which is transplanted later than HYVs (in July), the backscatter shows a dip as the fields are flooded during transplanting. It then shows an increasing trend as the plant growth takes place, similar to the trend in HYVs.

The fortnightly backscatter difference in VV and VH polarisations for short- and long-duration Basmati, and HYVs also shows that differentiation between the two groups of Basmati and HYVs can be made using temporal backscatter values in VV and VH polarisations (Figure 4.28). HYV rice is transplanted in the second fortnight of June, and the backscatter continues to increase as the plant grows from vegetative to flowering stage in September. On the other hand, Basmati varieties are transplanted in the second fortnight of July; hence, the backscatter is minimum at this time and continues to increase as the crop grows. The higher backscatter of Basmati varieties in June is because of the fallow fields.

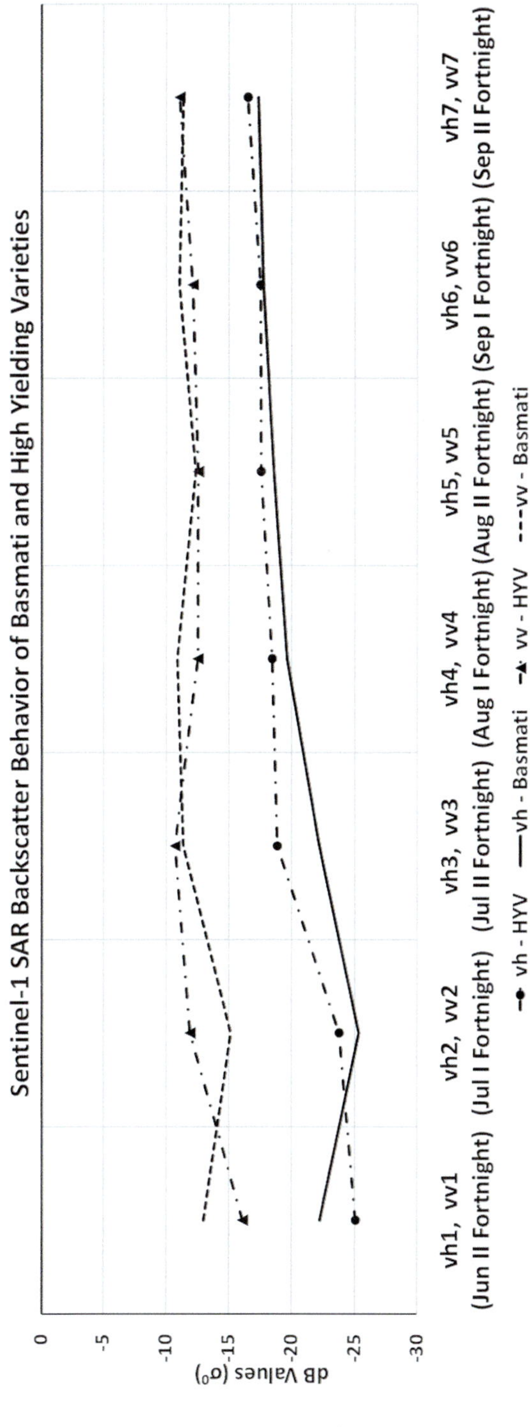

FIGURE 4.27 Temporal change in SAR backscatter coefficient (σ^0) of non-Basmati HYV and Basmati rice varieties.

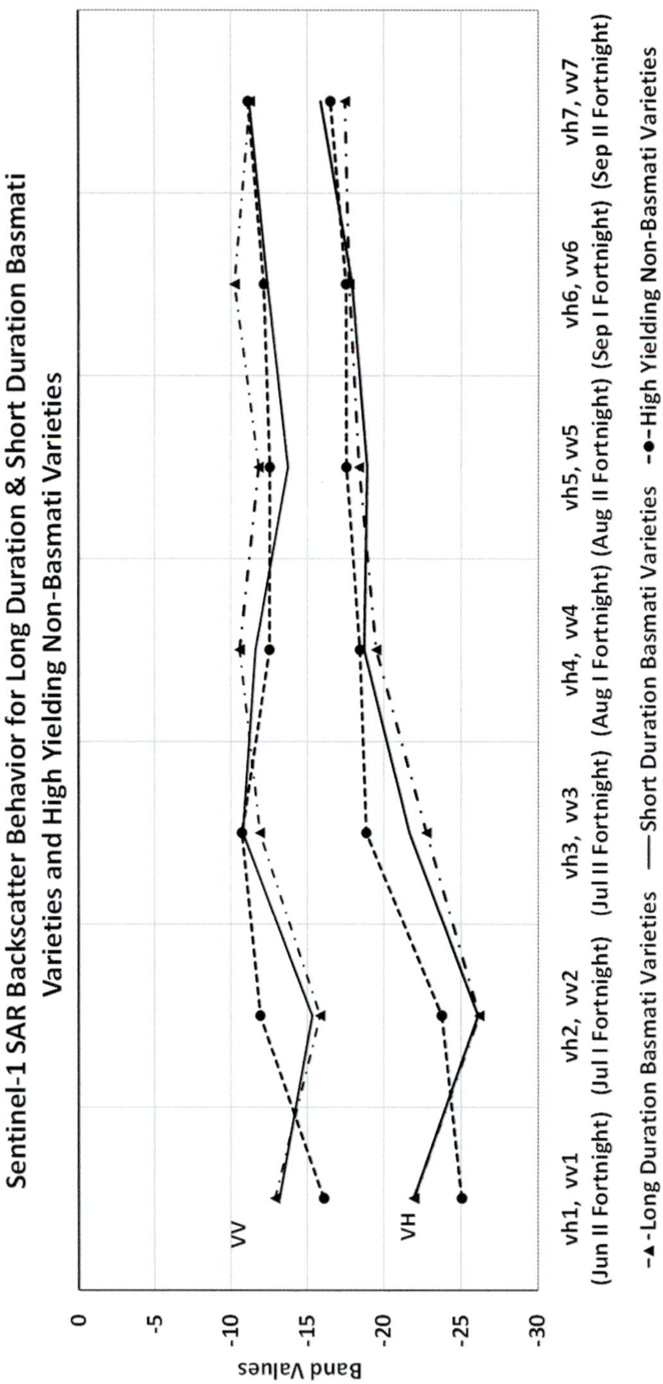

FIGURE 4.28 Temporal change in SAR backscatter coefficient $\left(\sigma^0\right)$ of short- and long-duration Basmati and non-Basmati HYV rice.

The NDVI profile of non-Basmati HYV and Basmati rice at different phenological stages in 2023 was also studied (Figure 4.29). In the case of HYV rice, NDVI increases continuously during August and until the third week of September when the crop is at the flowering stage and drops sharply afterwards as it attains maturity in October, and is harvested in the third week of October. In the case of Basmati rice, NDVI has slightly declined in the first week of October as the crop is senescing from flowering towards the grain filling stage. In the third week of October, NDVI has further reduced as the crop is towards maturity. Basmati and non-Basmati varieties can be easily differentiated during the first week of October based on the NDVI values.

Spectral separability of non-Basmati HYV and Basmati rice varieties in Sentinel-2 MSI bands (B3 green, B4 red, B5 and B7 red-edge, B8 NIR) during the first and the third week of October 2023 has also been studied (Figure 4.30). During the first week of October, HYV rice is at maturity, hence higher spectral values in the visible green, red, and red-edge bands (3, 4, and 5, respectively), compared to Basmati rice. During the third week of October, Basmati rice is at the flowering stage, hence spectral values are lower in the visible green, red, and red-edge, while it increases sharply in the NIR bands 7 and 8. It can thus be concluded that HYV and Basmati rice can be separated during the first week of October using both visible and NIR band spectral data.

4.8.2 Total Rice Mask Generation

The unique feature of rice fields is that they are flooded during transplanting for better seedling establishment and proper tillering. Rice is thus uniquely identified using its fortnightly temporal backscatter profile, which discriminates it from sugarcane, other crops, and other land uses. A rice map was generated to be used in further varietal discrimination. The total rice crop map generated for the study area during *kharif* 2023 is shown in Figure 4.31.

4.8.3 Flash Flood Mapping and Monitoring

During the monsoon season of 2023, the study area received rainfall higher than normal in the initial months (Figure 4.20), which resulted in flash floods in some parts of Karnal and Kaithal districts. The incessant rains in the second fortnight of July caused flash floods, resulting in complete damage to freshly transplanted rice, and it had to be retransplanted. Some rice fields ready for transplanting before flood occurrence also got inundated. SAR sensors with their all-weather and all-time imaging capabilities can provide timely information on flood hazards and have been extensively used for flood mapping (Amitrano et al., 2018; Mason et al., 2012). The deployment of ESA Copernicus Program Sentinel-1 satellites has further enhanced the scope of SAR image time series with high temporal resolution for flood assessment (Uddin et al., 2019). A flood map was prepared using Sentinel-1 A/B images of three available dates coinciding with the rain spell, namely, 16 July, 28 July, and 09 August 2023, and one date before the incessant rains, that is, 15 June 2023. The map was generated in the SNAP tool, an open-source image processing software by

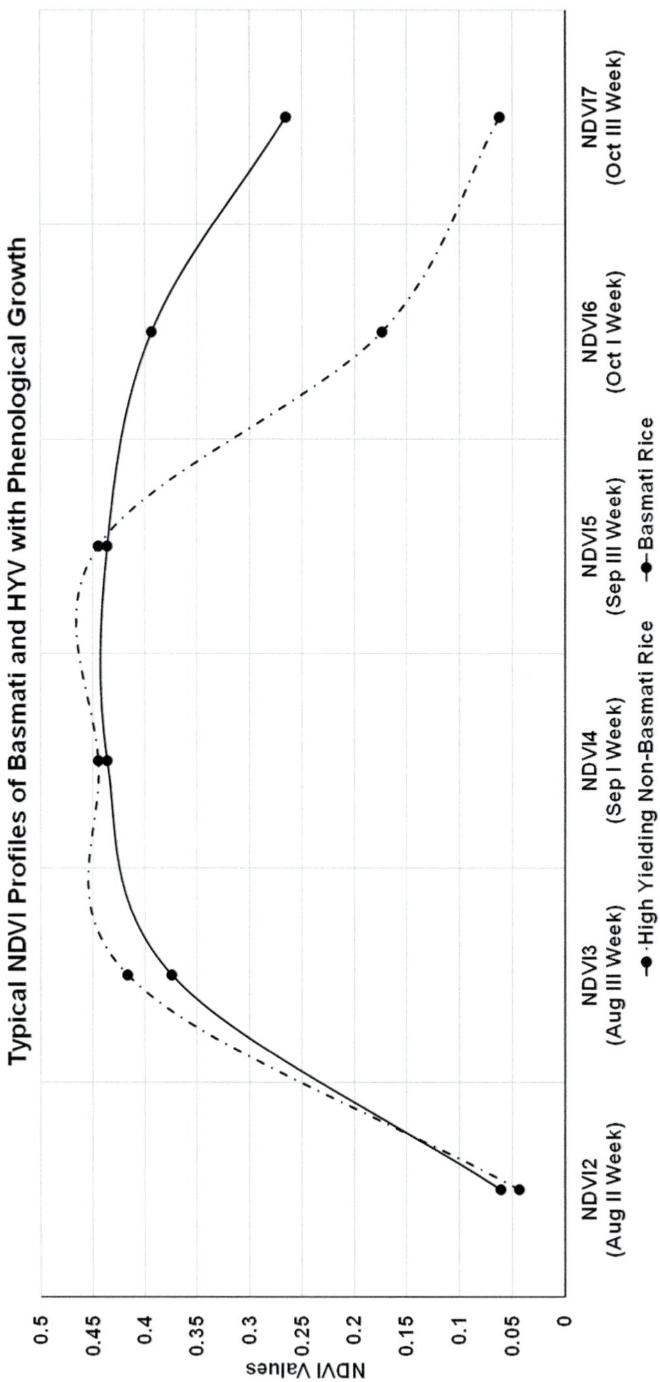

FIGURE 4.29 NDVI profile of non-Basmati HYV and Basmati rice at different phenological stages in 2023.

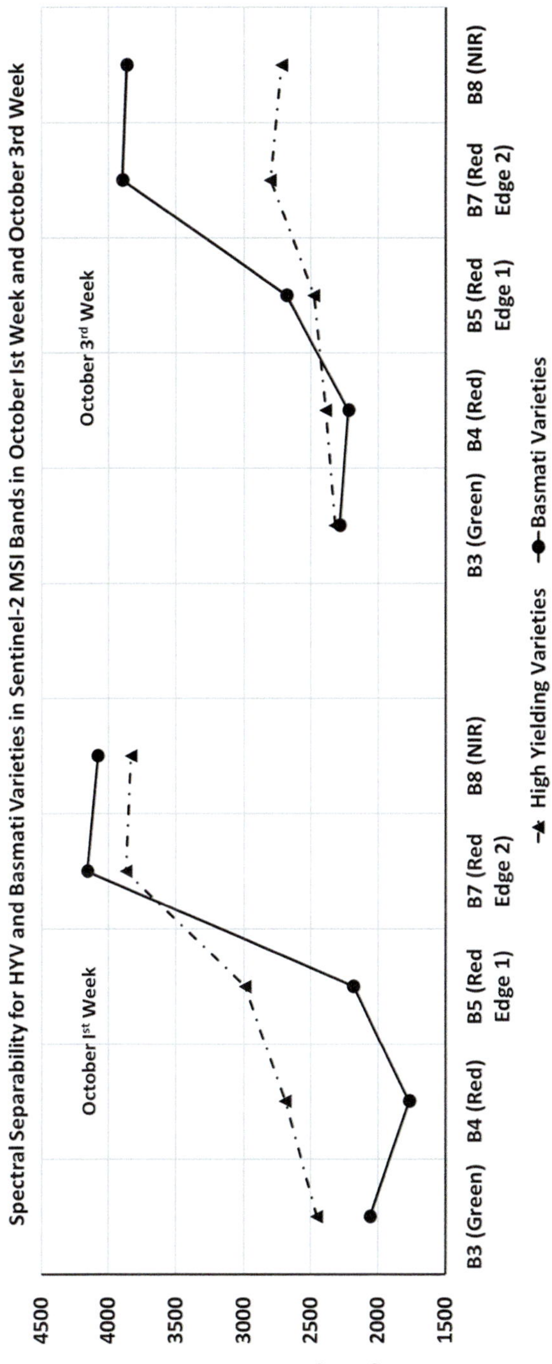

FIGURE 4.30 Spectral separability of non-Basmati HYV and Basmati rice varieties in Sentinel-2 MSI bands.

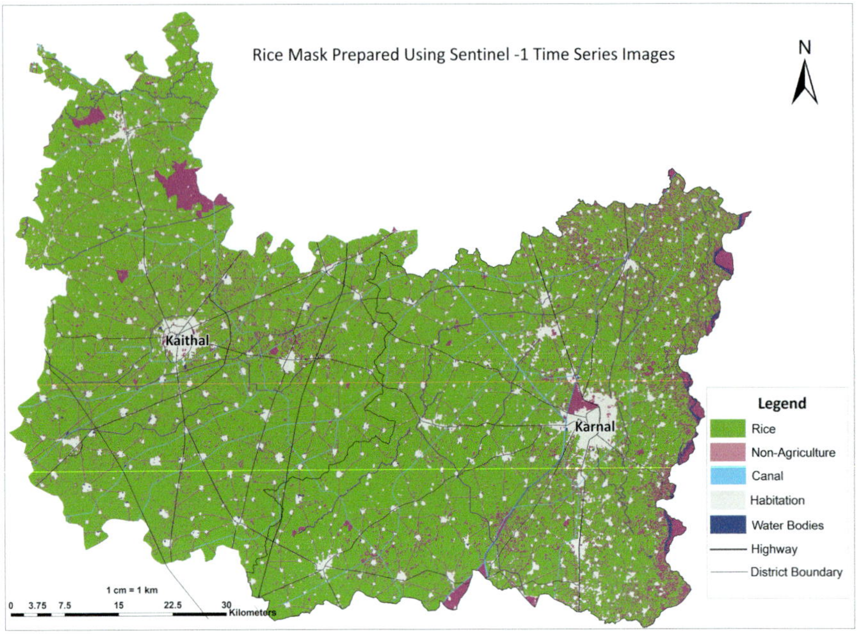

FIGURE 4.31 Total rice crop map during *kharif* season 2023 covering Karnal and Kaithal districts.

Copernicus. The flash flood inundation map covering Guha Block in Kaithal district is given in Figure 4.32.

4.8.4 Rice Transplanting Monitoring in Flood-inundated Areas

Retransplanting or delayed transplanting in flood-inundated areas after the recession of rainwater was done in the first fortnight of August. The flood map was imported as a GeoTIFF file and exported to a shape file in QGIS. The vector was imported into GEE to prepare a flood mask. The flood-inundated area was classified separately using a Random Forest classifier in GEE to generate a separate map of normal and late-planted rice. VV–amplitude of the SAR images was opted for separating water pixels from non-water pixels using histogram thresholding (Kocaman et al., 2018), which shows higher signal response in double bounce conditions and slightly higher thematic accuracy than VH polarisation (Twele et al., 2016; Xiao et al., 2002). The supervised classification of the Sentinel-1 image time series from the second fortnight of June to the first fortnight of August was carried out in GEE using a random forest classifier to generate a rice mask. The normal and late rice polygons over the pixels give different signatures in the RGB FCC of VH bands of three different date images. These polygons were used as training sites in the supervised classification. The overall accuracy was 99.7% with a Kappa coefficient of 0.99. The number of trees was kept at 150 with eight variables per split and seed size 25, giving OOB as low as 0.008. As the backscatter coefficient of the VH polarisation graph reveals, the early

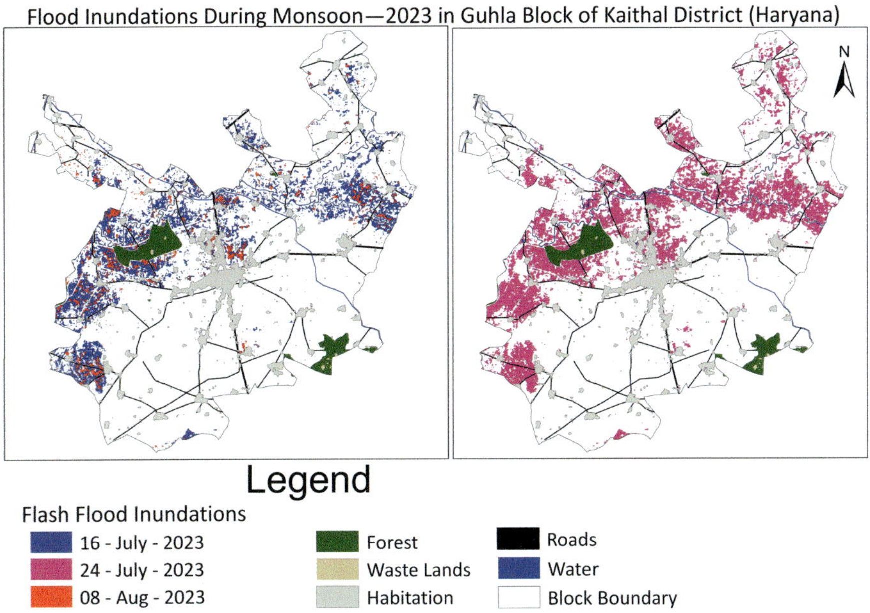

FIGURE 4.32 Flash flood inundated map covering Guha Block in Kaithal district.

rice was transplanted in the II week of June, transplanting of normal rice was carried out in the first fortnight of July, and the late rice was established through transplanting from the first week of August till the end of August. Initially, the backscatter values were very low due to flooding during transplanting in all three cases. But, during the growing season, the seedlings were established, and the booting stage (increasing foliage vigour) increased the backscatter values. The map showing rice transplanted during the second fortnight of June, first fortnight of July, and during the first fortnight of August in flood-inundated areas of Guha Block in Kaithal district, along with profiles of the SAR backscatter coefficient (σ^0), is presented in Figure 4.33.

The phenology-based classification approach leverages detailed knowledge of rice growth stages, which are reflected in the temporal dynamics of optical and microwave remote sensing data. During the transplanting phase, rice fields are characterised by inundated conditions with sparse vegetation, a combination that results in low VH backscatter values in Sentinel-1 SAR data. As the crop progresses through the vegetative and reproductive stages, the increasing canopy cover alters the backscatter response and spectral reflectance in Sentinel-2 time series. These phenological transitions, especially the presence of standing water during early growth, enable the discrimination of rice from other crops using VH polarisation and vegetation indices (Fatchurrachman et al., 2022). Kobayashi and Ide (2022) monitored rice crop growth using Sentinel-1 C-band imagery, revealing the backscattering behaviour corresponding to changes in crop phenology, water content, and surface condition. Similarly, Bazzi et al. 2019 analysed the temporal dynamics of SAR backscatter and demonstrated that rice cultivation exhibits a distinct Gaussian-like pattern in the VV/

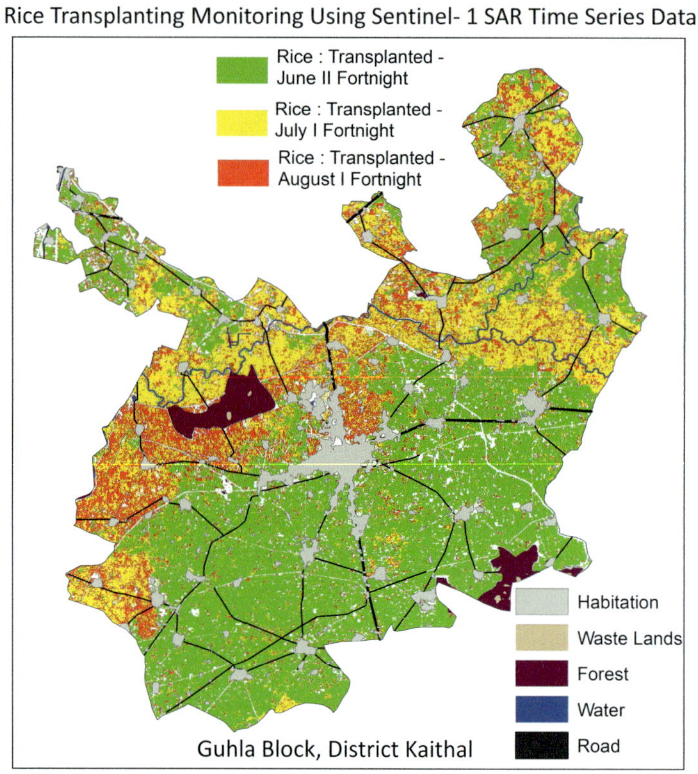

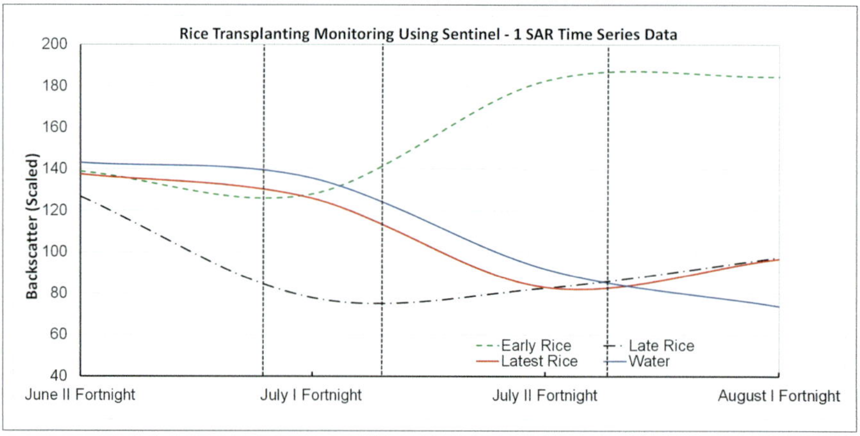

FIGURE 4.33 Monitoring of rice transplanting in flood-inundated areas of Guha Block in Kaithal district, and their backscatter profile.

VH ratio time series, along with a steadily increasing VH backscatter trend—both of which differ notably from the temporal profiles of other crop types.

4.8.5 Temporal Profile of VH–VV Difference Index in Rice Variety Discrimination

The Sentinel-1 polarisations (VV, VH, VH–VV) were studied for their temporal profile and discrimination of rice varieties at different phenological phases during their growth cycle. Figure 4.34 shows the temporal profiles during different dates, and the mean SAR backscatter $\left(\sigma^0\right)$ difference of VH and VV polarisations of different rice varieties. The observed increase in backscatter difference during the germination, tillering, and jointing stages can be attributed to rainfall events influencing surface and canopy conditions. Following these stages, a progressive decline in backscatter difference was noted, continuing up to the onset of crop maturity. The least backscatter difference indicates the harvested fields, which are higher in the case of long-duration varieties due to the prolonged vegetative phase and late maturity. The backscatter difference dropped early in the case of high yield non-Basmati rice, which matured earliest and was harvested by the end of the first week of October while the short-duration varieties have been harvested by the end of October. The harvesting of long-duration varieties started in the third week of October and continued till the second week of November (Verma et al., 2025).

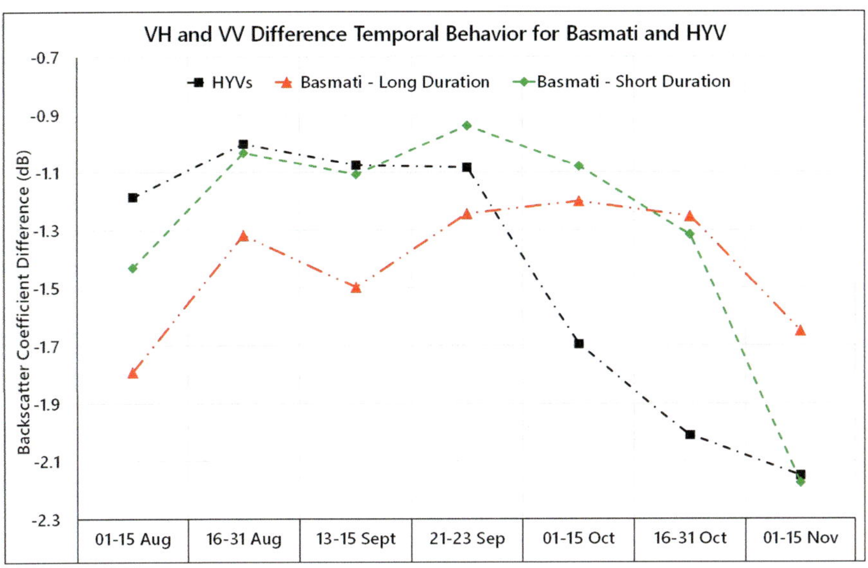

FIGURE 4.34 Temporal profiles of Sentinel-1 difference index VH–VV of HYV and Basmati rice varieties.

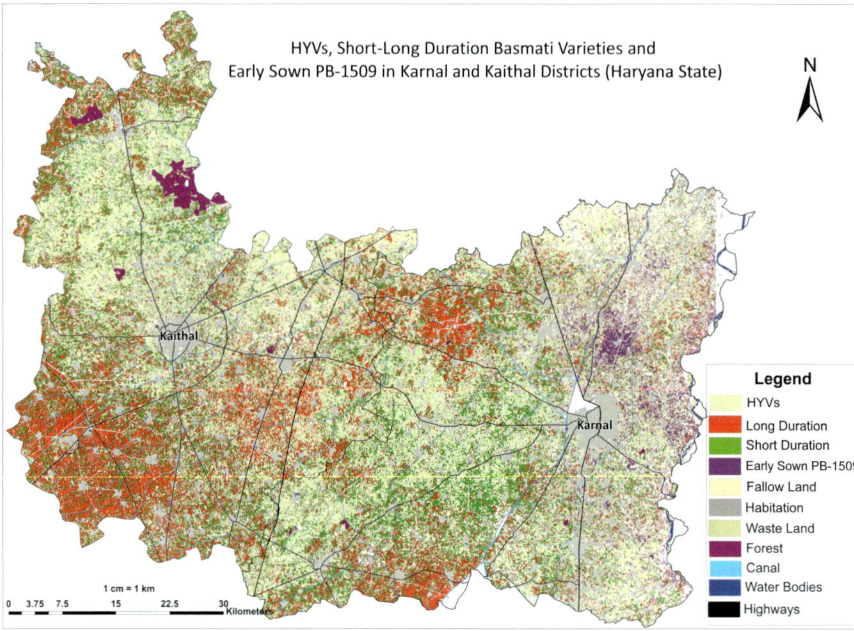

FIGURE 4.37 Map showing long and short-duration Basmati, and early Basmati (PB-1509) varieties in Karnal and Kaithal districts, Haryana.

TABLE 4.17
Area under Basmati and Non-Basmati Rice Varieties in Kaithal and Karnal Districts of Haryana in 2023

Rice Varieties	Area Estimates (000′ ha)		
	Kaithal	Karnal	Total
HYVs (non-Basmati)	78.95	76.89	155.84
Basmati: long-duration	54.67	34.28	88.94
Basmati: short-duration	49.49	56.65	106.14
Basmati early: Pusa Basmati-1509	3.10	11.72	14.82
Total Basmati	**107.25**	**102.65**	**209.90**
Total Rice	**186.20**	**179.54**	**365.74**

4.8.7.4 Classifier Accuracy and Performance

Most frequently, the overall accuracy (OA) and kappa coefficients are computed to evaluate the classifier's efficacy. The overall accuracy is the percentage of pixels/features classified correctly. In the case of satellite image classification, the evaluation of classification performance was based on overall accuracy, while class-specific reliability was assessed using producer's accuracy (sensitivity) and user's accuracy (precision) (Congalton, 1991; Foody, 2002). A greater OA value may not necessarily represent a more realistic map, particularly with multiple categories. Assessing classification performance or map outputs using only OA may lead to unreliable conclusions, as OA does not reflect the distribution of errors across individual classes (Shao et al., 2019). The evaluation of the RF classifier in this study included both OA and Cohen's kappa coefficient as performance metrics for all cases (Basmati–HYV separation, long-duration, and short-duration Basmati varieties separation), in both normal (non-flooded) and flooded area classification (Table 4.18). The consumer and producer accuracies of HYV and Basmati rice are summarised in Table 4.19.

TABLE 4.18
Accuracy Assessment of Classified Basmati Rice Using Sentinel-1 SAR and Sentinel-2 MSI Data

Sentinel-1 (S-1) /Sentinel-2 (S-2) Data Analysis	OA %	Kappa	No. of Trees	Variables Per Split	No. of Seeds	OOB Error
Early rice (PB-1509)	83.3	1.0	100	20	25	0.01
Total rice (S-1)	99.7	0.99	150	8	25	0.01
Basmati and HYVs(S-1 + S-2)	99.8	0.99	25	20	20	0.006
Basmati varieties(S-1 + S-2)	97.9	0.95	25	40	20	0.017
Basmati and HYVs (flooded area) (S-1 + S-2)	94.2	0.88	25	30	20	0.11
Basmati varieties (flooded area) (S-1 + S-2)	89.3	0.79	50	40	50	0.23

TABLE 4.19
Consumers and Producer Accuracy of Basmati and HYV Rice Varieties Analysed Using Sentinel-1 and -2 Data

	Consumer Accuracy		Producer Accuracy	
Sentinel-1 (S-1)/ Sentinel-2 (S-2) Data Analysis	HYVs	Basmati	HYVs	Basmati
Basmati and HYVs (S-1 + S-2)	100	99.2	99.7	98.1
Basmati – HYVs	92.3	96.2	96.0	92.6
Early Rice (PB-1509)	81	100	100	77.8
Total Rice (S-1)	96.8	99.9	99.9	98.1

(*Continued*)

TABLE 4.19 (Continued)

	Long-duration	Short-duration	Long-duration	Short-duration
Basmati Varieties (S-1 + S-2)	98.6	96.5	98.1	97.3
Basmati Varieties (Flooded Area (S-1 + S-2)	85.7	92.9	92.3	86.7

4.8.7.5 Receiver Operating Characteristics (ROC) and Area Under Curve (AUC)

Additionally, the analysis of model rates (model success and model prediction performance) for the receiver operating characteristics (ROC) and area under curve (AUC) was computed. In the context of RF classification, the area under the receiver operating characteristic curve (AUC-ROC) provides a comprehensive measure of the model's discriminatory power. AUC summarises the trade-off between the true positive rate (TPR) and false positive rate (FPR) across various classification thresholds. Higher AUC values indicate a better ability of the RF model to distinguish between classes, making it a robust metric for evaluating model performance, especially in imbalanced datasets (Fawcett, 2006). ROC analysis and AUC are standard tools for evaluating classifier performance, offering a threshold-independent measure of model discrimination capability (Bradley, 1997). Sun et al. (2020) evaluated 35 classifiers for built-up area intelligent classification using the ROC curve and AUC. Additional terms related to ROC are sensitivity (recall) and specificity (precision).

$$\textbf{\textit{Recall}} = \frac{\textbf{\textit{True Positives}}}{\textbf{\textit{False Negatives}} + \textbf{\textit{True Positives}}} \tag{4.1}$$

$$\textbf{\textit{False Positive Rate}}\left(\textbf{\textit{FPR}}\right) = \frac{\textbf{\textit{False Positives}}}{\textbf{\textit{False Positives}} + \textbf{\textit{True Negatives}}} \tag{4.2}$$

The area under the ROC curve (AUC) summarises the diagnostic accuracy and is a comprehensive metric that quantifies the overall discriminative ability of a diagnostic test. The value of AUC ranges between 0 and 1. In this case, a high AUC of 0.99 using both training and validation datasets, respectively, was observed in the case of the Basmati and HYVs model (Figure 4.38). The best ROC cutoff points were 0.39 and 0.12 in the case of training and validation, respectively. In the case of the long-duration and short-duration Basmati varieties classifier, the AUC was 0.99 in both the case of validation and training. The best ROC cutoff points were 0.31 and 0.51 in the case of model prediction and success, respectively. The high values of AUC validated the high efficiency of the RF classifier model. However, the hyper-tuning was also attempted to ensure the parameters of the classifier, namely, number of trees, number of variables, and the seed size, before finalising the model.

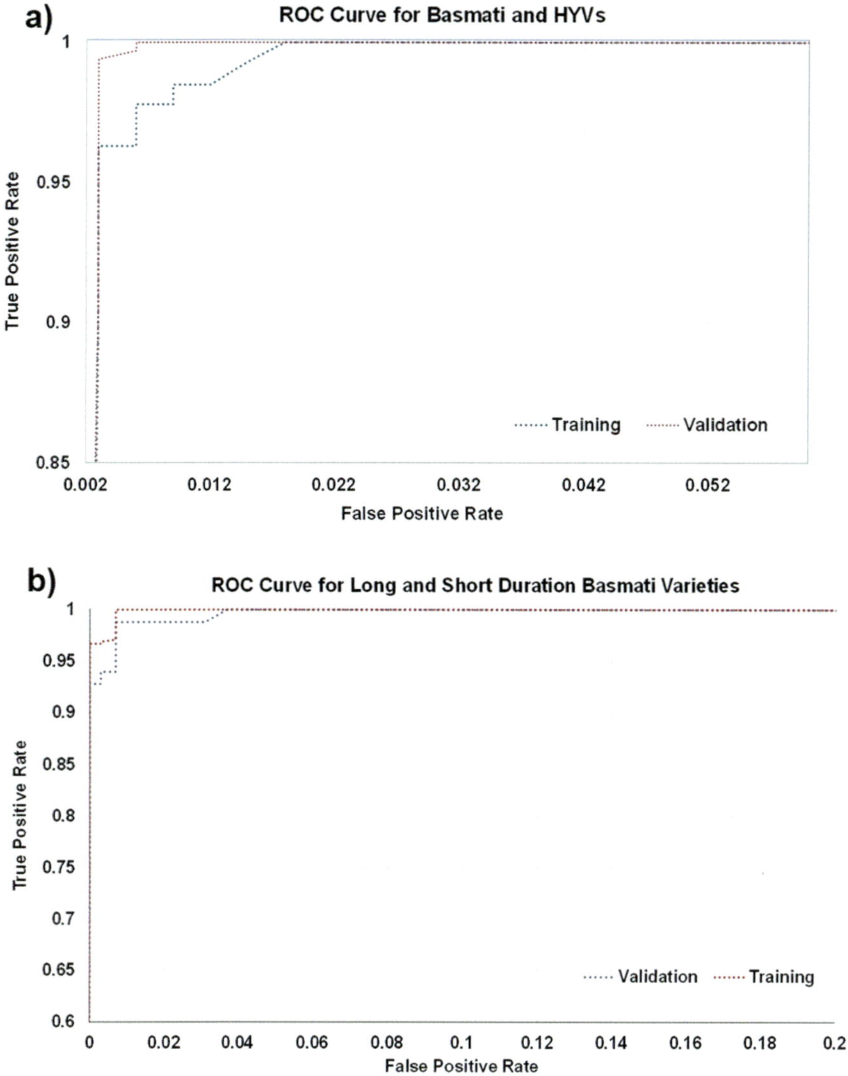

FIGURE 4.38 ROC curve for (a) map of Basmati and HYV rice varieties, and (b) map of long- and short-duration Basmati varieties.

4.9 CONCLUSIONS

The time-series Sentinel-1 SAR and Sentinel-2 MSI data were analysed for Basmati rice variety discrimination in two districts of Haryana, using an RF classifier in the Google Earth Engine (GEE) cloud computing platform. Time-series composites of VIs, like red-edge chlorophyll index (CIrededge), green chlorophyll index (CIgreen), and SAR VH–VV difference, were computed at different phenological growth stages, which showed significant differences in the spectral signatures of

HYVs, and long- and short-duration Basmati varieties. RF classifier was adopted in both cases, where 25 trees and 20 variables per split with seed size 20 were opted after hyper-tuning the parameters.

The performance of the RF classifier was evaluated for overall accuracy (OA) and kappa coefficient in all the cases (Basmati and HYVs separation, long-duration and short-duration Basmati separation). This resulted in an OOB error of 0.01 in the case of rice, and 0.006 in the case of Basmati and non-Basmati rice classification. The overall validation accuracy was 0.99, and Kappa was 0.99. The consumer accuracy was 1 for HYVs and 0.99 for Basmati, while the producer's accuracy was 0.99 and 0.98 for HYVs and Basmati, respectively. In order to check whether the model is overfitting to the training data, the sample data was split into training (70%) and validation (30%). The model was used separately to make predictions on the validation and training sets. The performance metrics, including validation accuracy and train accuracy, and OOB error, were compared.

Additionally, the analysis of the RF classifier model for ROC and AUC was computed to evaluate model success and model prediction performance. The results indicated a high AUC of 0.99 using both calibration and evaluation datasets in the case of Basmati and HYVs models. In the case of the long-duration and short-duration Basmati varieties classifier, the AUC was 0.99 in both the case of validation and training.

The area estimation of different varietal groups showed that Basmati varieties had a larger area coverage (57.4%) as compared to non-Basmati HYVs (42.6%) in the study districts. The area under all Basmati varieties was 2,09,905 ha, whereas HYVs were 1,55,841 ha. Among the Basmati varieties, the long-duration varieties covered 88,945 ha, and the short-duration varieties were 1,06,143 ha. Apart from these, an early Basmati group (mostly Pusa Basmati-1509) grown as direct-seeded rice, covering 14,817 ha, could also be separated. The results also revealed that the long-duration Basmati varieties are being replaced by shorter-duration Basmati varieties in the area, due to climate change, low irrigation requirement, and post-rice crop diversification possibilities. Non-Basmati HYVs (which are of short duration) and short-duration Basmati varieties together cover more than 75% area in these districts.

4.9.1 Observations on Basmati Varietal Shift in Recent Years

The two case studies, carried out using two different methodologies and in two timeframes (2007–2012) and 2023, have shown that a significant change in the Basmati area has taken place. The first is that the aromatic Basmati rice growing area in India is expanding, mainly due to economic reasons, as Basmati rice is considered more profitable to farmers as compared to non-Basmati varieties. However, year-to-year fluctuations in Basmati acreage also take place. In the years when farmers do not get a good price for their Basmati crop, the acreage is reduced in the following year. The other is that traditional Basmati varieties are now being grown only in a very small area, which is limited to a few districts. Among the evolved Basmati varieties that are presently being grown, the area under short-duration, recently-bred Basmati varieties has been increasing. The following may be the reasons for the change:

i. ***Emphasis on high-yielding traits coupled with pest and disease tolerance*:** Basmati farmers are shifting from earlier available Basmati varieties (both traditional and evolved) to new, higher-yielding Basmati varieties. Some farmers growing HYVs earlier due to their higher yield have shifted to growing Basmati, as some recently released Basmati varieties have yield potential almost similar to HYVs. In areas prone to specific pests and diseases, recently released pest/disease-resistant varieties are replacing older Basmati varieties.
ii. ***Preference for shorter duration varieties*:** Previously, farmers following rice-potato-wheat rotation in some areas used HYVs due to their shorter duration to fit in the rotation. Now, with the availability of Basmati varieties of duration similar to HYVs, some farmers have shifted from HYV rice to Basmati.
iii. ***Rice market price*:** Longer duration Basmati varieties usually fetch a higher market price (about 20% higher) than the shorter duration Basmati varieties (possibly due to their better taste when cooked). Hence, some farmers still prefer to grow longer-duration Basmati varieties. Manually harvested Basmati has been reported to fetch a higher market price compared to the combine-harvested grain, due to higher moisture content in the latter. Manually harvested Basmati crop is left to dry in the field for a few days after its harvest, for the ease of manual threshing.

REFERENCES

Agrinet Solutions. 2007. *Satellite image and field survey-based acreage assessment and yield estimation for Basmati and non-Basmati rice (Kharif 2007)*. Report submitted to Basmati Export Development Foundation, APEDA, New Delhi. November 2007.

AIREA, 2023. Basmati rice varieties. *All India Rice Exporters' Association*. https://airea.net

Amitrano, D., G. D. Martino, A. Iodice, D. Riccio, and G. Ruello. 2018. Unsupervised rapid flood mapping using Sentinel-1 GRD SAR images. *IEEE Transactions on Geosciences and Remote Sensing* 56: 3290–3299. https://doi.org/10.1109/TGRS.2018.2797536

ASD, 2006. *FieldSpec Pro*. Boulder: Analytical Spectral Devices. https://www.asdi.com/products-spectroradiometers.asp.

Bazzi, H., N. Baghdadi, M. El Hajj et al. 2019. Mapping paddy rice using Sentinel-1 SAR time series in Camargue, France. *Remote Sensing* 11: 887.

Belgiu, M. and L. Dragut. 2016. Random forest in remote sensing: A review of applications and future directions. *ISPRS Journal of Photogrammetry and Remote Sensing* 114: 24–31.

Bending, J., K. Yu, H. Aasen et al. 2015.Combining UAV-based plant height from crop surface models, visible, and near-infrared vegetation indices for biomass monitoring in barley. *International Journal of Applied Earth Observation* 39: 79–87. https://doi.org/10.1016/j.jag.2015.02.012

Birth, G. S. and G. McVey. 1968. Measuring the colour of growing turf with a reflectance spectrophotometer. *Agronomy Journal* 60: 640–643.

Boschetti, M., L. Busetto, G. Manfron, A. Laborte, S. Asilo, S. Pazhanivelan, and A. Nelson. 2017. PhenoRice: A method for automatic extraction of spatio-temporal information on rice crops using satellite data time series. *Remote Sensing of Environment.* 194: 347–365.

Boschetti, M., F. Nutini, G. Manfron, P. A. Brivio, and A. D. Nelson. 2014. Comparative analysis of normalised difference spectral indices derived from MODIS for detecting surface water in flooded rice cropping systems. *PLoS ONE 9*, e88741.

Bouvet, A. and T. Le Toan. 2011. Use of ENVISAT/ASAR wide-swath data for timely rice field mapping in the Mekong River Delta. *Remote Sensing of Environment* 115: 1090–1101.

Bouvet, A., T. Le Toan, and N. Lam-Dao. 2009. Monitoring of the rice cropping system in the Mekong Delta using ENVISAT/ASAR dual polarization data. *IEEE Transactions on Geoscience and Remote Sensing* 47: 517–526.

Bradley, A. P. 1997. The use of the area under the ROC curve in the evaluation of machine learning algorithms. *Pattern Recognition* 30: 1145–1159. https://doi.org/10.1016/S0031-3203(96)00142-2

Breiman, L. 2001. Random forests. *Machine Learning* 45: 5–32.

Broge, N. H. and E. Leblanc. 2001. Comparing prediction power and stability of broadband and hyperspectral vegetation indices for estimation of green leaf area index and canopy chlorophyll density. *Remote Sensing of Environment 76*: 156–172.

Carter, G. A. 1991. Primary and secondary effects of water content on the spectral reflectance of leaves. *American Journal of Botany* 78: 916–924.

Carter, G. A. 1993. Responses of leaf spectral reflectance to plant stress. *American Journal of Botany* 80: 239–243.

Carter, G. A., W. G. Cibula, and R. L. Miller. 1996. Narrowband reflectance compared with thermal imagery for early detection of plant stress. *Journal of Plant Physiology* 148: 515–522.

Carver, K. R. 1988. *SAR Synthetic Aperture Radar – Earth Observing System*, Vol. II f, Washington: NASA Instrument Report, 233.

Casanova, D., G. F. Epema, and J. Goudriaan. 1998. Monitoring rice reflectance at the field level for estimating biomass and LAI. *Field Crop Research* 55: 83–92.

Cen, H., L. Wan, J. Zhu et al. 2019. Dynamic monitoring of biomass of rice under different N treatments using a lightweight UAV with dual-frame snapshot cameras. *Plant Methods* 15:32–52.

Chandrasekar, K., M. V. R. Sesha Sai, P. S. Roy, and R. S. Dwivedi. 2010. Land Surface Water Index (LSWI) response to rainfall and NDVI using the MODIS Vegetation Index product, *International Journal of Remote Sensing* 31: 3987–4005. https://doi.org/10.1080/01431160802575653

Chang, K. W., Y. Shen, and J. C. Lo. 2005. Predicting rice yield using canopy reflectance measured at the booting stage. *Agronomy Journal* 97: 872–878.

Chen, C., H. Lin and Z. Pei. 2007. Application of ENVISAT ASAR data in mapping rice crop growth in Southern China. *IEEE Geoscience and Remote Sensing Letters* 4: 431–435.

Chen, C. and H. McNairn. 2006. A neural network integrated approach for rice crop monitoring. *International Journal of Remote Sensing* 27: 1367–1393.

Cheng, Q. 2006. Multi-sensor comparisons for validation of MODIS vegetation indices. *Pedosphere* 16: 362–370.

Choudhury, I. and M. Chakraborty. 2004. Analysis of temporal SAR and optical data for rice mapping. *Journal of the Indian Society of Remote Sensing* 32: 373–385.

Choudhury, I., M. Chakraborty, and J. S. Parihar. 2007. Estimation of rice growth parameter and crop phenology with conjunctive use of RADARSAT and ENVISAT. *Proceedings of the Envisat Symposium*, Montreux, Switzerland, April 23–27, 6 p.

Cian, F., M. Marconcini, and P. Ceccato. 2019. Normalized difference flood index for rapid flood mapping: Taking advantage of EO Big Data. *Remote Sens. Environ.* 2019: 712–730.

Cibula, W. G. and G. A. Carter. 1992. Identification of a far-red reflectance response to *Ectomycorrhizae* in slash pine. *International Journal of Remote Sensing.* 13: 925–932.

Clark, R. N., T. V. King, C. Ager, and G. A. Swayze. 1995. Initial vegetation species and senescence/stress indicator mapping in the San Luis Valley, CO, using imaging spectrometer data. *Proceedings, Summitville Forum*, ed. H. Posey, J. Pendelton, and D. Van Zyl. Colorado Geological Survey Publication # 38, 64–69. https://speclab.cr.usgs.gov

Clauss, K., H. Yan, and C. Kuenzer. 2016.Mapping paddy rice in China in 2002, 2005, 2010, and 2014 with MODIS time series. *Remote Sensing* 8: 434. https://doi.org/10.3390/rs8050434

Clevers J. G., and A. A. Gitelson. 2013. Remote estimation of crop and grass chlorophyll and nitrogen content using red-edge bands on Sentinel-2 and 3. *International Journal of Applied Earth Observation Geoinformation* 23: 344–351.

Collins, W. 1978. Remote sensing of crop type and maturity. *Photogrammetric Engineering and Remote Sensing* 44: 43–55.

Colwell, R. N. 1973. Some recent developments in photographic interpretation. *Photonirvachak, Journal of Indian Society of Photo-interpretation* 1: 1–20.

Congalton, R. G. 1991. A review of assessing the accuracy of classifications of remotely sensed data. *Remote Sensing of Environment* 37: 35–46

Dadhwal, V. K., R. P. Singh, S. Dutta, and J. S. Parihar. 2002. Remote sensing-based crop inventory: A review of Indian experience. *Tropical Ecology* 43: 107–122.

Delegido, J., J. Verrelst, L. Alonso, and J. Moreno. 2011. Evaluation of Sentinel-2 red edge bands for empirical estimation of green LAI and chlorophyll content. *Sensors* 11: 7063–7081. https://doi.org/10.3390/s110707063

Dong, J. and X. Xiao. 2016. Evolution of regional to global paddy rice mapping methods: A review. *The ISPRS Journal of Photogrammetry and Remote Sensing (P&RS)* 119: 214–227.

Dong, J., X. Xiao, M. A. Menarguez et al. 2016. Mapping paddy rice planting area in northeastern Asia with Landsat 8 images, phenology-based algorithm, and Google Earth. *Engine. Remote Sensing of Environment* 185: 142–154.

Dong, Y., G. Sun, and Y. Pang. 2006. Monitoring of rice crop using ENVISAT ASAR data. *Science in China: Series D: Earth Sciences* 49: 755–763.

Du, M. and N. Noguchi. 2017. Monitoring of wheat growth status and mapping of wheat within-field variations using colour images acquired from a UAV-camera system. *Remote Sensing* 9: 14. https://doi.org/10.3390/rs9030289

Fatchurrachman, R., N. C. Soh, R. M. Shah, S. G. E. Giap, B. I. Setiawan, and B. Minasny. 2022. High resolution mapping of paddy rice extent and growth stages across Peninsular Malaysia using a fusion of Sentinel-1 and -2 time series data in Google Earth Engine. *Remote Sensing* 14: 1875. https://doi.org/10.3390/rs14081875

Fawcett, T. 2006. An introduction to ROC analysis. *Pattern Recognition Letters* 27: 861–874.

Fiorillo, E., E. D. Giuseppe, G. Fontanelli, and F. Maselli. 2020. Lowland rice mapping in the Sedhiou region (Senegal) using Sentinel-1 and Sentinel-2 data and Random Forest. *Remote Sensing* 12: 3403. https://doi.org/10.3390/rs12203403

Foody, G. M. 2002. Status of land cover classification accuracy assessment. *Remote Sensing of Environment* 80: 185–201. https://doi.org/10.1016/S0034-4257(01)00295-4

Gao, B. C. 1996. NDWI: A normalized difference water index for remote sensing of vegetation liquid water from space. *Remote Sensing of Environment* 58: 257–266.

Gausmann, H.W., W. A. Allen, and R. Kardenas. 1969. Reflectance of cotton leaves and their structure. *Remote Sensing of Environment* 1: 110–122.

Gitelson, A., Y. Gritz, and M. Merzlyak. 2003. Relationships between leaf chlorophyll content and spectral reflectance and algorithms for non-destructive chlorophyll assessment in higher plant leaves. *Journal of Plant Physiology* 160: 271–282.

Gitelson, A. A. and M. N. Merzlyak. 1996. Signature analysis of leaf reflectance spectra: Algorithm development for remote sensing of chlorophyll. *Journal of Plant Physiology* 148: 494–500.

Okamoto, K. and H. Kawashima. 1999. Estimation of rice planted area in the tropical zone using a combination of optical and microwave satellite sensor data. *International Journal of Remote Sensing* 20: 1045–1048.

Onojeghuo, A. O., G. A. Blackburn, Q. Wang, P. M. Atkinson, D. Kindred, and Y. Miao. 2018. Mapping paddy rice fields by applying machine learning algorithms to multi-temporal Sentinel-1A and Landsat data. *International Journal of Remote Sensing* 39: 1042–1067.

Orynbaikyzy, A., U. Gessner, B. Mack, and C. Conrad. 2020. Crop type classification using fusion of Sentinel-1 and Sentinel-2 data: Assessing the impact of feature selection, optical data availability, and parcel sizes on the accuracies. *Remote Sensing* 12: 2779, https://doi.org/10.3390/rs12172779

Oyoshi, K., L. Samarakoon, S. Sobue et al. 2022. Rice monitoring in Southeast Asia using Earth observation satellite data. In *Remote Sensing of Agriculture and Land Cover/Land Use Changes in South and Southeast Asian Countries*. ed. K. P. Vadrevu, T. Le Toan, S.S. Ray and C. Justice, 217–233. https://doi.org/10.1007/978-3-030-92365-5_12

Panigrahy, R. K., S. S. Ray, and S. Panigrahy. 2009. Study on the utility of IRS-P6 AWiFS SWIR band for crop discrimination and classification. *Journal of the Indian Society of Remote Sensing*. 37: 325–333. https://doi.org/10.1007/s12524-009-0026-6

Panigrahy, S., M. Chakraborty, S. A. Sharma, N. Kundu, S. C. Ghosh, and M. Pal. 1997. Early estimation of rice area using temporal ERS-1 synthetic aperture radar data: A case study for the Howrah and Hughly districts of West Bengal, India. *International Journal of Remote Sensing* 18: 1827–1833.

Panigrahy, S., K. R. Manjunath, M. Chakraborty. M. Kundu, and J. S. Parihar. 1999. Evaluation of RADARSAT standard beam data for identification of potato and rice crops in India. *ISPRS Journal of Photogrammetry and Remote Sensing* 54: 254–262.

Panigrahy, S., J. S. Parihar, N. K. Patel et al. 1991. Rice acreage estimation for Orissa using remotely sensed data. *Journal of the Indian Society of Remote Sensing*, 19: 16–26.

Parihar, J. S. and V. K. Dadhwal. 2002. Crop production forecasting using remote sensing data: Indian experience. *Proceedings of the ISPRS Comm. VII Symposium 'Resources and Environmental Monitoring' held at Hyderabad, India. Dec. 3-6, 2002*. IAPRS &GIS vol. 34, Part 7. pp. 313–318.

Parihar, J. S. and K. R. Manjunath. 2013. Agricultural applications: Evolutions during the last 25 years. In *Indian Remote Sensing Satellite Series and Applications: A Saga of 25 Years. NNRMs Bulletin 37*. Bangalore, India: NNRMS, Department of Space. pp. 58–75.

Parihar, J. S. and M. P. Oza. 2006. FASAL: An integrated approach for crop assessment and production forecasting. In *Agriculture and Hydrology Applications of Remote Sensing*, ed. R. J. Kuligowski, J.S. Parihar, and G. Saito. Proceedings of SPIE Vol. 6411, 641101. https://doi.org/10.1117/12.713157

Patel, N. K., T. T. Medhavy, C. Patnaik, and A. Hussain. 1995. Multi-temporal ERS-1 SAR data for the identification of rice crop. *Journal of the Indian Society of Remote Sensing* 23: 33–39.

Patel, N. K., T. P. Singh, B. Sahai, and M. S. Patel. 1985. Spectral response of rice crop and its relation to yield and yield attributes. *International Journal of Remote Sensing* 6: 657–664.

Pateriya, B., S. Subramanian, D. Kar, and R. Rajan. 2009. Reliable remote sensing- Test case: A comparison with supply chain *mandis*/millers. *Geospatial World*. https://geospatialmedia.net

Phan, H., T. Le Toan, A. Bouvet, L. D. Nguyen, T. P. Duy, and M. Zribi. 2018. Mapping of rice varieties and sowing date using X-band SAR data. *Sensors* 18: 316. https://doi.org/10.3390/s18010316

Qiu, B., W. Li, Z. Tang, C. Chen, and W. Qi. 2015. Mapping paddy rice areas based on vegetation phenology and surface moisture conditions. *Ecological Indicators* 56: 79–86.

Ramadhani, F., P. Reddy, G. Kereszturi, and J. Proctor. 2020. Automatic mapping of rice growth stages using the integration of Sentinel-2, MOD13Q1, and Sentinel-1. *Remote Sensing* 12: 3613. https://doi.org/10.3390/rs12213613

Rauf, U., W. S. Qureshi, H. Jabbar et al. 2022. A new method for pixel classification for rice variety identification using spectral and time series data for Sentinel-2 satellite imagery. *Computers and Electronics in Agriculture* 193: 196731. https://doi.org/10.1016/j.compag.2022.106731

Ray, S. S. and S. Neetu 2017. Crop area estimation with Remote Sensing. In *Handbook on Remote Sensing for Agricultural Statistics*, J. Delincé ed., Rome: Global Strategy for Improving Agricultural and Rural Statistics (GSARS), FAO Statistics Division (ESS), FAO. pp. 131–183.

Ray, S. S., S. M. Neetu, and S. Gupta. 2014. Use of remote sensing in crop forecasting and assessment of impact of natural disasters: Operational approaches in India. *Proceedings of the Expert Meeting, Vientiane, Lao PDR*. 17 February 2014. Rome, and ADB, Manila: Organised by FAO.

Ribbes, F. and T. Le Toan. 1996. Use of ERS-1 SAR for rice field mapping and rice crop parameter retrieval. *IGARSS* 4: 1983–1985.

Ribbes, F. and T. Le Toan. 1999. Rice field mapping and monitoring with RADARSAT data. *International Journal of Remote Sensing* 20: 745–765.

Rouse J. W., R. H. Haas, J. A. Schell, and D. W. Deering. 1974. Monitoring vegetation systems in the Great Plains with ERTS. *Proceedings of Third Earth Resources Technology Satellite Symposium 1*, Greenbelt, USA: NASA SP-351, 309–317.

Running, S. W., C. O. Justice, V. Solomonson et al. 1994. Terrestrial remote sensing science and algorithms planned for EOS/MODIS. *International Journal of Remote Sensing* 15: 3587–3620.

Saad El Imanni, H., A. El Harti, and J. Panimboza. 2022. Investigating Sentinel-1 and Sentinel-2 data efficiency in studying the temporal behavior of wheat phenological stages using google earth engine. *Agriculture* 12: 1605. https://doi.org/10.3390/agriculture12101605

Saadat, M., S. T. Seydi, M. Hasanlou, and S. A. Homayouni. 2022. Convolutional neural network method for rice mapping using time series of Sentinel-1 and Sentinel-2 imagery. *Agriculture* 12: 2083. https://doi.org/10.3390/agriculture12122083

Sahai, B. 1985. Agricultural remote sensing in the Indian context. *Proceedings of U.S.-INDIA Symposium-cum- Workshop on Remote Sensing Fundamentals and Applications*. March 11–15. Ahmedabad, India: Space Applications Centre.

Sahai, B. and V. K. Dadhwal. 1990. Remote sensing in agriculture. In *Technology Blending and Agrarian Prosperity* eds. J. P. Verma and A. Verma. New Delhi: Malhotra Publishing House, pp. 83–98.

Salam, M. A. and H. Rahman. 2014. Application of remote sensing and geographical information system (GIS) techniques for monitoring *boro* rice expansion area in Bangladesh. *Asian Journal of Geo-informatics* 14: 11–17.

Shao, G., L. Tang, and J. Liao. 2019. Overselling the overall map accuracy misinforms about research reliability. *Landscape Ecology* 34: 2487–2492. https://doi.org/10.1007/s10980-019-00916-6

Shao, Y., X. Fan, H. Liu et al. 2001. Rice monitoring and production estimation using multi-temporal RADARSAT. *Remote Sensing of Environment* 76: 310–325.

Singh, A. N. and V. P. Singh 2003. Extent, distribution, and growing environments of aromatic rices of India. In *A treatise on the scented rices of India*, ed. R. K. Singh and U. S. Singh, 211–229. Ludhiana: Kalyani Publishers.

Singh, R. P., S. R. Oza, and M. R. Pandya. 2006. Observing long-term changes in rice phenology using NOAA-AVHRR and DMSP-SSM/I/I/satellite sensor measurements in Punjab, India. *Current Science* 91: 1217–1221.

5.2.2 Meteorological Data-based Models

Yield estimation models using meteorological data rely either on meteorological factors (e.g., precipitation, temperature, solar radiation) or agrometeorological indices derived from these factors, such as drought indices (Mathieu and Aires, 2018). Agrometeorological models utilise these meteorological parameters, alongside yield data, to estimate crop yield variations and their correlation with weather conditions (Sheehy et al., 2006). Crop-weather models, typically using statistical techniques, offer simplified representations of the complex relationships between climate factors and crop growth, yield, and its components (Bair, 1977). The least-squares regression method is often employed to explore these relationships, regressing yield (the dependent variable) against meteorological variables (independent variables) (Jones, 1982). For example, in Madurai district, Tamil Nadu, India, eight machine learning models, such as linear regression (LR), Random Forest regression (RFR), support vector regression (SVR), cat boost regression (CBR) and hybrid machine learning with variance inflation factor (VIF) LR-VIF, RFR-VIF, SVR-VIF, and CBR-VIF techniques were used to predict rice yield based on weather data spanning over 15 years (Sakthipriya and Chandrakumar, 2024). Among these, CBR-VIF showed the best performance, followed by CBR, and it was found that variables like maximum temperature, minimum temperature, normal rainfall, and actual rainfall significantly affected the rice yield in the district.

5.2.3 Remote Sensing and Meteorological Data-based Indices Used in Yield Modelling

Among remote sensing-based indicators for crop yield estimation, the Normalised Difference Vegetation Index (NDVI) remains one of the most widely utilised, particularly when assessed across distinct phenological stages. Alongside NDVI, a range of alternative vegetation indices, such as Leaf Area Index (LAI), Enhanced Vegetation Index (EVI), Perpendicular Vegetation Index (PVI), Green Area Index (GAI), Vegetation Condition Index (VCI), Fraction of Photosynthetically Active Radiation (FPAR), and Wide Dynamic Range Vegetation Index (WDRVI), are frequently incorporated as explanatory variables in statistical regression frameworks to improve yield prediction accuracy (Zhu et al., 2021). Yield estimation models that rely on meteorological data typically use input variables such as maximum and minimum temperatures (T_{max}, T_{min}), relative humidity (RH%), accumulated rainfall (ARF), and solar radiation indicators like sunshine hours (SSH). In some cases, these primary weather parameters are further transformed into agrometeorological indices to better capture crop-weather interactions (Mathieu and Aires, 2018). Various remote sensing and meteorological data-based indices normally used in yield prediction, along with their formulation, are given in Table 5.1.

TABLE 5.1
Meteorological and Remote Sensing Data-based Indices and Their Mathematical Formula for Yield Modelling

Index	Formula	References
	Meteorological Data–based Indices	
GDD: Growing Degree Days	$\Sigma\left[\left\{\left(T_{max}+T_{min}\right)/2\right\}-Tb\right]$	Ketring, and Wheless, (1989)
TD = Temperature Difference	$TD=\left(T_{max}-T_{min}\right)$	Kothawale, and Rupa Kumar (2005)
HTU: Heliothermal units	Σ GDD x n $(n = \text{actual Sunshine Hours})$	Rajput (1987)
PTU: Photothermal units	$\Sigma\left(GDD^{*}\ \text{Day Length}\right)$ Day Length = Sunshine Hours	Nuttonson (1948)
HUE = Heat Use Efficiency	Grain Yield x Heat Units	Nandini, and Sridhara (2019)
CWSI: Crop Water Stress Index	$\left(T_C-T_{wet}\right)/\left(T_{dry}-T_{wet}\right)$	Jones (1992)
HI: Harvest index (%)	Grain Yield / Biological Yield* 100	Donald (1962)
LAI: Leaf Area Index	Total Leaf Area (m^2) / Ground Area (m^2)	Watson (1958)
	Remote Sensing Data-based Spectral Vegetation Indices	
NDVI: Normalised Difference Vegetation Index	$\rho NIR-\rho Red/\rho NIR+\rho Red$	Rouse et al. (1974)
GNDVI: Green Normalised Difference Vegetation Index	$(NIR-Green)/(NIR+Green)$	Gitelson et al. (1996)
EVI: Enhanced Vegetation Index	$2.5 \text{ x } \rho NIR-\rho Red/\rho NIR+6\,\rho Red-7.5\,\rho Blue+1$	Jiang, et al. (2008)
SAVI: Soil Adjusted Vegetation Index	$1+L^{*}\,\rho NIR-\rho Red/\rho NIR+\rho Red+L\ (L=0.5)$	Huete (1988)
MSI: Moisture Stress Index	$\rho SWIR\,1/\rho NIR$	Welikhe, et al. (2017)
LSWI: Land Surface Water Index	$NIR-SWIR/NIR+SWIR$	Zhou et al. (2017)
RGVI2: Rice Growth Vegetation Index2	$1.05-(Blue+Red)(NIR+SWIR1+0.5)$	Choudhary, Shi, and Dong (2021)
VCI (%): Vegetation Condition Index	$(NDVI-NDVI\,min)/(NDVI\,max-NDVI\,min)^{*}\,100$	Kogan (1990)
TCI (%): Temperature Condition Index	$\left(BT_{max}-BT\right)/\left(BT_{max}-BT_{min}\right)^{*}\,100$	Kogan (1990)
PVI: Perpendicular Vegetation Index	$\left(NIR-a^{*}Red-b\right)/\left(sqrt\left(1+a2\right)\right)$	Richardson and Wiegand (1977)
TVI: Transformed Vegetation Index	$\left\{SQRT\left[(NIR-Red)/(NIR+Red)\right]\right\}+0.5$	Deering et al. (1975)

(Continued)

TABLE 5.1 (Continued)

Index	Formula	References
GAI: Green Area Index	Green leaf & stem area/area of ground	Baret, et al. (2010)
fPAR: Fraction of Photosynthetically Active Radiation	fraction of incoming solar radiation in the spectrum of 400–700 nm	Moreau, and Li (1996)
WDRVI: wide Dynamic Range Vegetation Index	$WDRVI = (a^{*}\rho NIR - \rho red) / (a^{*}\rho NIR + \rho red)$	Gitelson (2004)
Red Edge Vegetation Indices Using Sentinel-2 MSI Data		
Chl Index Red (Chl-r)	$(Band7)/(Band5) - 1$	Gitelson et al. (2005)
Chl Index Green (Chl-g)	$(Band7)/(Band3) - 1$	Gitelson et al. (2005)
GCI: Green Chlorophyll Index	$(NIR)/(Green) - 1$	Gitelson, Gritz, and Merzlyak (2003)
RERI: Red-edge Rice Index	$(2x\,\rho NIR + \rho Red\text{-}edge\text{-}2 - 3x\,\rho Red) / (1 + \rho Red)$	Wan et al. (2024)

Note: ρ is the surface reflectance values for blue (B), red (R), near-infrared (NIR), Shortwave infrared (SWIR1 and SWIR2) bands, BT: brightness temperature, a = slope of the soil line, b = gradient of the soil line, T_{max} = maximum Temperature, T_{min} = minimum Temperature, Tb = Base Temperature (10°C).

5.3 AGROMETEOROLOGICAL RICE YIELD MODELLING STUDIES

During the crop growing season, crop yields are estimated by conducting crop-cutting experiments (Ahmad et al., 2021) and adopting standard statistical approaches such as the multivariate linear regression (MLR). However, these methods often fail to deliver satisfactory accuracy. As a result, machine learning approaches, capable of uncovering complex patterns between climatic factors and crop yield through data-driven algorithms, are gaining widespread adoption in agricultural modelling. Numerous studies have employed machine learning techniques for crop yield prediction, particularly for crops like rice (Rai and Chandrahas, 2000; Satpathi et al., 2023) and wheat (Aravind et al., 2022). For instance, a district-level Agromet rice yield model was developed using data from the Karnal district in Haryana, India, spanning 1981–1994 (Kalubarme and Ahuja, 1996). This model incorporated meteorological variables such as minimum and maximum temperatures and sunshine hours. Key factors like GDD, TD, and SSH were integrated across three crop growth stages: active vegetative phase (AVP), reproductive phase (RP), and maturity phase (MP). The optimal model, which included TD during AVP and RP, GDD during RP, and ASH during AVP and RP, achieved an R^2 of 0.842 and a standard error of estimate (SEOE) of 0.663, explaining 84% of the variability in district-level rice yields.

5.3.1 Effect of Rainfall and Temperature on Rice Yield

The impact of weather conditions on rice yield in Karnal District of Haryana state was analysed using weekly meteorological data from June 1 to harvest during October. The historical time series data on rice yield of the past 35 years from 1985–86 to 2019–20 have been analysed to examine how weather variables influence rice productivity. The impact of climate change on rice yields at various growth phases of the rice crop has been investigated, and different statistical models have been developed (Chetna et al., 2023). The results of the effect of maximum and minimum temperatures indicated that a 1°C increase over the average weekly maximum temperature had a positive effect on rice yield during the initial growth and vegetative growth stages, but it had a negative effect during the reproductive stage. However, during the ripening and maturity stages, increases in maximum temperature were found to be advantageous. The results also indicated that a 1-mm increase in rainfall over the normal weekly rainfall has had a favourable effect up to the vegetative growth stage (Chetna et al., 2023).

5.3.2 Effect of Sunshine Duration on Rice Productivity

A study by Horie et al. (1997) emphasised the significant impact of spatial and temporal variation in solar radiation on rice productivity. Building on this, Garima et al. (2021) analysed the influence of sunshine hours on the yearly variability of rice yield in Ludhiana district, Punjab, using 15 years of data (2003–2017). The analysis revealed that increases in average sunshine hours during the vegetative, reproductive, and entire growth phases were positively correlated with grain yield. A significant relationship was observed between the reproductive phase, panicle initiation, anthesis, and grain yield. In years with below-average yields, overcast skies and reduced sunshine hours from transplanting to maturity were identified as the primary cause of yield reduction. Chetna et al. (2023) also observed the beneficial effect of an increase of one hour above the average weekly sunshine hours during the crop sowing and vegetative growth stages.

5.4 STUDIES ON RICE YIELD MODELLING

Various modelling approaches have been employed to estimate rice yield, ranging from empirical and semi-empirical methods to mechanistic crop simulation models. Empirical models, also known as statistical or correlation-based models, are especially effective for large-area applications, such as regional or national assessments. These models typically rely on long-term yield records and climatic trends to establish correlations, as discussed by Lobell and Burke (2010), thereby offering a valuable approach for yield estimation over expansive areas.

5.4.1 Empirical Models

Empirical models are relatively simple and scalable, particularly for uniform sites cultivated with the same rice variety. These models are commonly used for trend analysis to explore the relationships between various variables. Through regression analysis, factors such as meteorological and economic variables are incorporated into these models. Empirical models have been applied to estimate rice yields across different regions globally. As a part of empirical modelling, machine learning has introduced innovative approaches to estimate rice productivity and yields of other crops (Palanivel and Surianarayanan, 2019). Random forests, for example, have been successfully applied with Sentinel-1 data, showing a strong correlation ($r^2 = 0.86–0.93$) for three rice-growing seasons within a single year, despite varying transplanting and harvesting dates (Clauss et al., 2018).

5.4.2 Semi-Empirical Models

Semi-empirical models bridge the gap between purely statistical models and crop simulation models. Several rice-specific semi-empirical remote sensing (RS)–based crop models have been developed, with GRAMI, introduced by Maas (1992), being one of the earliest. A revised version of the GRAMI model incorporates an in-season adjustment component that serves as a data assimilation mechanism (Yeom, Ko, and Kim, 2015). Such models frequently make use of remote sensing-derived inputs, either by directly representing canopy conditions or by refining internal model states when observational data become available. In a recent advancement, a novel approach was introduced to map rice-growing areas and estimate yield by leveraging NDVI values extracted from a series of nine Sentinel-2 satellite acquisitions. The Leaf Area Index (LAI) was derived using the Surface Energy Balance Algorithm for Land (SEBAL), and both NDVI and LAI were employed within an empirical yield estimation framework. This model achieved a mean prediction error of 0.66 tons per hectare. The SEBAL model predicted LAI with a Mean Absolute Percentage Error (MAPE) of 6.76%, and the combination of NDVI at the heading stage and LAI data demonstrated effective yield estimation (Ali et al., 2021).

5.4.3 Process-based Rice Crop Models

Process-based crop models, also known as crop growth simulators, are mathematical frameworks designed to replicate how plant development and yield respond to environmental factors such as climate conditions, soil properties, water availability, physiological stress, and agricultural management practices (Batchelor et al., 2002). These models are more complex than empirical models, offering a deeper understanding of the mechanisms underlying rice growth and yield. As crop yield modelling technology has advanced, numerous models have been developed, with rice-specific models such as ORYZA (Li et al., 2017), RiceGrow (Tang et al., 2009), and WARM (Confalonieri, Rosenmund, & Baruth, 2009) standing out. The ORYZA model, created by the International Rice Research Institute (IRRI), simulates rice crop performance across varying conditions, including optimal growth scenarios as well as environments constrained by nitrogen or water availability. The model has been utilised with optical and SAR data, using the LAI as the state variable, to accurately estimate rice yield (Setiyono et al., 2018).

5.4.4 Rice Yield Modelling Using Simulation Models

Crop simulation models quantify the interactions between plant physiological functions and environmental variables over time. Well-known models like WOFOST (World Food Studies), APSIM (Agricultural Production Systems Simulator), and DSSAT (Decision Support System for Agrotechnology Transfer) operate on a daily timestep to replicate crop phenology, biomass accumulation, and yield formation. These models serve as essential tools for evaluating agricultural outcomes under diverse climatic conditions and management strategies (Cao et al., 2021). Despite their utility, these models demand extensive and site-specific input data, such as local soil profiles, crop management practices, and detailed weather records, and are often computationally intensive.

To overcome these limitations, integrating remote sensing data with process-based crop models has emerged as an effective approach. Remote sensing provides timely, spatially continuous information on actual crop conditions, which can complement traditional model inputs and improve estimation accuracy (dela Torre et al., 2021). For instance, Huang et al. (2002) developed the Rice-SRS model by modifying the original ORYZA1 framework to assimilate NDVI data from various satellite sources. Their study demonstrated that integrating NOAA AVHRR (LAC) NDVI data with the simulation model led to accurate rice yield predictions in the Shaoxing region of China. Similarly, Sailaja et al. (2013) applied the ORYZA2000 model in conjunction with spatial technologies, including remote sensing and GIS, to estimate paddy rice yields in the irrigated rice-growing region of Nalgonda district, Andhra Pradesh, India. In this study, NDVI derived from IRS-1D LISS-III satellite imagery and multiple GIS layers representing soil and climate parameters were incorporated into the model. The resulting yield estimates were comparable with those reported by the Department of Agriculture, validating the integration approach. A broader pilot study conducted by Setiyono et al. (2014) across several countries in South and Southeast Asia demonstrated the effectiveness of combining crop growth modelling with both real-time and historical weather data, alongside key SAR–derived variables such as the start of the growing season and leaf area index (LAI). This system incorporated two modules: MAPscape-Rice, which processes satellite observations into parameters like phenological status and start of season (SOS); and ORYZA2000, which assimilates these variables to generate yield forecasts. The study concluded that integrating SAR products into process-based models significantly enhances yield estimation accuracy and has practical implications for applications such as crop insurance programs.

5.4.5 Rice Yield Prediction Using Deep Learning Models

Deep learning, a subfield of machine learning (ML), leverages artificial neural network (ANN) architectures to model complex data patterns. ANNs consist of interconnected layers of artificial neurons that collaboratively process and learn from input data. Unlike traditional ML models, deep learning techniques typically require large-scale datasets to effectively capture intricate relationships (De Clercq and Mahdi, 2025). Both ML and deep learning fall under the broader domain of artificial intelligence (AI). Among deep learning methods, Long Short-Term Memory (LSTM) and Bidirectional LSTM (BiLSTM) are particularly effective in modelling temporal dependencies and uncovering complex, nonlinear interactions in data (Divakar et al., 2022). A comprehensive overview of precision agriculture proposes novel approaches to rice yield prediction using deep learning models such as LSTM and Gated Recurrent Units (GRU) (Chang, 2024), emphasising their ability to autonomously extract meaningful patterns from diverse agricultural datasets. Within precision agriculture, Convolutional neural networks (CNNs) have proven effective in processing satellite and UAV imagery by automatically extracting spatial patterns essential for identifying crop types and evaluating plant health. Meanwhile, recurrent neural networks (RNNs), including LSTM, are well-suited for time-series applications such as yield forecasting, irrigation planning, and assessing the impacts of weather variability on crop growth.

Sathya and Gnanasekaran (2023) proposed a hybrid predictive framework combining MLR with LSTM for rice yield estimation in the Tamil Nadu Delta region. The results showed that the combined MLR-LSTM approach delivered higher predictive accuracy compared to the individual models used separately. Similarly, Mia et al. (2023) developed a multimodal deep learning model that integrates UAV–derived multispectral imagery and weather data to predict rice grain yield at the heading stage. Their results highlighted the effectiveness of using weekly weather data and a simplified CNN architecture, suggesting that even basic CNN–based feature extractors are capable of generating accurate yield predictions while maintaining computational efficiency.

5.5 RICE YIELD MODELLING ADOPTED IN CAPE/FASAL PROJECTS IN INDIA

The foundation of crop yield prediction in India was established with the initiation of the Indian Remote Sensing (IRS) Utilisation Programme in 1983, which focused on preparatory activities such as crop stress detection and yield modelling in anticipation of the IRS-1A satellite launch in 1988 (Navalgund and Kasturirangan, 1983), and was further advanced through early experiments that employed Landsat and IRS satellite data (IRS-1A, IRS-1B, IRS-P2) for estimating crop acreage and monitoring crop conditions (Sahai and Dadhwal, 1990). A significant milestone was the launch of the *Crop Acreage and Production Estimation (CAPE)* programme in 1988 by the Ministry of Agriculture, Government of India. CAPE aimed to provide pre-harvest, district-level production forecasts for major crops, such as wheat, rice, rabi sorghum, groundnut, cotton, and rapeseed-mustard, using stratified random sampling of RS data and agrometeorological variables to develop district-level yield models. Building on CAPE's success, the Ministry initiated the *Forecasting Agricultural output using Space, Agrometeorology and Land-based observations (FASAL)* programme (Parihar and Oza, 2006). FASAL continues to deliver multiple pre-harvest forecasts at national and state levels for five key crops: *kharif* and rabi rice, jute, rapeseed and mustard, winter potato, and wheat.

At the national scale, crop yield prediction incorporates diverse methodologies, including weather-yield models, crop simulation models, and RS–driven statistical approaches (Jain et al., 2019). Recent developments in remote sensing, characterised by improved spectral, spatial, radiometric, and temporal resolutions, have enabled semi-physical yield estimation approaches. During the 2018–19 *kharif* season, a model combining RS–derived parameters with physiological concepts such as photosynthetically active radiation (PAR) and the fraction of PAR absorbed by the crop was used. The Net Primary Productivity (NPP) was estimated via the Monteith model, and rice yield was derived using NPP, radiation use efficiency, and harvest index (Dwivedi et al., 2019). Furthermore, RS–guided Crop Cutting Experiments (CCEs) are now conducted to enhance yield estimates at district and state levels. For rice, crop density classification maps are generated using Resourcesat-2 AWiFS NDVI composites from September to mid-October. Based on the NDVI frequency distribution, rice fields are categorised into three density classes (A, B, C), from which a 10% random sample in each class is selected for CCE. The Mahalanobis National Crop Forecasting Centre (MNCFC) in the Ministry of Agriculture and Farmers' Welfare, Government of India, is the main organisation implementing the FASAL project, collaborating with ISRO and ICAR-IARI for operational yield estimates using semi-physical and crop simulation models (MNCFC website: www.mncfc.gov.in, accessed on 16 June 2025). The correlation-weighted empirical agrometeorological model-based yield is generated by the India Meteorological Department (IMD) (Ray et al., 2017). Approaches adopted in yield modelling during CAPE/ FASAL projects in India have been summarised in Table 5.2.

5.6 STUDIES ON AROMATIC BASMATI RICE YIELD MODELLING

5.6.1 Use of Agrometeorological Indices and Phenology of Basmati Rice

Field experiments were conducted during the *kharif* seasons of 2015 and 2018 to study the phenological development of Basmati rice varieties under different transplanting dates and agro-meteorological conditions. At the Rice Research Station, Haryana Agricultural University, a 2015 study (Abhilash Dagar et al., 2017) evaluated the performance of Basmati rice varieties, CSR-30, HB-2, PB-1121, and PB-1509, under transplanting dates ranging from June 25 to July 25. Key indices such as GDD, helio thermal units (HTU), photo thermal units (PTU), radiation use efficiency (RUE), and heat use efficiency (HUE) were assessed, revealing a progressive decline in heat unit requirements for various phenological stages with delayed transplanting across all varieties. Complementing these findings, a 2018 study at Sher-e-Kashmir University of Agricultural Sciences and Technology, Jammu (Sultan Singh et al., 2021), examined Basmati-370 and Pusa-1121, predominant in the region, under varying sowing dates and agro-climatic conditions. Minimum temperature, rainfall, and evening relative humidity exhibited a positive association with the panicle initiation and flowering phases of the crop. Early sowing on June 1st consistently outperformed later dates in terms of grain yield, largely due to enhanced heat use efficiency. Among the

TABLE 5.2
Yield Model Approaches Adopted in CAPE/FASAL Projects in India

Year	Satellite/Sensor Used	Methodology for Rice Yield/Production Forecast	States Covered/ References
CAPE Phase I (1988–1991)	Landsat MSS and IRS-1 LISS- I	District-level historical yield trend and rainfall-based agrometeorological model	2 rice-growing states (Parihar and Dadhwal, 2002)
CAPE Phase II (1992–2006)	IRS LISS-I/ LISS-II (1992–95), ERS-1 (1996), RADARSAT-1 (1998-2006)	District-level historical yield trend and rainfall-based agrometeorological model used for yield prediction. Two forecasts, the first in October and the second in the following January	11 major rice-growing states (Parihar and Dadhwal, 2002)
FASAL (implementation in 2007, by ISRO-SAC and collaborating institutions)	RADARSAT ScanSAR	District-level historical yield trend and rainfall-based agrometeorological model	Pilot implementation in Orissa (Parihar and Oza, 2006)
2012 (FASAL Taken over by MNCFC, Dept. Agric. & FW	RADARSAT-2 ScanSAR	District-level yield using Correlation weighted agro-meteorological model (empirical and crop simulation)	14 major rice-growing states (Ray and Neetu, 2017; Jain et al., 2019)
2013–2014	RISAT-1 ScanSAR HH	District level yield using Correlation weighted agro-meteorological model + yield estimate using SAR data-based biomass modelling	14 major rice-growing states (Jain et al., 2019)
2015	RISAT-1 ScanSAR HH	District level yield using Agromet + crop simulation models + RS input-based CCE yield	14 major rice-growing states (Jain et al., 2019)
2016	Sentinel-1, VV	District level yield using correlation weighted agrometeorological model+ VCI remote sensing-based model	14 major rice-growing states (Jain et al., 2019)
2017 onwards	Sentinel-1, VV	Correlation weighted agrometeorological model+ crop simulation models + VCI remote sensing-based model	14 major rice-growing states (Jain et al., 2019)

varieties, Basmati-370 exhibited greater adaptability and a higher benefit–cost ratio than Pusa Basmati-1121, highlighting its suitability under diverse sowing methods and climatic conditions.

5.6.2 Use of Remote Sensing Data-based Indices in Basmati Yield

Singh et al. (2022) developed predictive models for estimating grain yield and nitrogen (N) uptake in different Basmati rice (*Oryza sativa* L.) cultivar groups using NDVI measurements. Separate regression models were formulated for CSR-30 (a tall cultivar) and the Pusa Basmati group varieties (semi-dwarf cultivars). Validation results demonstrated that NDVI measurements taken at the panicle initiation stage provided reliable estimates of both grain yield and N uptake, closely aligning with observed field data. These models offer a non-destructive, timely, and cultivar-specific approach to support precision nutrient management and yield forecasting in Basmati rice cultivation.

5.6.3 Use of Agromet-Spectral Model in Basmati Yield

A comprehensive study was conducted to estimate acreage of aromatic Basmati rice varieties and predict yields in North-Western India using remotely sensed data (Verma et al. 2022). The differentiation was achieved by exploiting phenological differences that influence spectral reflectance patterns captured in satellite imagery. Yield prediction models were developed using data from CCE, agro-meteorological variables, and spectral data based on NDVI and NDWI. Agromet-spectral-yield models were constructed using multiple regression techniques. CCEs were carried out across selected plots in the study districts to support model development and production forecasting (Verma et al., 2022). Notably, strong correlations ($R^2 > 0.84$) were observed between NDVI and CCE yields for major Basmati cultivars such as Pusa Basmati-1 and Pusa-1121, and a non-Basmati variety, Sharbati, in Haryana. The integration of satellite-derived acreage estimates with yield models enabled the generation of timely district-level production forecasts, well ahead of harvest.

5.6.4 Use of Crop Growth Simulation Model in Basmati Rice Yield

Crop growth simulation models serve as powerful tools to quantify the effects of climatic, edaphic, and management factors on crop development, productivity, and sustainability. The CERES (Crop Environment Resource Synthesis) model has been effectively utilised to simulate phenology, biomass accumulation, and yield in wheat and barley under typical growing conditions (Nain and Kersebaum, 2007). Expanding on this, Nyang'au et al. (2014) applied the CERES-Rice model (version 4.5 of the DSSAT system) to assess the impact of climatic variability on Basmati-370 and IR-2793-80-1 rice varieties cultivated under the System of Rice Intensification (SRI) across four irrigation schemes in Kenya. Incorporating daily weather inputs, such as maximum and minimum temperatures, precipitation, and solar radiation, alongside atmospheric CO_2 concentration scenarios, the study showed that an increase in temperature adversely affected grain yield, while elevated CO_2 levels and improved solar radiation positively influenced the productivity. These findings highlight the potential of crop simulation models like CERES-Rice in evaluating the implications of climate change and informing adaptive strategies in rice-based production systems.

5.7 MODELLING BASMATI RICE YIELD: A CASE STUDY IN HARYANA

This study aimed to assess the performance of various multivariate statistical techniques and machine learning models, with a particular focus on MLR, RF, and Extreme Gradient Boosting (XGBoost), for pre-harvest yield prediction of aromatic Basmati rice varieties in Karnal district, Haryana. Seventeen years (2007–2023) of historical time-series weather data across key rice growth stages during the *kharif* season were analysed in conjunction with observed yield data to establish robust weather–yield relationships. Analysis was carried out using R within the RStudio environment.

The major objectives of the study were:

i. To develop a simplified and interpretable modelling approach for predicting the yield of aromatic Basmati rice based on agro-meteorological parameters such as maximum temperature (T_{max}), minimum temperature (T_{min}), accumulated rainfall (ARF), and sunshine hours (SSH).
ii. To apply and assess the predictive capabilities of MLR, RF, and XGBoost models in forecasting rice yield.
iii. To evaluate and validate the predictive accuracy of these models using standard forecast error metrics, including root mean square error (RMSE), mean absolute error (MAE), and mean absolute percentage error (MAPE).

This research aims to contribute to the development of climate-resilient and data-driven yield forecasting frameworks for region-specific Basmati rice production management.

5.7.1 Study Area

This study was conducted in Karnal district, a key contributor to Basmati rice acreage and production in Haryana, India. Haryana, located in northern India between latitudes 27°39′N to 30°55′N and longitudes 74°27′E to 77°36′E, spans 4.421 million hectares—approximately 1.35% of India's total geographical area. Karnal district, situated between 29°26′16″ to 29°59′36″N and 76°29′43″ to 77°13′46″E, falls within the Eastern Plain Zone of the state and covers 1,967 square kilometres. The region receives an average of 766 mm of rainfall during the monsoon season (June to October) and features deep alluvial soils with medium to medium-heavy texture, making it highly suitable for *kharif*-season rice cultivation. The location map of the Karnal district in Haryana State is given in Figure 5.1.

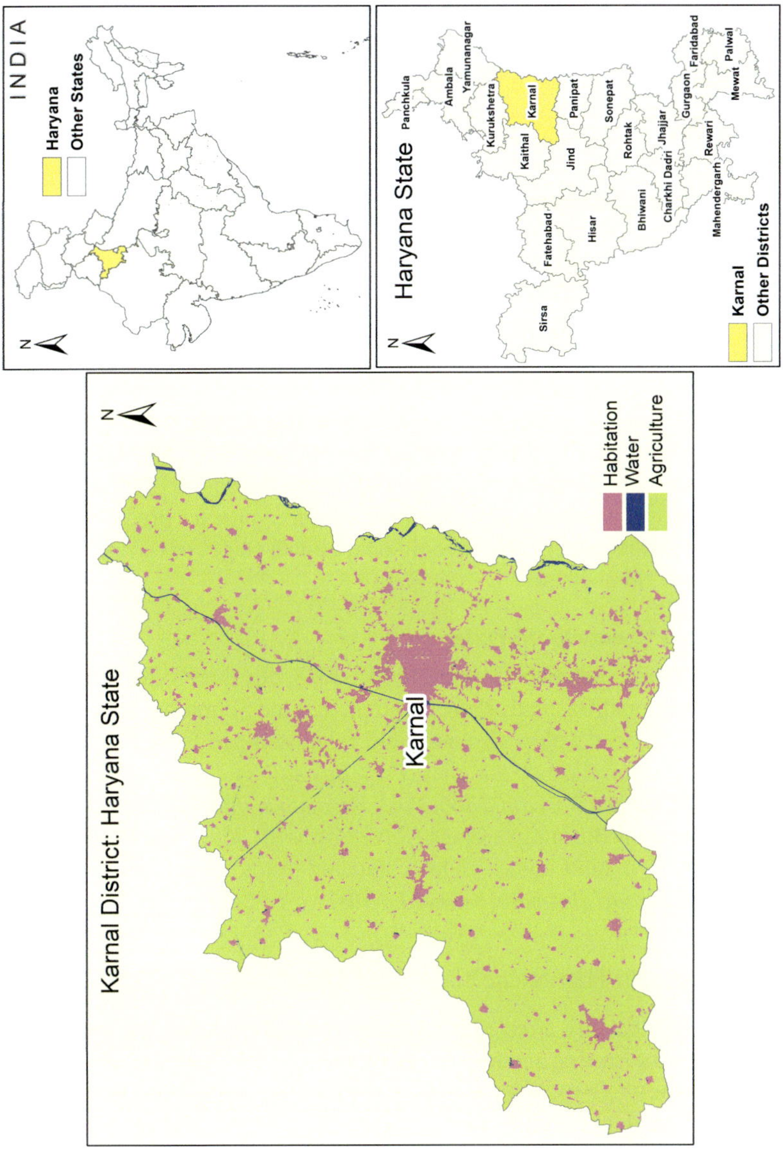

FIGURE 5.1 Location map of Karnal District in Haryana State.

5.7.2 Climate of the Study District

Haryana experiences a humid subtropical climate with dry winters, classified under the Köppen climate classification as ranging from semi-arid to sub-humid. The state is divided into three climatic zones: hot semi-arid, hot desert, and hot sub-humid. It is characterised by extreme summer heat and cold winters. During the 2023 monsoon season (June 1 to September 30), Haryana received varied rainfall patterns. In June, the state recorded 80.7 mm of rainfall, which was 48% above the normal of 54.7 mm. July witnessed significant excess rainfall, receiving 236.9 mm, 59% above the normal 149.1 mm, with five districts recording substantial departures. In contrast, August experienced a significant deficit, with only 58.5 mm of rainfall, 60% below the normal 146.1 mm (India Meteorological Department, 2023).

5.7.2.1 Weather Data

Seventeen years (2007–2023) of historical time-series weather data for the *kharif* rice season in Karnal district, Haryana, were obtained from the Central Soil Salinity Research Institute (CSSRI), Karnal. The dataset covers key meteorological parameters across the rice growth period, corresponding to Standard Meteorological Weeks (SMWs) 25 to 42 (June 18 to October 21), encompassing 18 weeks from transplanting to physiological maturity. Daily weather records were aggregated into weekly averages for five critical variables: T_{max}, T_{min}, RH%, ARF, and SSH. These variables were used to develop weather–yield relationships and construct predictive models for in-season yield forecasting. The phenological growth stages of different Basmati varieties and SMW numbers are given in Figure 5.2.

5.7.2.2 Basmati Variety Yield Data

Time series yield data spanning 17 years (2007–2023) for three groups of Basmati rice varieties, traditional Basmati, short duration, and long duration, were collected from Karnal district to develop weather-yield relationships (Table 5.2). The data reveal that long-duration Basmati varieties (*Pusa Basmati-1121*, *Pusa Basmati-1718*, *Pusa Basmati-1885*, *Pusa Basmati-1886, Pusa Basmati-1, Pusa Basmati-6* and *CSR-30*), and short-duration Basmati varieties (*Pusa Basmati-1509*, *Pusa Basmati-1692*, *Pusa Basmati-1847*) exhibit higher yields compared to traditional cultivars such as *Basmati-370 and Taraori Basmati (Karnal Local*).

5.7.3 Weather and Historical Yield Data Used

Rice yields are significantly affected by weather conditions during the crop growth period. In-season forecasting models for Basmati rice yield were developed using weather parameters.

MLR and two machine learning methods, RF and XGBoost, were employed, with Basmati rice yield as the dependent variable (response) and weekly weather variables, namely T_{max}, T_{min}, RH %, ARF, and SSH as independent variables. Weekly weather data spanning from the 25th to 43rd SMW were used (refer to Figure 5.4) in stepwise regression to select key independent variables for the model. Data from the first 17 years (2007–2023) were used for model development, with predictions extended to forecast district-level yields for the 2024 cropping season.

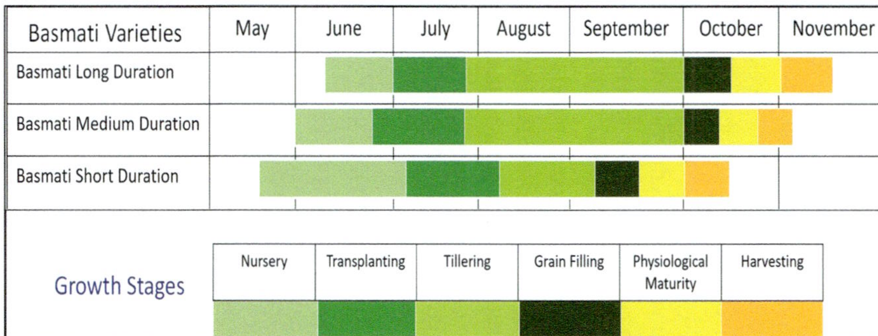

Standard Meteorological Weeks

Week No.	Dates	Week No.	Dates
25	18 Jun – 24 Jun	34	20 Aug – 26 Aug
26	25 Jun – 01 Jul	35	27 Aug – 02 Sep
27	02 Jul –08 Jul	36	03 Sep – 09 Sep
28	09 Jul – 15 Jul	37	10 Sep – 16 Sep
29	16 Jul – 22 Jul	38	17 Sep – 23 Sep
30	23 Jul – 29 Jul	39	24 Sep – 30 Sep
31	30 Jul – 05 Aug	40	01 Oct – 07 Oct
32	06 Aug – 12 Aug	41	08 Oct – 14 Oct
33	13 Aug – 19 Aug	42	15 Oct – 21 Oct
		43	22 Oct – 28 Oct

FIGURE 5.2 Phenological growth stages of different Basmati varieties and Standard Meteorological Week (SMW) numbers.

5.8 MULTIPLE LINEAR REGRESSION (MLR) ANALYSIS

MLR is a multivariate statistical method employed to explore how a set of independent variables collectively influences a single response variable. In this context, let **Y** represent the dependent variable, specifically the yield of Basmati rice varieties, which is assumed to have a linear relationship with **k** predictor variables $\mathbf{x_1, x_2, \ldots, x_k}$, corresponding to weekly weather parameters. The general form of the MLR model can be expressed as:

$$Y = \beta_0 + \beta_1 x_1 + \beta_2 x_2 + \cdots\cdots + \beta_k x_k \,\ldots\ldots\ldots \tag{5.1}$$

Here, $\beta_1, \beta_2, \ldots, \beta_k$ denote the regression coefficients corresponding to the independent variables, which include weekly weather parameters such as T_{max}, T_{min}, RH%, ARF, and SSH. A stepwise regression approach was utilised to identify the most relevant predictors, resulting in the final specification of the multiple linear regression model (Nain et al., 2021).

5.8.1 Aromatic Basmati Rice Yield Models using Multiple Linear Regression (MLR)

The study employed MLR to investigate how various weather parameters influence Basmati rice yield in Haryana's Karnal district. Weekly weather data from SMW 25 (June 18–24) to SMW 43 (October 23–28) across 19 weeks, spanning 17 years (2007–2023), were used as independent variables. These included T_{max}, T_{min}, RH%, ARF, and SSH. The dependent variable was the yield data for three distinct groups of Basmati rice varieties: traditional, long duration, and short duration. Yield prediction models were developed for each of these groups, and the corresponding regression equations and coefficients for the multivariate models were derived. These models provide valuable insights into the weather-yield relationship, enabling better prediction of Basmati rice yields based on meteorological data.

5.8.1.1 Group A: Traditional Basmati Rice Varieties

$$\begin{aligned}\mathbf{Y = 3.148 + 0.256^{*}Tmax40 + 0.016^{*}SSH30 + 0.033^{*}SSH33}\\ \mathbf{+\,0.128^{*}SSH39 - 0.054^{*}RH36 + 0.068^{*}RH39}\end{aligned}$$

$$\begin{aligned}R^2 = 0.884, \text{Adj.}R^2 = 0.814, \text{F-Statistics} = 12.65, \text{P-Value}\\ = 0.00037, \text{RMSE} = 0.1412, \text{Std Dev.} = 0.427\end{aligned}$$

In this study, the top 6–7 most important weather variables affecting the yields of Basmati rice varieties were identified through RF regression analysis. These variables were subsequently used in MLR analysis to further assess their impact on rice yields. For traditional Basmati rice varieties, key weather parameters influencing yields included Tmax, SSH, and RH%. The impact of sunshine hours was particularly significant during the grain filling and maturity stages. Specifically, SSH during

SMW 36, 37, and 39 (spanning from dough stage to physiological maturity) was found to be crucial in determining the yields of traditional Basmati varieties.

5.8.1.2 Group B: Long-duration Basmati Rice Varieties

$$\begin{aligned}\mathbf{Y} &= \mathbf{4.181 - 0.047^{*}ARF27 + 0.142ARF40 - 0.229^{*}T_{min}27 + 0.184T_{max}33} \\ &\quad \mathbf{+ 0.318^{*}T_{min}34 - 0.1SSH33}\end{aligned}$$

$$\begin{aligned}R^2 &= 0.916, \text{Adj.}R^2 = 0.865, \text{F Statistics} : 18.06, \text{P - Value} = 0.000, \text{RMSE} \\ &= 0.11, \text{Std Dev.} = 0.39\end{aligned}$$

5.8.1.3 Group C: Short-duration Basmati Rice Varieties

$$\begin{aligned}\mathbf{Y} &= \mathbf{4.81 + 0.027^{*}ARF26 - 0.058^{*}ARF29 + 0.138^{*}T_{max}39 + 0.339^{*}T_{min}27} \\ &\quad \mathbf{- 0.284^{*}T_{min}39 + 0.13^{*}RH30 - 0.08^{*}RH31 - 0.061^{*}RH35}\end{aligned}$$

$$\begin{aligned}&R^2 = 0.985, \text{Adj.}R^2 = 0.926, \text{F Statistics} : 16.68, \\ &\quad \text{P-Value} = 0.058, \text{RMSE} = 0.15, \text{Std Dev.} = 0.581\end{aligned}$$

For short-duration Basmati rice varieties, only 10 years of observed data were available for MLR analysis, as these varieties were recently introduced for cultivation. Key weather parameters influencing grain yield included total rainfall (ARF36) during the first week of September (grain-filling stage), maximum temperature (Tmax29 and Tmax31) during the second fortnight of July and early August (transplanting and tillering stages), and sunshine hours (SSH33 and SSH35) during the second fortnight of August (tillering and flag leaf initiation stages).

5.8.1.4 Multiple Linear Regression (MLR) Model Performance

This study employed stepwise MLR analysis to develop yield prediction models for three groups of Basmati rice varieties: traditional, long duration, and short duration. The accuracy of these models was assessed using the coefficient of correlation $\left(R^2\right)$, which explains the percentage of variability in the dependent variable (rice yield) accounted for by the independent weather variables. Additionally, the adjusted R^2 was calculated to measure the extent of variance in the yield predictions explained by the model inputs, providing a refined evaluation of model fit (Aravind et al., 2022; Wangkheimayum and Paliwal, 2023). To further evaluate forecasting accuracy, the study also utilised RMSE and MAPE, both of which are generally used metrics to evaluate model performance. RMSE indicates the average intensity of errors in predictions, while MAPE measures the percentage deviation between predicted and observed values.

The results, as shown in Table 5.3, indicate that the MLR model accurately predicted yields for traditional and short-duration Basmati rice varieties, with MAPE consistently within 10% for most years. The MAPE values reinforce these observations, underscoring areas where model refinement could improve predictions for

TABLE 5.3
Observed and Predicted Yields (Tons/ha) of Basmati Varieties in Karnal

Year	Traditional Basmati			Long Duration Varieties			Short Duration Varieties		
	Observed	Predicted	MAPE	Observed	Predicted	MAPE	Observed	Predicted	MAPE
2007	2.53	2.69	−6.3	4.50	4.24	5.8			
2008	3.00	3.11	−3.7	4.00	4.11	−2.8			
2009	2.98	3.13	−5.0	4.19	4.35	−3.8			
2010	2.75	2.52	8.4	3.75	3.64	2.9			
2011	3.01	2.96	1.7	3.99	3.87	3.0			
2012	3.40	3.49	−2.6	3.93	3.88	1.3			
2013	3.31	3.01	9.1	3.98	3.99	−0.3	4.89	4.86	0.6
2014	3.81	3.54	7.1	4.45	4.36	2.0	5.56	5.62	−1.1
2015	3.71	3.72	−0.3	4.32	4.21	2.5	4.94	4.89	1.0
2016	3	3.08	−2.7	3.83	3.87	−1.0	4.92	4.93	−0.2
2017	3.6	3.58	0.6	4.12	4.21	−2.2	4.85	4.82	0.6
2018	2.8	2.91	−3.9	4.05	4.06	−0.2	4.5	4.51	−0.2
2019	2.3	2.41	−4.8	4.30	4.33	−0.7	3.6	3.77	−4.7
2020	3.5	3.48	0.6	3.30	3.44	−4.2	4.77	4.75	0.4
2021	3	2.94	2.0	5.10	5.1	0.0	4.45	4.34	2.5
2022	3.4	3.45	−1.5	3.98	4.09	−2.8	4.1	4.05	1.2
2023	3.58	3.67	−2.5	4.50	4.56	−1.3	5.83	5.87	−0.7
2024		2.98			4.96			5.28	

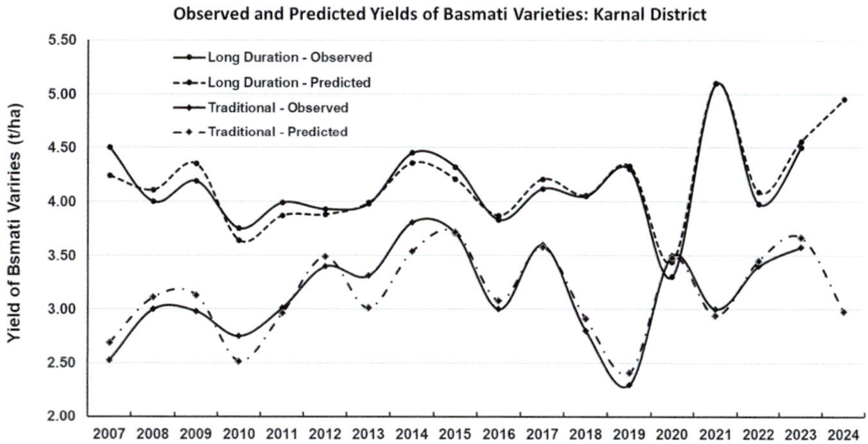

FIGURE 5.3 MLR model predicted and observed yields of traditional and long duration Basmati Varieties in Karnal District.

long duration varieties. These findings affirm the suitability of MLR for predicting yields of traditional and short-duration Basmati rice, while also highlighting the need for further model adjustments to enhance prediction accuracy for long-duration varieties. The MLR model predicted and observed yields of traditional and long duration Basmati rice varieties are given in Figure 5.3.

5.8.2 Comparison of Observed and Predicted Yields

The results of predicted yields of different Basmati varieties obtained from the MLR analysis were compared with the observed yields by visualisations of these yield values in the scatter plots given in Figures 5.4 and 5.5.

5.9 RANDOM FOREST IN BASMATI RICE YIELD PREDICTION

5.9.1 Beyond Traditional Models: AI–driven Approaches for Climate-resilient Yield Prediction

Climate change threatens global food security by disrupting rainfall patterns and causing more frequent extreme weather events like droughts, heat waves, and floods. These changes reduce agricultural output and result in financial hardships for farmers. Accurate crop yield prediction has become increasingly complex due to the combined effects of climate variability, soil degradation, water scarcity, and shifting pest dynamics. Traditional statistical models, reliant on historical yield data, often fall short in accounting for these nonlinear and evolving factors, especially under the influence of climate change, marked by erratic precipitation, extreme temperatures, and prolonged droughts. AI models offer a robust alternative by processing large, heterogeneous datasets to capture complex interactions affecting yield. Climate-induced

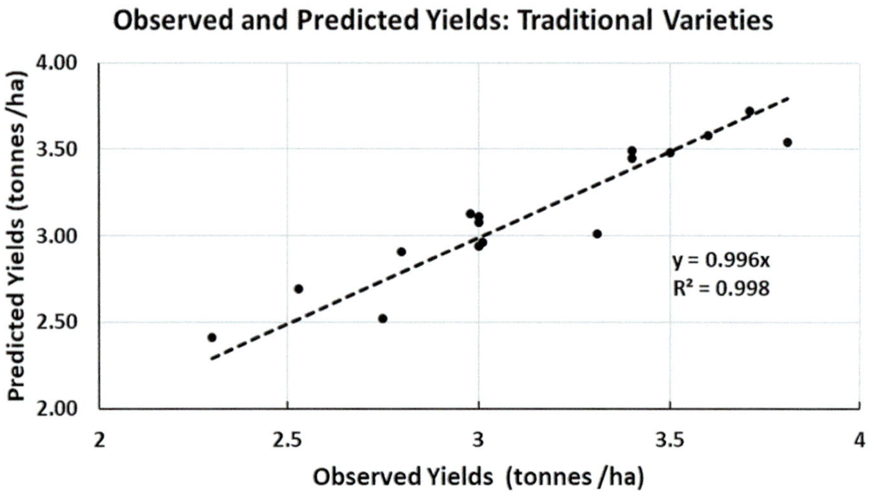

FIGURE 5.4 Scatter plot between observed and model-predicted yields of traditional Basmati varieties in Karnal District.

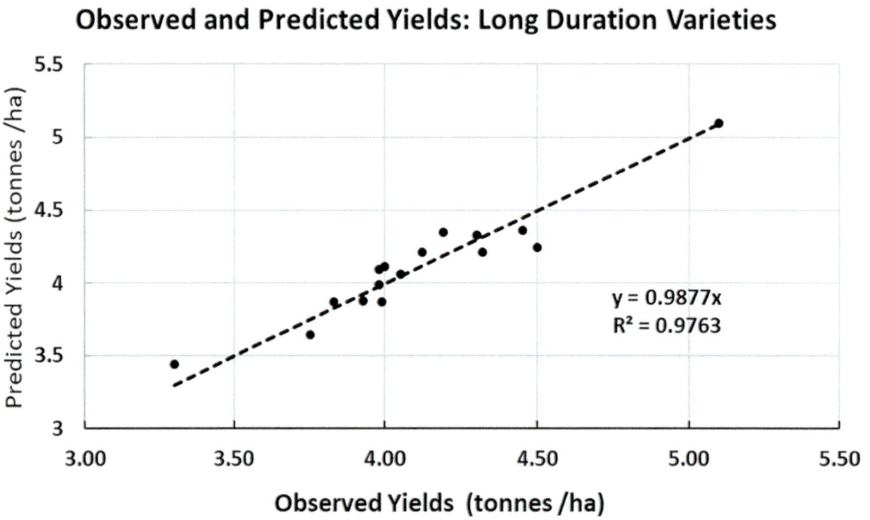

FIGURE 5.5 Scatter plot between observed and model-predicted yields of long-duration Basmati varieties in Karnal District.

shifts in crop phenology and pest seasonality further complicate forecasts. AI can model these dynamics more effectively, offering scalable, adaptive solutions for yield forecasting under changing environmental conditions.

Traditionally, crop growth models have been used to simulate and predict crop yields based on soil type, climate, management practices, and crop genetics (Blanc,

2017). While these models offer insights into biophysical processes, they often face limitations in parameter estimation and adaptability to complex or extreme conditions (Roberts et al., 2017). Earlier yield prediction efforts were constrained by narrow scopes, limited crop varieties, and isolated environmental variables, reducing generalisability (Bishop et al., 2014). To overcome these limitations, linear regression models were introduced but achieved limited success due to their simplicity and inability to capture nonlinear relationships (Westcott and Jewison, 2013; Shook et al., 2021). Time-series models such as ARIMA have also been applied (Alam et al., 2018), but these primarily perform well within stable time series and often fail under changing climate conditions (Petrică et al., 2016). ML techniques, especially RF, have emerged as a robust alternative for crop yield modelling. Introduced by Breiman (2001), RF combines multiple decision trees built on bootstrapped data with random feature selection, effectively reducing variance and overfitting. It can manage high-dimensional, noisy datasets and missing values, making it ideal for agricultural applications. RF has been implemented widely in R and Python, ensuring accessibility across disciplines. The review by Belgiu and Drăguţ (2016) has revealed that RF classifier can successfully handle high data dimensionality and multicollinearity, being both fast and insensitive to overfitting.

Recent studies demonstrate RF's utility in yield modelling due to its proficiency to represent complex, non-linear interactions among weather, soil, and crop variables (Sangeetham and Narayana Reddy, 2023; Shawon et al. (2025); Jhajharia et al., 2023; Reddy and Kumar, 2021; Padma and Sinha, 2023). Its interpretability, through feature importance scores, adds value for agronomic decision-making and climate adaptation (Jeong et al., 2016; Shankar and Manjula, 2021). The RF algorithm has proven useful for analysing and predicting crops like rice (Venkata Narasimhamurthy and Pavan Kumar 2017), wheat, potato (Tamayo-Vera et al., 2025), and sugarcane across geographies, with enhanced accuracy, unlike traditional methods (Kumar et al., 2020; Everingham et al., 2016; Bowden et al., 2023). Studies by Wibisono and Saiful (2025) and Padma and Sinha (2023) further highlighted that fine-tuning RF parameters improves performance significantly, often outperforming other ML models.

5.9.2 Data Preprocessing—Quantile Normalisation in Crop Yield Prediction

When applying machine learning to crop yield prediction, normalising the data is an essential preprocessing step that ensures each input variable has a balanced influence on the training process. Quantile normalisation, in particular, aligns the statistical distributions of variables with differing scales, making it effective for datasets with inherent distributional differences due to technical or environmental variability (Hicks and Irizarry Quantro 2015). In this study, we applied quantile normalisation after first preprocessing the dataset by converting the 'Year' column to numeric values. The data for the year 2024, which contained an unknown target value, was excluded from normalisation, and the remaining data (df_non2024) was used for reference quantile calculation. Each feature's values were ranked and scaled between 0 and 1 and mapped to their corresponding quantiles from the reference distribution

(Bolstad et al., 2003). Though no missing values were present, na.rm = TRUE was included as a safeguard during quantile computation. To further support analysis, observed crop yield values were binned into four quantile-based categories (quartiles) using the cut() function. This binning facilitated pattern recognition across yield ranges and potentially enhanced model performance by emphasising categorical comparisons over raw continuous values, thereby enabling better handling of nonlinear relationships. Additionally, we explored "class-specific" quantile normalisation, as recommended by Zhao et al. (2020), where data is normalised within predefined classes (e.g., by year or label group) to correct for batch effects and preserve meaningful signal. However, in this case, class-specific normalisation offered no additional improvement, suggesting that the training and test data distributions were already well-aligned, with minimal year-specific variation or signs of substantial climate shift.

5.9.3 Stratified Sampling for Model Validation

To ensure proportional representation of crop yield categories in the test dataset, stratified sampling was employed based on the `obs_bin` column, which grouped the 'Observed' values into four quantile-based bins. The `lapply()` function was used to iterate over the bin numbers (1 through 4), extracting one random row from each bin in the df_non2024 dataset (Steven and Sanderson, 2025). Each subset (`bin_data`) represented a different range of observed yield values. The individual subsets were then combined using `do.call(rbind, ...)`, resulting in a single stratified dataset, test_stratified, containing one randomly selected row from each yield category. This method ensured that all value ranges were represented equally in the test set, which is critical for model evaluation. By maintaining balanced representation across yield segments, the stratified sample provided a more robust and generalisable estimate of model performance across the entire distribution of crop yield outcomes.

5.9.4 Train-test Split and Inclusion of 2024 for Prediction

To evaluate model generalisation on unseen data, a clear separation between training and test datasets was established, with each containing distinct and non-overlapping years. The year 2024, which had an unknown 'Observed' value (NA), was added to the test dataset to assess the model's predictive capability for future or missing data. The test dataset was created by combining df_2024 with the previously prepared `test_stratified` dataset, which included one row from each bin of the observed yield distribution to ensure balanced representation. Duplicate years were removed using a distinct operation based on the 'Year' column, with `.keep_all = TRUE` preserving all associated feature columns. This approach ensured that each year appeared only once, either in the training or test set, effectively preventing data leakage. By incorporating both stratified sampling and future prediction (2024), the test dataset facilitated a thorough assessment of the model's effectiveness across different yield levels and its capacity to generalise to previously unseen temporal conditions.

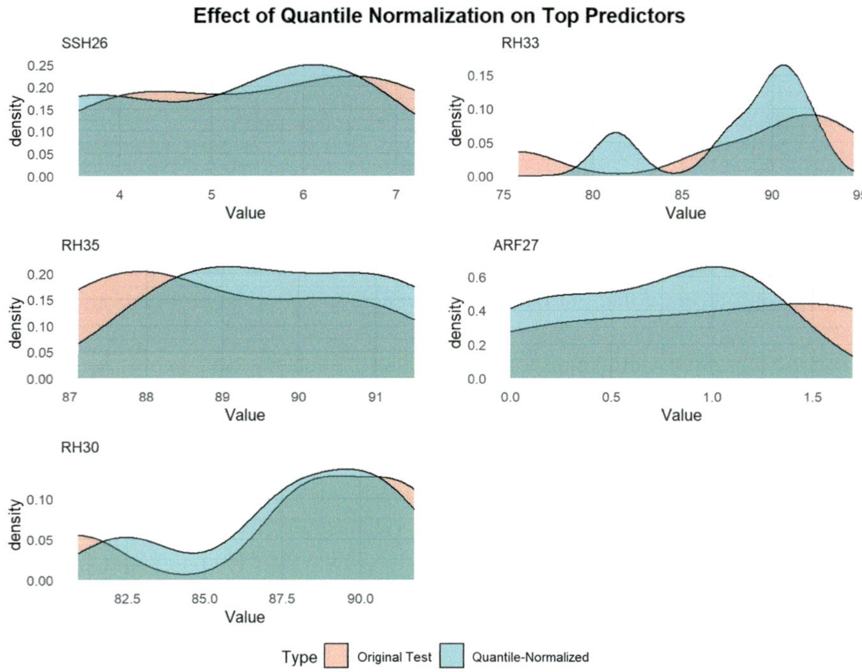

FIGURE 5.6 Quantile normalised values vs. original values of the top 5 predictors in long-duration varieties.

5.9.4.1 Feature Engineering and Data Normalisation for Model Training

As part of the feature engineering process, an `obs_bin` column was created by discretising the continuous observed variable into quantile-based bins. This aided in stratified sampling, ensuring balanced test representation across the target value range. While binning can help improve sampling strategies and model robustness in imbalanced data settings (Hastie et al., 2009; Kuhn and Johnson, 2013; Friedman, 2001), `obs_bin` was excluded from the model training to avoid data leakage. Only variables with genuine predictive power were retained to support better generalisation and avoid overfitting (Figure 5.6).

Non-numeric variables were excluded to prepare the predictors, and the 'Observed' column was removed from the feature set. The resulting list of numeric predictors was used for both training and testing the model. Columns with near-zero variance or insufficient unique values were identified and excluded, as they offered little to no information gain. Next, global quantile normalisation was applied separately to the training and test datasets. This non-parametric method aligned the distributions of the variables, addressing skewness commonly present in agro-climatic features such as rainfall and humidity. By ranking and averaging feature values, this technique mitigated scale mismatches and made the datasets more comparable, reducing the risk of distributional shift (Kouw and Loog, 2021; Moreno-Torres et al., 2012). As a result, the model training became more stable and fairer, enhancing performance on unseen data.

5.9.5 Time-based Dataset Splitting and Data Integrity

To maintain consistency in the dataset, any year values with decimal points, possibly introduced during averaging or interpolation, were rounded to align with standard year formats. The dataset was subsequently partitioned into training and testing sets using non-overlapping Year values to ensure temporal separation. This temporally stratified split ensured that the model was trained on historical observations (train_years) and validated on future or unseen periods (test_years), including 2024, for which the Observed yield value was not available. Such chronological separation is a standard practice in time series modelling to prevent data leakage and assess generalisation on forward-looking data (Yan et al., 2025). To facilitate traceability and performance evaluation, a split column was added to both datasets using the mutate() function (Wickham and Grolemund, 2017), clearly marking the origin of each observation (Table 5.3). In case of three Basmati varietal segments, the train and test years were selected using split and mutation so that the data representation is well balanced in both train and test data (Tables 5.4 and 5.5).

TABLE 5.4
Reduction in Observed Value Standard Deviation Post-Adjustment

	Train Observed SD	Test Observed SD
Initial	0.4370472	0.2194691
After Adjustment	0.1179513	0.05549775

TABLE 5.5
Distribution of Training and Test Years across Basmati Rice Duration Segments

Segment	Training Year Rows	Test Year Rows
Long Duration	2007, 2008, 2009, 2010, 2012, 2015, 2016, 2018, 2019, 2020, 2021, 2022, 2023	2011, 2013, 2014, 2017, 2024
Traditional	2007, 2008, 2009, 2012, 2013, 2014, 2016, 2017, 2018, 2019, 2021, 2022, 2023	2010, 2011, 2015, 2020, 2024
Short Duration	2013, 2015, 2017, 2018, 2019, 2021, 2022, 2023	2014, 2016, 2020, 2024

5.9.6 Aligning Train and Test Distributions Using Quantiles

To ensure balanced and representative sample distributions between the train and test datasets, quantile-based ranges were applied. This method divides the data into consistent intervals, reducing bias from extreme outliers or skewed distributions (Joaquin et al., 2022). The model was trained on the central tendency of the data, focusing on typical values unless a different distribution was specifically desired. Quantile variables were used to define ranges for the 'Observed' column, including key percentiles such as the 0th (minimum), 25th (Q1), median (Q2), and beyond. For instance, the training data concentrated on the middle 60% (between quantiles (Alam et al., 2018) and quantiles (Batchelor et al., 2002)) to avoid extreme outliers. The test data focused on a slightly broader range (e.g., between the 50th and 75th percentiles) to assess the model's generalisability.

By comparing quantile distributions across the train and test datasets, we ensured they followed similar distributions. Significant discrepancies in quantiles could indicate data shift, leading to biased predictions or reduced model accuracy. Large quantile differences might also signal the need for additional scaling or normalisation (Quiñonero-Candela et al., 2008; Moreno-Torres et al., 2012), especially for models like linear regression that are sensitive to feature scales. Quantile calculations provide valuable insights into data distribution, revealing key metrics such as range, median, and percentile spread. This helps diagnose skewness, outliers, and potential misalignment between datasets, ensuring reliable model performance. To operationalise this, the 'split' column was checked in both datasets and added if absent, labelling rows as either 'train' or 'test'. This column is crucial for tracking the origin of each row and simplifying model evaluation. After confirming that both datasets contained the same feature columns, the response variable ('Observed') was separated from the predictors, and y_train and y_test were extracted accordingly. A final visual comparison of the 'Observed' distributions, with the median marked, helped evaluate alignment between the train and test data (Figures 5.7, 5.8, and 5.9).

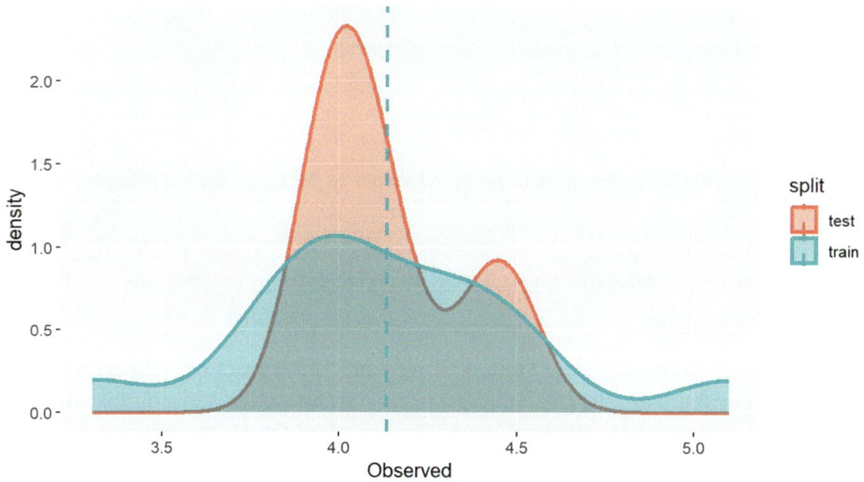

FIGURE 5.7 Density distribution of observed values in the train and test datasets in long-duration varieties.

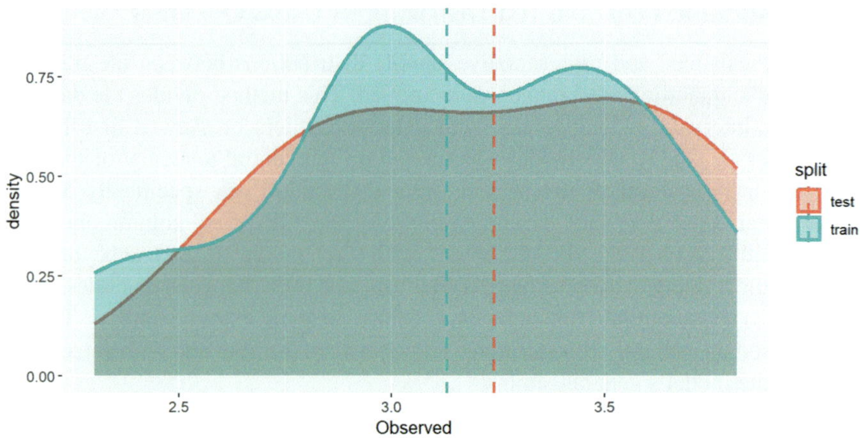

FIGURE 5.8 Density distribution of observed values in the train and test datasets in traditional varieties.

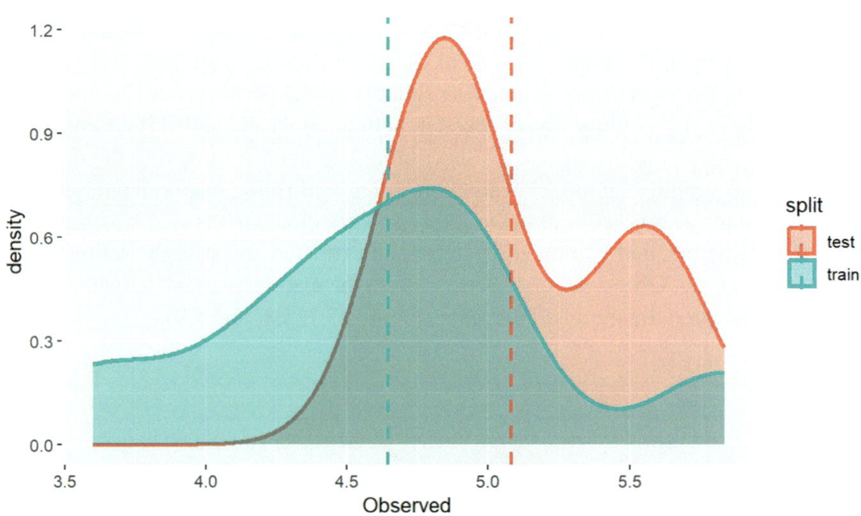

FIGURE 5.9 Density distribution of observed values in the train and test datasets in short-duration varieties.

5.9.7 Handling Missing Values and Data Cleaning for Model Preparation

Missing values in the response variable can negatively impact model performance, as many algorithms do not handle missing target values. To address this, missing values were imputed using the median, a robust method for continuous variables that minimises the influence of outliers (Boris and Hoaglin, 1993; Hoaglin et al., 1986;

Hodge and Austin, 2004). It is essential to handle missing values before applying normalisation or scaling to ensure that the imputed data integrates smoothly into the dataset for further processing. To ensure data integrity, the `train_values` were cleaned by removing any NA values using `na.omit()`, and the `test_values` were also checked for missing data. This step ensures that only valid data points are used for subsequent normalisation and model training.

5.9.8 Handling the Too-small Dataset

In scenarios involving sparse training data lacking sufficient variability, quantile normalisation may be ineffective due to the inability to match the quantiles of the test data. As a fallback, the function substitutes missing values in the test dataset with the mean value of the training data, ensuring the function remains robust and doesn't fail due to limited data. This is particularly useful in group-wise modelling, cross-validation, or nested training loops (Bengio et al., 2013). To further improve the robustness of preprocessing, quantiles for both training and test data were computed using a fixed probability sequence (probs_seq <- seq (0, 1, length.out = 100)). This ensures that distribution diagnostics are consistent across different data subsets, regardless of sample size, skewness, or missing values. This approach is essential for small or dynamic datasets, where data may be incomplete (Quiñonero-Candela et al., 2008). It also facilitates reproducible validation checks, automated plotting, and debugging during feature engineering and handling missing values. Additionally, columns with only one unique value were removed from both the training and test datasets. These columns did not contribute useful variation for modelling and were discarded to improve model performance, reduce dimensionality, and ensure the dataset remains clean for analysis. Such columns, with no differentiation between observations, offer no value to the model and were eliminated to prevent inefficiencies during model training (Hastie et al., 2009).

5.9.9 Data Cleaning and Re-splitting

After removing columns with a single unique value, the train and test datasets were re-split based on the 'Year' column to ensure that the model is trained on specific time periods. This approach is particularly useful in time series forecasting tasks, where the model needs to generalise to future or unseen time periods. Additionally, categorical variables in the dataset were identified using a vector of TRUE or FALSE values (`factor_cols`), indicating which columns in the `train_data` are factors. To enhance model performance and prevent overfitting, columns with constant values, which do not contribute meaningful information, were dropped. This step resulted in a cleaner version of the training data, `train_data_clean`, which improves model interpretability and performance by retaining only the variables with sufficient variability.

5.9.10 Top Predictor Selection via Random Forest Importance

To identify the most effective features, a random forest model was trained on the cleaned `train_data` using the `train()` function from the 'caret' package in

R. The model utilised all columns in the dataset except the 'Observed' column. The training process involved splitting the data into subsets, performing cross-validation, and building multiple decision trees (Liaw and Wiener, 2002). The `train()` function simplifies the training process by automating tasks such as cross-validation, hyperparameter tuning, and model evaluation. Variable importance was then calculated using the `varImp()` function, which identifies the most influential features for predicting the outcome. The importance scores were accessed and converted into a data frame, with feature names added as a column. The features were sorted in descending order based on their importance score, and the top 20 most important features were selected to reduce dimensionality, improve model performance, and enhance interpretability (Figures 5.10–5.12).

5.9.11 Feature Selection, Data Preparation, Imputation, and Standardisation

To ensure compatibility with model training, `train_data_final` was constructed using the original train_data rows but restricted to the top 20 selected predictor features. Similarly, `test_data_final` was prepared by retaining only the test years and the same top features. The Year and Observed columns were excluded from the feature set, as they are not valid predictors in the modelling process.

To enhance model reliability and interpretability, redundant features exhibiting high pairwise correlation (absolute Pearson correlation > 0.7) were identified using a safely computed correlation matrix and subsequently removed. This step reduced multicollinearity, simplified the model, and mitigated the risk of overfitting (Table 5.5). Missing values were imputed using column-wise medians from the training data (ref_df), a robust method less sensitive to outliers compared to mean imputation. Numeric columns were selectively imputed, and the cleaned data frames

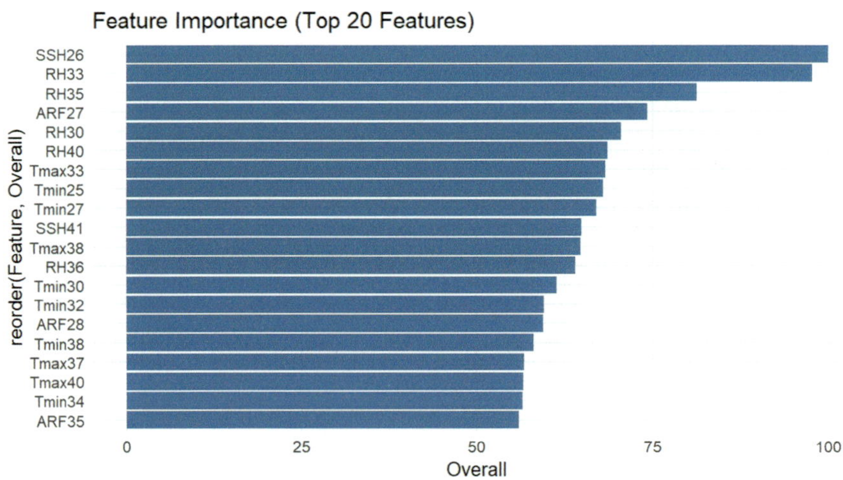

FIGURE 5.10 Importance order of the top 20 selected features in long-duration varieties.

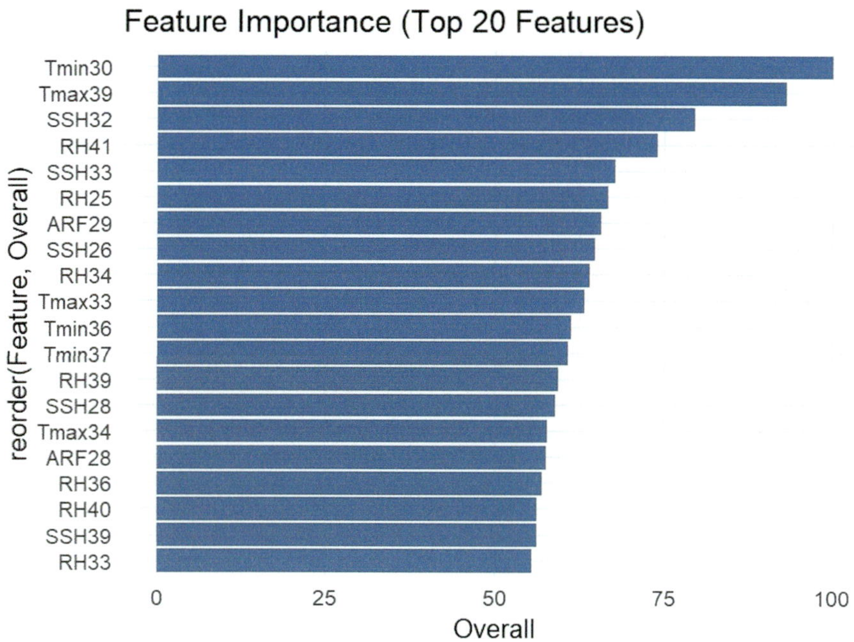

FIGURE 5.11 Importance order of the top 20 selected features in traditional varieties.

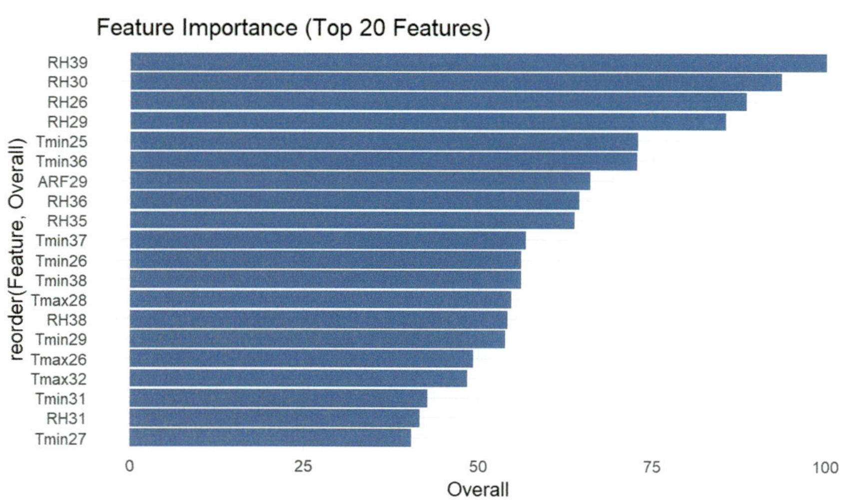

FIGURE 5.12 Importance order of the top 20 selected features in short-duration varieties.

preserved the original structure and column names. A split column was temporarily added to facilitate operations like imputation, scaling, and validation. Duplicate column names were also removed to prevent silent errors during modelling. Both `x_train` and `x_test` were filtered to include only the final selected features and were free from non-predictive columns like Year and Observed. Character columns were converted to factors and then to integers to ensure compatibility with machine learning algorithms such as Random Forest and XGBoost, which require numeric input. The standardisation process utilised the mean and standard deviation calculated exclusively from the training set to avoid any risk of information leakage into the model evaluation phase. The datasets were scaled to a consistent range and converted back to data frames. The row for the year 2024, initially excluded, was reintroduced into the test set to allow prediction for its unknown Observed value. To ensure correct alignment between predictions and actual values, the target variable (Observed) was separated from `test_data_updated`, just as with `y_train`. This allowed for accurate model evaluation post-prediction. Lastly, the number of final selected features was further reduced from 25 to 11 to simplify the model and eliminate potentially irrelevant predictors. This reduction helps minimise overfitting and improves model generalisation by focusing on the most impactful features. Out of the selected 25 features, after removing highly correlated features, only 11 top features were selected from the remaining for use in modelling (Table 5.7).

5.9.12 Cross-validation Modelling

To develop a robust and well-generalised Random Forest model, a systematic hyperparameter tuning process was carried out. The key parameters tuned included `mtry` (the number of features considered at each split), `min.node.size` (minimum samples in terminal nodes), `num.trees` (number of trees), and `splitrule` (splitting criterion, set to "variance" for regression tasks). Hyperparameter tuning was conducted to optimise the trade-off between model flexibility and predictive

TABLE 5.6
Highly Correlated Variables for Different Basmati Rice Varieties

1.	Long Duration Varieties	T_{min}27
2	Traditional Varieties	RH34, T_{min}37, T_{max}33, SSH33
3.	Short Duration Varieties	T_{min}26, RH35, T_{min}36, RH36, RH30, T_{min}27, T_{min}37, T_{max}26, RH31, T_{min}25, RH26, RH39, T_{min}38

TABLE 5.7
Top 11 Weather Features for Basmati Rice Varietal Segments

1.	Long Duration Varieties	SSH26, RH33, RH35, ARF27, RH30, RH40, Tmax33, T_{min}25, T_{min}27, SSH41, T_{max}38
2.	Traditional Varieties	T_{min}30, T_{max}39, SSH32, RH41, SSH33, RH25, ARF29, SSH26, RH34, T_{max}33, T_{min}36
3.	Short Duration Varieties	RH39, RH30, RH26, RH29, T_{min}25, T_{min}36, ARF29, RH36, RH35, T_{min}37, T_{min}26

performance, a critical consideration when working with limited datasets susceptible to overfitting. Given the limited number of observations, Leave-One-Out Cross-Validation (LOO-CV) was adopted for evaluating model performance during tuning. This method trains the model on all rows except one and tests on the excluded row, repeating the process for every observation. LOO-CV offers the advantage of using nearly all the data for training in each iteration and provides an unbiased estimate of model error, making it a sound choice for small datasets where traditional k-fold cross-validation might be unstable.

$$LOOCVRMSE = \sqrt{\frac{1}{n}\sum_{i=1}^{n}\left(yi - yi\left(-i\right)\right)2}$$

where: yi is the actual value for the ith observation, n: total number of observations.

$yi\left(-i\right)$: the predicted value using a model trained on all data except ith row.

This RMSE is based on predictions made only on the data points (test), not used during the training. LOO-CV gives an unbiased estimate of model performance, since each prediction is made on a data point not seen by the model during training.

To ensure reproducibility, `set.seed(123)` was applied at the beginning of the script. This fixed the random number generator sequence, allowing for consistent results across repeated executions, a critical aspect during model training and validation. The hyperparameter grid was defined using `expand.grid()` to include combinations of key tuning parameters: mtry (number of features considered at each split), `min.node.size` (minimum number of samples required in a terminal node), and `splitrule` (set to "variance", appropriate for regression tasks). Hyperparameter tuning was conducted using the caret package (Kuhn, 2008), which simplifies the model training process and integrates various resampling and tuning strategies efficiently (Probst et al., 2019). Model performance was further explored by iterating over different values for `num.trees` (100, 200, and 300) to observe its impact on accuracy. As supported in the literature, gains in model performance typically plateau after approximately 100 trees (Oshiro et al., 2012; Probst and Boulesteix, 2017), making this a balanced choice between performance and computational efficiency. Leave-One-Out Cross-Validation (LOO-CV) is especially suitable for small datasets as it maximises training data usage in each iteration. The model is trained n times, each time leaving out a single observation for testing, thus providing nearly unbiased performance estimates and minimising overfitting risk. During each iteration, 75% of the training data was randomly sampled (`sample.fraction = 0.75`) with replacement. This bagging strategy introduces variability across trees, enhancing model robustness and generalisability. The model's forecasting performance was assessed using RMSE, a preferred metric for regression problems due to its sensitivity to large errors. The tuning process identified the hyperparameter combination that minimised RMSE, which was then used to train the final model.

The final Random Forest regression model was built using the ranger package (Wright and Ziegler, 2017), known for its speed and ability to parallelise computation. The model formula y_train ~ . included all predictor variables to model the target response. Based on tuning results, the optimal parameters selected were `mtry`

= 4, `min.node.size = 1`, and `splitrule = "variance"`, with `num.trees = 100`. Deep tree growth (enabled by `min.node.size = 1`) was considered appropriate for capturing detailed patterns in the small dataset. All trees were trained on the full dataset with bootstrap resampling enabled (`sample.fraction = 1`), leveraging ensemble learning effectively. To evaluate feature relevance, permutation-based importance (`importance = "permutation"`) was used. This model-agnostic method randomly shuffles the values of each feature and measures the resulting increase in RMSE. Features that cause a greater drop in performance are deemed more important. Compared to traditional impurity-based methods, permutation importance is less biased and better suited to datasets with multicollinearity or variables on differing scales. The choice of the "variance" split rule ensured that each decision tree aimed to minimise the variability of the response variable within the resulting nodes, which enhances prediction accuracy in regression contexts. The overall approach, combining seed control, careful hyperparameter tuning, robust cross-validation, bagging, and unbiased importance evaluation, ensured that the final Random Forest model was interpretable, reliable, and well-optimised for the available data.

5.9.13 Model Performance Evaluation Using R^2, RMSE, and MAE

The model predictions were first generated for the training dataset using the `predict final_ranger_model, data = x_train$ predictions` function, which provided the predicted values based on the input features. These predictions were stored as `train_predictions`. Similarly, predictions for the test dataset were made with `predict(final_ranger_model, data = x_test_clean_updated)$predictions` and stored as `test_predictions`. To address missing values in the target variable for the test set (such as the NA for 2024), a logical index (`valid_test_idx<- !is.na(y_test_clean_updated)`) was created to include only rows with valid observed values for evaluation. The performance of the model was assessed using several metrics. The R^2 (coefficient of determination) was computed to quantify the model's ability to explain variability in both the training and test datasets (Montgomery et al., 2012). For the training set, `train_r2 <-cor(train_predictions, y_train)^2` calculated the squared correlation between the predicted and actual values, indicating the proportion of variance explained. For the test set, the R^2 was computed similarly, but only for rows with valid observations, using `test_r2 <-cor(test_predictions[valid_test_idx], y_test_clean_updated[valid_test_idx])^2`, ensuring an accurate measure free from the influence of missing values.

$$\mathbf{R^2} = \left(\mathbf{cor}\left(\mathbf{y}, \hat{\mathbf{y}}\right)\right)^2$$

where: y is the actual yield and $\hat{y}$ is the predicted yield.

Root Mean Squared Error (RMSE) was used to measure the average intensity of errors in predictions, with greater weight given to larger errors (Hastie et al., 2009;

James et al., 2021). The training RMSE was calculated using `train_rmse <- Metrics::rmse(y_train, train_predictions)`, and the test RMSE was calculated as `test_rmse <- Metrics::rmse(y_test_clean_updated[valid_test_idx], test_predictions[valid_test_idx]`. RMSE provides a clear indication of the typical prediction error intensity.

$$RMSE = \sqrt{\frac{1}{n}\sum_{i=1}^{n}\left(yi - \hat{y}i\left(-i\right)\right)2}$$

where, yi is the actual observed yield, ŷi is the predicted yield and n = number of observations.

Mean Absolute Error (MAE) was also computed to assess the model's accuracy by measuring the average absolute differences between predicted and actual values. The training MAE was calculated via `train_mae <- Metrics::mae(y_train, train_predictions)`, and the test MAE was computed using `test_mae <- Metrics::mae(y_test_clean_updated[valid_test_idx], test_predictions[valid_test_idx])`. MAE treats all errors equally, making it easier to interpret compared to RMSE, and serves as an additional layer of performance evaluation (Hastie et al. 2009; James et al., 2021).

$$MAE = \frac{1}{n}\sum_{i=1}^{n}\left|yi - \hat{y}i\right|$$

Together, R^2, RMSE, and MAE offer a detailed assessment of model performance metrics providing insights into variance explained and prediction accuracy. To facilitate comparison, both training and test results were merged into a single data frame, `combined_predictions_full`, which included columns for Year, set (Train/Test), Observed, and Predicted values. The data was sorted by year using order(`combined_predictions_full$Year`), making it well-suited for time series analysis and allowing for effective visualisation of trends, over- or under-predictions, and potential anomalies in the predictions.

5.10 XGBOOST MODEL IN BASMATI RICE YIELD PREDICTION

Despite the growing success of ML models, XGBoost, a gradient boosting algorithm known for its high accuracy, efficiency with structured data, and robustness to missing values, remains underexplored in crop yield modelling. Its adoption could mark the next step in developing scalable, high-performing models for yield prediction under climate variability.

XGBoost constructs an ensemble of decision trees in a sequential manner, where each subsequent tree is trained to minimise the residual errors of the preceding trees, thereby enhancing overall model performance. It offers superior accuracy when fine-tuned, establishing it as widely adopted for predictive modelling. In this study, XGBoost was applied to three datasets: long duration, traditional, and short duration varietal segments. The model uses a standard objective function that combines a differentiable training loss (e.g., mean squared error) with a regularisation component incorporated to penalise model complexity and enhance generalisation. The objective function is defined as:

$$L(\Phi) = \sum_i l(\hat{y}i, yi) + \sum_k \Omega(\mathbf{f}_k) \tag{5.2}$$

where, $L(\Phi)$ the overall objective function to be minimised during training, combining both the training loss and regularisation terms. $l(\hat{y}i, yi)$ represents the loss function measuring the difference between the predicted value $\hat{y}i$ and the true label yi for each training example i. $\Omega(\mathbf{f}_k)$ is the regularisation term applied to each tree f_k, penalising model complexity to prevent overfitting.

and $\Omega(f)$ is the regularisation term for tree complexity, given by:

$$\Omega(f) = \gamma T + \frac{1}{2}\lambda \sum \omega_j^2 \tag{5.3}$$

Here, T is the number of leaves, ω_j^2 represents the leaf weights (score), and γ and λ are regularisation parameters. This formulation enables XGBoost to achieve high predictive accuracy while mitigating overfitting, making it particularly effective for structured agricultural data (Chen and Guestrin, 2016).

5.10.1 Data Formatting

To initiate the modelling process, the training and test datasets were converted into the optimised `xgb.DMatrix` format, which is designed for efficient computation in XGBoost. The training data was structured with `dtrain <- xgb.DMatrix(data = as.matrix(x_train), label = y_train)`, where the feature set `xtrain` was converted to a numeric matrix and the target variable `ytrain` was specified. This format allowed XGBoost to efficiently compute gradients and hessians during model training. Similarly, the test dataset was transformed using `dtest <- xgb.DMatrix(data = as.matrix(x_test_clean_updated))`, including only the predictors, as target values were not needed for inference. This transformation optimised memory usage and computational efficiency, crucial for large datasets and repeated evaluations during model training.

5.10.2 Cross-validation

The optimal number of boosting iterations was identified through 5-fold cross-validation implemented via the `xgb.cv()` function. The data was divided into five folds, with training and validation rotated across these folds. Up to 1000 boosting iterations were allowed (`nrounds = 1000`), and early stopping was enabled (`early_stopping_rounds = 100`) to prevent overfitting by halting training if the validation RMSE did not improve for 100 consecutive rounds. The evaluation metric used was root mean squared error (RMSE), which is appropriate for regression tasks. The process identified the optimal number of boosting iterations (`best_nrounds`) based on the lowest validation error.

5.10.3 Setting Hyperparameters and Final Model Training

Following cross-validation, the final XGBoost model was trained using the full training dataset (`dtrain`) and the ideal count of boosting iterations (best_nrounds). The model was trained with the `xgboost()` function, utilising the predefined parameters and iteration count. Verbose output was suppressed (verbose = 0) to streamline training logs. The trained model successfully learned intricate non-linear patterns within the dataset and was subsequently applied to generate predictions on the test set (`dtest`).

5.10.3.1 Long-duration Varieties

The XGBoost algorithm was trained with hyperparameters fine-tuned through cross-validation to enhance model performance. The model was configured with the best number of boosting rounds (`nrounds = best_nrounds`), using the "`reg:squarederror`" objective to predict a continuous variable. A learning rate (eta) of 0.2 was chosen to balance model training and prevent overfitting. The max_depth was set to 2 to limit tree complexity, while `min_child_weight = 3` ensured adequate sample sizes for node splits. Randomness was introduced with `subsample = 0.8` and `colsample_bytree = 0.8` to enhance generalisation. Regularisation was applied with `alpha = 0.5` for L1 and `lambda = 0.1` for L2, reducing model complexity. The verbose parameter was set to 0 for cleaner output. This combination of hyperparameters was selected to optimise performance and generalisation on the smaller dataset, resulting in a robust and effective model.

5.10.3.2 Traditional Varieties

A more complex and regularised XGBoost model was configured for traditional varietal data, using a different set of hyperparameters. The booster was set to "`gbtree`" for tree-based boosting, which is suited for regression tasks. The objective was set to "`reg:squarederror`," employing squared error loss for continuous value prediction. A lower learning rate (eta = 0.05) was used to promote gradual learning and minimise overfitting. A `max_depth` of 6 allowed for deeper trees, capturing more complex relationships in the data. To enhance generalisation and prevent overfitting, subsample and `colsample_bytree` were both set to 0.7, limiting the data and features used for each tree. L2 (ridge) and L1 (lasso) regularisation were applied with `lambda = 0.05` and `alpha = 0.05`, controlling model complexity by penalising large coefficients and encouraging feature sparsity. These hyperparameters were selected to balance model complexity and generalisation, ensuring effective pattern capture without overfitting.

5.10.4 XGBoost Model Optimisation Through Grid Search

To optimise the XGBoost model's performance, a grid search was conducted over a range of hyperparameter combinations. This approach allowed for systematic exploration of the configuration space to find the optimal model settings. The model used the "`gbtree`" booster for gradient-boosted decision trees, suitable for regression tasks, with the objective function set to "`reg:squarederror`" to minimise mean squared error. The learning rate (`eta`) was tested at 0.005 and 0.01 for gradual convergence, while tree complexity was controlled by varying `max_depth` between 2 and 3. Stochastic elements, including row subsampling (subsample) and column subsampling (`colsample_bytree`), were varied between 0.7 and 0.8 to enhance generalisation and prevent overfitting. Regularisation was applied with L2 (`lambda`) and L1 (`alpha`) penalties at values of 0.05 and 0.2 to reduce model complexity. Each combination was evaluated using 5-fold cross-validation with early stopping to minimise overfitting. The optimal parameter set, determined by the lowest test RMSE, was used to train the final model on the full dataset.

5.10.5 Model Evaluation and Prediction Workflow Using XGBoost

After training the XGBoost model on the processed dataset, predictions were generated for both the training and test sets using the `predict()` function. Training set predictions were compared with actual data to assess fit, while test set predictions were used to evaluate model generalisation. Observed values from the test set, excluding missing data, were aligned with predicted values for performance evaluation. Missing values, particularly for the year 2024, were handled by appending NA values to ensure structural consistency. Predictions were stored separately for clarity, and both observed and predicted values were merged into a comprehensive data frame for year-wise comparison and analysis. The predicted value for 2024 was manually inserted into the dataset to ensure accurate tracking. The model's predictive accuracy was assessed through metrics including the coefficient of determination $\left(R^2\right)$, MAE, and RMSE. R^2 was calculated for both the training and test sets to assess model fit and generalisation, while MAE and RMSE provided additional insights into prediction accuracy. This thorough evaluation process allowed for a comprehensive understanding of the model's performance, particularly in forecasting future periods such as 2024.

5.11 IMPORTANCE OF DIFFERENT WEATHER VARIABLES IN RF AND XGBOOST

An analysis of the top 11 most influential weather predictors used in both Random Forest and XGBoost, based on the Random Forest model, highlights the pivotal role of specific climatic variables during critical crop growth stages.

***Relative Humidity (RH)*:** Relative humidity emerged as a consistently influential factor throughout the growing season. For short-duration varieties, RH was particularly important during the transplanting, tillering, and physiological maturity stages. In contrast, both long-duration and traditional varieties demonstrated a strong dependence on RH during the grain filling stage, with RH40 and RH41 showing substantial predictive importance.

***Rainfall (ARF)*:** Initial rainfall events during the transplanting phase (ARF27 and ARF29) were found to be beneficial for maintaining optimum field moisture. These early spells support root establishment and contribute positively to overall crop performance.

***Sunshine Hours (SSH)*:** Sunshine hours also showed significant influence, particularly during distinct phenological stages. In long-duration and traditional varieties, SSH during the nursery stage was critical, while SSH41, representing sunshine during physiological maturity, played a crucial role in determining yield in long-duration varieties. Notably, SSH32 and SSH33, representing sunshine during the vegetative stage, were consistently important in both long-duration and traditional varieties, contributing over 60% importance in the Random Forest model, highlighting their strong influence on photosynthetic activity and biomass accumulation.

Temperature (T_{max} and T_{min}): Temperature variables were also key yield determinants. Maximum temperatures during the tillering stage $\left(T_{max}33\right)$ and advanced vegetative stages ($T_{max}38$ and $T_{max}39$) were significant for crop productivity. Additionally, minimum temperatures during the advanced vegetative phase ($T_{min}36$ and $T_{min}37$) were found to influence yield outcomes across all varietal groups. These findings suggest that both day and night temperatures during mid-season growth stages are crucial for optimal crop development.

5.12 HYPERPARAMETER TUNING IN RF AND XGBOOST

5.12.1 Random Forest

The model was applied to all three varietal segments of Basmati rice: long duration, traditional, and short duration. Hyperparameter tuning was performed to optimise model performance, and the top 11 most important features were selected for training. These features were used in both the cross-validation phase and the final Random Forest models implemented using the ranger package. The most suitable hyperparameter values were identified through cross-validation, resulting in the selection of the best-performing models for each varietal group (Table 5.8).

To optimise performance for small datasets, the "`variance`" split rule was adopted across all varietal segments. This criterion is well-suited for regression tasks, as it

TABLE 5.8
Optimised Random Forest Hyperparameters for Basmati Varietal Segments

Segment	mtry	Min. node.size	Split rule	Num.trees	Sample. fraction	Importance
Long Duration	6	1	Variance	100	0.75	Permutation
Traditional	2	2	Variance	100	0.75	Permutation
Short Duration	4	1	Variance	1000	0.75	Permutation

minimises the variance of the target variable in the child nodes following a split, thereby aligning with the objective of reducing RMSE and MAE. Unlike the "`extratrees`" splitrule, which evaluates a random subset of split points, the "variance" criterion considers all possible splits, making it more data-efficient and reliable for smaller datasets (i.e., limited number of years). Furthermore, it exhibited greater stability and consistency across cross-validation folds and maintained compatibility with variable importance metrics. For assessing feature importance, the permutation method was employed. Permutation importance evaluates the contribution of each predictor by observing the rise in prediction error after randomly permuting the values of that feature, thereby reflecting its influence on the model's accuracy. It is less biased by variable type or scale and treats all features uniformly. This approach was consistently applied across all three Basmati rice segments: long duration, traditional, and short duration, ensuring fair and interpretable importance rankings. The `sample.fraction` parameter was set to 0.75, indicating that each tree in the random forest was built using 75% of the training data. This subset was sampled either with or without replacement, allowing a balance between model diversity and data efficiency, which is especially important in scenarios involving limited training instances.

5.12.2 XGBoost Hyperparameters

We applied to cross validate XGBoost before applying XGBoost model with fine-tuned parameters in order to find the best number of rounds (iterations), which were applied to the xgb_model (Table 5.9).

5.13 RF AND XGBOOST MODEL PERFORMANCE ON THREE BASMATI VARIETAL SEGMENTS

5.13.1 Long-Duration Basmati Varieties

Model performance varied considerably across algorithms for Long Duration Basmati varieties. The cross-validated Random Forest model (cv_RF_Model) achieved a Train R^2 of 0.85 and a Test R^2 of 0.69, indicating moderate predictive ability. While

TABLE 5.9
Optimised XGBoost Hyperparameters for Basmati Varietal Segments

Segment	nrounds	Objective	eta	Max. depth	Min. child.wt	Subsample	Cosample_ bytree
Long Duration	35	reg:squarederror	0.2	2	3	0.8	0.8
Traditional	35	reg:squarederror	0.05	6	-	0.7	0.7
Short Duration	988	reg:squarederror	0.005	2	-	0.8	0.8

the model captured the overall trends, its relatively lower test performance suggests mild underfitting, likely due to constrained model complexity. Nevertheless, the RMSE and MAE values remained within reasonable bounds, reflecting consistent, albeit suboptimal accuracy.

In contrast, the final tuned Random Forest model exhibited a marked improvement, with a Train R^2 of 0.95 and a Test R^2 of 0.85, indicating a stronger fit and improved generalisation. The high training performance reflects a good proportion of explained variance (Draper and Smith, 1998), though the noticeable gap between training and test metrics points to mild overfitting, where the model may have begun to capture noise in the training data. Despite this, the final model showed better stability and predictive strength compared to the cross-validated version.

The XGBoost model, while showing a slightly lower Train R^2 of 0.84, delivered a Test R^2 of 0.78 and achieved the lowest Test MAE (0.13) among all models (Table 5.10). This indicates a high level of prediction accuracy and generalisation, with minimal overfitting. Its performance suggests that The XGBoost algorithm exhibited enhanced performance in identifying complex data patterns, making it the most robust and reliable model for yield prediction in long-duration Basmati varieties (Table 5.11).

Despite the limited size of the test dataset (only five years), the XGBoost model demonstrated strong generalisation capability, with predicted values closely aligning with observed yields for at least three out of the five test years. Notably, although the observed yield for the year 2024 was unavailable, the model's prediction remained consistent with established trends, underscoring its robustness in handling unseen data. In contrast, the Random Forest model, while producing accurate predictions for the training years, exhibited greater variance in the test set, frequently over- or underestimating yields. This suggests a tendency toward overfitting, with reduced ability to generalise beyond the training data. XGBoost achieved a more effective balance between model fit and generalisation, particularly evident in its stable performance across both historical test years and the unobserved 2024 scenario. Its visual alignment with observed trends and lower prediction error makes it a more robust and climate-resilient tool for yield forecasting, especially under scenarios of increasing climatic variability (Figure 5.13).

TABLE 5.10
Model Performance Metrics for Predicting Long Duration Basmati Yield: Training vs. Test Sets

Model	Train R^2	Test R^2	Train RMSE	Test RMSE	Train MAE	Test MAE
Cv_RF_Model	0.85	0.69	0.16	0.17	0.14	0.16
Final RF_Model	0.95	0.85	0.18	0.17	0.15	0.17
XGBoost	0.84	0.78	0.18	0.16	0.13	0.13

TABLE 5.11
Year-wise Comparison of Observed and Predicted Yields Using RF and XGBoost Models: Long-duration Varieties

Sl. No.	Year	Set	Observed Yield (t/ha)	RF_Model Predicted	XGBoost Predicted
1	2007	Train	4.50	4.36	4.52
2	2008	Train	4.00	4.01	3.99
3	2009	Train	4.19	4.27	4.29
4	2010	Train	3.75	3.81	3.60
5	2011	Test	3.99	4.02	4.15
6	2012	Train	3.93	4.07	4.08
7	2013	Test	3.98	4.06	4.07
8	2014	Test	4.45	4.34	4.47
9	2015	Train	4.32	4.24	4.30
10	2016	Train	3.83	3.90	3.72
11	2017	Test	4.12	4.32	4.38
12	2018	Train	4.05	4.05	4.08
13	2019	Train	4.30	4.23	4.21
14	2020	Train	3.30	3.66	3.55
15	2021	Train	5.10	4.70	4.58
16	2022	Train	3.98	4.10	4.09
17	2023	Train	4.50	4.34	4.45
18	2024	Test	NA	4.16	4.28

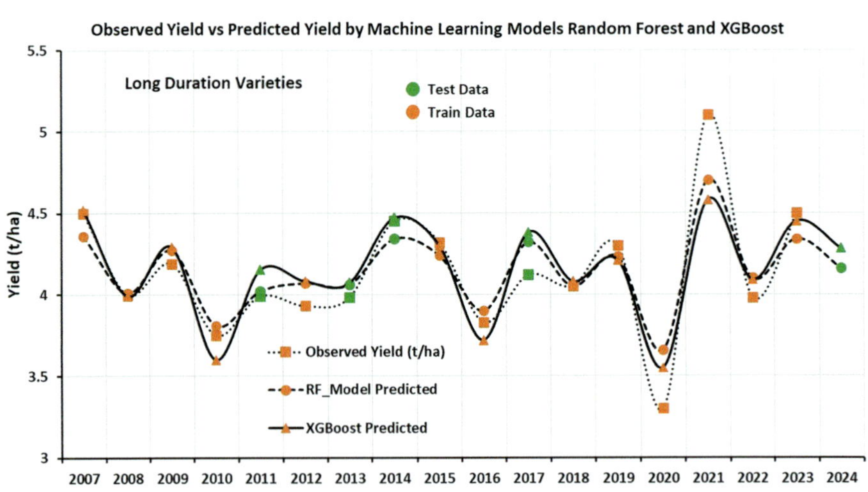

FIGURE 5.13 Observed vs. predicted yield for long duration Basmati varieties using RF and XGBoost Models (2007–2024).

5.13.2 Model Performance for Traditional Varieties

5.13.2.1 CV_RF_model

For traditional Basmati rice varieties, the cross-validated Random Forest model (cv_RF_Model) achieved a moderate Train R^2 of 0.85 and a lower Test R^2 of 0.69, indicating a reasonable fit to training data but limited generalisation to unseen data. Corresponding RMSE and MAE values for both training and testing sets were moderate, suggesting acceptable but suboptimal predictive accuracy.

In comparison, the final tuned Random Forest model (Final RF_Model) demonstrated notable improvement, with a high Train R^2 of 0.95 and a Test R^2 of 0.85, reflecting enhanced model fit and significantly improved generalisation. The Test RMSE and MAE remained consistent with the cross-validated model, indicating stable and reliable performance. The XGBoost model demonstrated superior performance, attaining a training R^2 of 0.95 and an ideal test R^2 of 1.00, demonstrating an exceptional capacity to capture underlying patterns and generalise effectively. It also recorded the lowest Train RMSE (0.12) and Train MAE (0.08), indicating high precision on the training set. Although the Test RMSE (0.28) and Test MAE (0.24) were relatively higher, they remained within acceptable bounds, suggesting that minor overfitting or test set variability may have influenced performance (Table 5.12). Nonetheless, XGBoost proved to be the most robust and high-performing model for yield prediction in traditional Basmati varieties (Table 5.13 and Figure 5.14).

5.13.3 Model Performance for Short-duration Varieties

Short-duration Basmati varieties, introduced only in the past decade, presented a limited dataset spanning 2013 to 2024. Three modelling approaches were evaluated: a cross-validated Random Forest (CV_RF_Model), a final tuned Random Forest (Final_RF_Model), and XGBoost.

Among them, XGBoost demonstrated superior performance, achieving the highest Test R^2 of 0.73 and an exceptionally high Train R^2 of 0.986, indicating strong predictive power and effective learning without evident overfitting. Fine-tuning

TABLE 5.12
Model Performance Metrics for Predicting Traditional Basmati Yield: Training vs. Test Sets

Model	Train R^2	Test R^2	Train RMSE	Test RMSE	Train MAE	Test MAE
Cv_RF_Model	0.85	0.69	0.16	0.17	0.14	0.16
Final RF_Model	0.95	0.85	0.18	0.17	0.15	0.17
XGBoost	0.95	1.0	0.12	0.28	0.08	0.24

TABLE 5.13

Year-wise Comparison of Observed and Predicted Yields Using RF and XGBoost Models: Traditional Varieties

Sl. No.	Year	Set	Observed Yield (t/ha)	RF_Model Predicted	XGBoost Predicted
1	2007	Train	2.53	2.82	2.56
2	2008	Train	3.00	3.08	3.01
3	2009	Train	2.98	2.98	2.96
4	2010	Test	2.75	3.21	3.22
5	2011	Test	3.01	3.19	3.30
6	2012	Train	3.40	3.59	3.40
7	2013	Train	3.31	3.32	3.30
8	2014	Train	3.81	3.19	3.76
9	2015	Test	3.71	3.59	3.76
10	2016	Train	3.00	3.11	3.02
11	2017	Train	3.60	3.48	3.59
12	2018	Train	2.80	2.83	2.79
13	2019	Train	2.30	2.71	2.31
14	2020	Test	3.50	3.48	3.59
15	2021	Train	3.00	3.15	3.01
16	2022	Train	3.40	3.27	3.37
17	2023	Train	3.58	3.43	3.57
18	2024	Test	NA	3.19	3.15

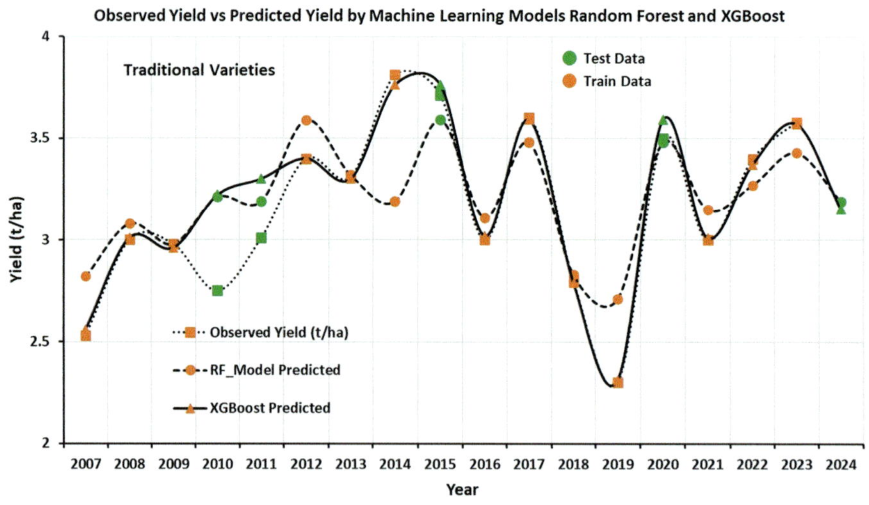

FIGURE 5.14 Observed vs. predicted yield for traditional Basmati varieties using RF and XGBoost Models (2007–2024).

TABLE 5.14
Model Performance Metrics for Predicting Short Duration Basmati Yield: Training vs. Test Sets

Model	Train R^2	Test R^2	Train RMSE	Test RMSE	Train MAE	Test MAE
Cv_RF_ Model	0.82	−0.32	0.26	0.39	0.17	0.31
Final RF_Model	0.96	0.60	0.26	0.40	0.18	0.32
XGBoost	0.99	0.73	0.13	0.41	0.07	0.35

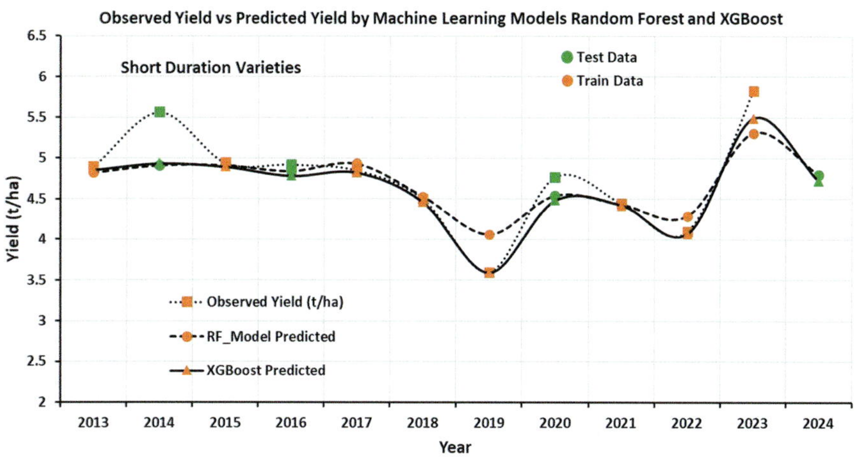

FIGURE 5.15 Observed vs. predicted yield for short duration Basmati varieties using RF and XGBoost Models (2013–2024).

through grid search and tracking of optimal iterations significantly enhanced its performance. The Final_RF_Model also performed reasonably well, with a Train R^2 of 0.96 and Test R^2 of 0.60, and slightly better Test RMSE and MAE than XGBoost. In contrast, the CV_RF_Model yielded a negative Test R^2 (−0.32), reflecting the limitations of resampling-based validation in very small datasets. Given the constrained data availability and the need to reliably forecast yields for 2024, XGBoost emerged as the most robust and generalisable model for short-duration Basmati varieties (Table 5.15 and Figure 5.15).

For short-duration Basmati varieties, introduced over the last 10–12 years, the dataset was limited to the period between 2013 and 2024. Three modelling

TABLE 5.15
Year-wise Comparison of Observed and Predicted Yields Using RF and XGBoost Models: Short-duration Varieties

Sl. No.	Year	Set	Observed Yield (t/ha)	RF_Model Predicted	XGBoost Predicted
1	2013	Train	4.89	4.82	4.85
2	2014	Test	5.56	4.91	4.93
3	2015	Train	4.94	4.91	4.89
4	2016	Test	4.92	4.84	4.78
5	2017	Train	4.85	4.93	4.82
6	2018	Train	4.50	4.53	4.46
7	2019	Train	3.60	4.06	3.59
8	2020	Test	4.77	4.54	4.48
9	2021	Train	4.45	4.43	4.42
10	2022	Train	4.10	4.29	4.07
11	2023	Train	5.83	5.31	5.49
12	2024	Test	NA	4.80	4.72

approaches were compared: a cross-validated Random Forest (CV_RF_Model), a final tuned Random Forest (Final_RF_Model), and XGBoost. Of the models evaluated, XGBoost delivered the best results, recording a test R^2 of 0.73 and a training R^2 of 0.99, indicating robust predictive capability and efficient learning of data patterns while avoiding overfitting. The Final_RF_Model showed reasonable performance with a Train R^2 of 0.96 and Test R^2 of 0.60, while providing slightly better Test RMSE and MAE than XGBoost (Table 5.14). Fine-tuning through grid search and tracking optimal rounds significantly boosted XGBoost's performance.

5.14 CONCLUSIONS

This study evaluated yield prediction across three Basmati rice varietal groups: long-duration, short-duration, and traditional, by integrating statistical and machine learning approaches. Model performance varied by varietal groups, reflecting their distinct agro-climatic responses and the complexities of yield modelling.

For long-duration Basmati varieties, the cross-validated Random Forest (cv_RF_Model) achieved a Train R^2 of 0.85 and a Test R^2 of 0.69, indicating a good fit to training data but only moderate generalisation. The Final_RF_Model improved this to a Train R^2 of 0.95 and a Test R^2 of 0.85, suggesting better generalisation, albeit with signs of overfitting. XGBoost outperformed both, with a Test R^2 of 0.78 and the lowest Test MAE (0.13), demonstrating superior generalisation capacity. In comparison, Multiple Linear Regression (MLR) achieved an R^2 of 0.92 and an Adjusted R^2 of 0.87. While it offered reasonable predictive power, its RMSE and MAE were higher than those of the machine learning models, limiting its ability to fully capture data complexity, especially in high-variability years.

For short-duration Basmati, data scarcity posed modelling challenges. Despite this, XGBoost performed best, with a Train R^2 of 0.99 and a Test R^2 of 0.73, showing strong generalisation. The final RF model posted a lower Test R^2 of 0.60, while the cv_RF_Model underperformed with a negative Test R^2 of −0.32. MLR achieved an R^2 of 0.96 and an Adjusted R^2 of 0.93, its highest across all varietal groups. Although MLR had higher RMSE and MAE compared to XGBoost, it offered reliable predictions under constrained data conditions.

For traditional Basmati varieties, the cv_RF_Model yielded a Train R^2 of 0.85 and a Test R^2 of 0.69, again showing moderate generalisation. The Final_RF_Model improved to a Train R^2 of 0.95 and a Test R^2 of 0.85. XGBoost achieved a perfect Test R^2 of 1.00 with the lowest RMSE and MAE, indicating exceptional predictive accuracy. MLR attained an R^2 of 0.88 and an Adjusted R^2 of 0.81. Though less accurate than XGBoost, MLR effectively captured underlying trends, despite limitations in modelling non-linear patterns.

Across all varietal groups, XGBoost consistently outperformed Random Forest and MLR, demonstrating strong generalisation and robustness. Random Forest models performed well in long-duration and traditional groups, but XGBoost's adaptability to nonlinear interactions and variable climatic patterns made it the most reliable model for forecasting Basmati rice yields. MLR, while useful as a baseline, struggled with capturing complex relationships, especially in long-duration varieties. Its strongest performance was observed in the short-duration group (R^2: 0.96) and traditional group (R^2: 0.88). Importantly, the Mean Absolute Percent Error (MAPE) remained below 10% for all three varietal segments, highlighting the overall reliability of the models.

In the case of MLR, 6–7 weather variables were selected for each varietal group based on correlation analysis. These variables contributed significantly to improvements in R^2 and Adjusted R^2 values and helped enhance the model's overall predictability.

These findings underscore the value of machine learning, particularly XGBoost, in modelling agricultural yields under climate variability. Such models offer actionable insights to researchers, farmers, and policymakers for managing uncertainties introduced by shifting rainfall patterns and temperature regimes, ultimately contributing to more resilient agricultural practices.

REFERENCES

Abhilash Dagar, C. S., R. S. Premdeep, and R. Sharma. 2017. Agrometeorological Indices and Phenology of Basmati Rice (*Oryza sativa* L.) under different dates of Transplanting. *International Journal of Current Microbiology and Applied Sciences* 6(3): 212–222. https://doi.org/10.20546/ijcmas.2017.603.022

Ahmad, T., P. M. Sahoo, M. Singh, and A. Biswas. 2021. *Crop Cutting Experiment techniques for the determination of yield rates of field crops*. New Delhi: Division of Sample Surveys, ICAR- Indian Agricultural Statistics Research Institute.

Alam, W., K. Sinha, R. R. Kumar, M. Ray, S. Rathod, K. N. Singh, and P. Arya. 2018. Hybrid linear time series approach for long-term forecasting of crop yield. *Indian Journal of Agricultural Sciences*, 88, 1275–1279.

Ali, A. M., S. Igor, P. Anton, A. Mohamed, S. Nasser, A. Khaled, E.-S. Mohammed, and D. Peter. 2021. Integrated method for rice cultivation monitoring using Sentinel-2 data and Leaf Area Index. *The Egyptian Journal of Remote Sensing and Space Sciences* 24(2021): 431–441.

Aravind, K., A. Vashisth, P. Krishnan, and B. Das. 2022. Wheat yield prediction based on weather parameters using multiple linear, neural network, and penalized regression models. *The Journal of Agrometeorology* 24(1): 18–25. https://doi.org/10.54386/jam.v24i1.1002

Bair, W. 1977. Crop weather models and their use in yield assessments. Tech. Note No.151, WMO, Geneva, pp.48

Baret, F., B. De Solan, and R. Lopez-Lozano, M. Kai, and M. Weiss. 2010. GAI estimates of row crops from downward-looking digital photos taken perpendicular to rows at 57.5° zenith angle: Theoretical considerations based on 3D architecture models and application to wheat crops. *Agricultural and Forest Meteorology*, 150(11): 13931401, 15 October 2010,

Batchelor, W. D., B. Basso, and J. O. Paz. 2002. Examples of strategies to analyze spatial and temporal yield variability using crop models. *European Journal of Agronomy* 18(1–2): 141–158.

Belgiu, M. and L. Drăguţ. 2016. Random forest in remote sensing: A review of applications and future directions. *ISPRS Journal of Photogrammetry and Remote Sensing* 114(April): 24–31. https://doi.org/10.1016/j.isprs jprs.2016.01.011

Bengio, Y., Léonard, N., & Courville, A. 2013. *Estimating or Propagating Gradients Through Stochastic Neurons for Conditional Computation*. arXiv preprint arXiv:1308.3432.

Bishop, K. A., A. D. Leakey, and E. A. Ainsworth 2014. How seasonal temperature or water inputs affect the relative response of C3 crops to elevated [CO2]: A global analysis of open-top chamber and free-air CO2 enrichment studies. *Food and Energy Security* 3(1): 33–45. https://doi.org/10.1002/fes3.44

Blanc, É. 2017. Statistical emulators of maize, rice, soybean, and wheat yields from global gridded crop models. *Agricultural and Forest Meteorology*, 236: 145–161. https://doi.org/10.1016/j.agrformet.2016.12.022

Bolstad, B. M., R. A. Irizarry, M. Astrand, and T. P. Speed. 2003. A comparison of normalization methods for high-density oligonucleotide array data based on variance and bias. *Bioinformatics* 19: 185–193.

Boris, I. and D. C. Hoaglin. 1993. *Volume 16: How to Detect and Handle Outliers*. ASQ Quality Press, ISBN: 0873892607

Bowden, C., T. Foster, and B. Parkes. 2023. Identifying links between monsoon variability and rice production in India through machine learning. *Scientific Reports* 13, 2446. https://doi.org/10.1038/s41598-023-27752-8

Breiman, L. 2001. Random forests. *Machine Learning*, 45(1): 5–32. https://doi.org/10.1023/A:1010933404324

Cao, J., Z. Zhang, F. Tao, L. Zhang, Y. Luo, J. Zhang, J. Han, and J. Xie. 2021. Integrating multi-source data for rice yield prediction across china using machine learning and deep learning approaches. *Agricultural and Forest Meteorology* 297: 108275.

Chang, X.. 2024. Rice yield prediction based on deep learning. In *Artificial Intelligence Technologies and Applications*, ed. C. Chen, pp. 490–496. https://doi.org/10.3233/FAIA231333

Chen, T., and C. Guestrin. 2016. XGBoost: A scalable tree boosting system. *Proceedings of the 22nd ACM SIGKDD International Conference on Knowledge Discovery and Data Mining* (pp. 785–794). Association for Computing Machinery. https://doi.org/10.1145/2939672.2939785

Chetna, M., D. Aditi, and A. Sharma. 2023. A study on the impact of climate change on rice yield. *International Journal of Statistics and Applied Mathematics 2023* SP-8(3): 23–29.

Choudhary, K., Shi, W., and Dong, Y. 2021. Rice growth vegetation index 2 for improving the estimation of rice plant phenology in coastal ecosystems. *Computer Optics*, 45(3) https://doi.org/10.18287/2412-6179-CO-827

Clauss, K., M. Ottinger, P. Leinenkugel, and C. Kuenzer. 2018. Estimating rice production in the Mekong Delta, Vietnam, utilizing time series of Sentinel-1 SAR data. *International Journal of Applied Earth Observation and Geoinformation* 73: 574–585. https://doi.org/10.1016/j.jag.2018.07.022

Confalonieri, R., A. S. Rosenmund, and B. Baruth. 2009. An improved model to simulate rice yield. *Agronomy for Sustainable Development* 29(3): 463–474. https://doi.org/10.1051/agro/2009005

De Clercq, D., and A. Mahdi. 2025. Modern computational approaches for rice yield prediction: A systematic review of statistical and machine learning – based methods. *Computers and Electronics in Agriculture* 231: 109852, ISSN 0168-1699. https://doi.org/10.1016/j.compag.2024.109852

Deering, D. W., J. W. Rouse, R. H. Haas, and J. A. Schell. 1975. Measuring Forage Production of Grazing Units from Landsat MSS Data. *Proceedings of the 10th International Symposium on Remote Sensing of Environment*, II: 1169–1178. ERIM, Ann Arbor, MI. https://ntrs.nasa.gov/api/citations/19800024311/downloads/19800024311.pdf

dela Torre, D. M. G., J. Gao, and M.-N. Cate. 2021. Remote sensing-based estimation of rice yields using various models: A critical review. *Geo-Spatial Information Science* 2021, 24(4): 580–603. https://doi.org/10.1080/10095020.2021.1936656

Divakar, M.S., M.S. Elayidom, and Rajesh, R. 2022. Forecasting crop yield with deep learning based ensemble model. *Materials Today: Proceedings* **2022**(58): 256–259.

Donald, C. M. 1962. In search of yield. *The Journal of the Australian Institute of Agricultural Science* 1962(28), 171–178.

Draper, N. R. and H. Smith. 1998. *Applied Regression Analysis*. Chichester, England, UK: John Wiley & Sons, Ltd..

Dwivedi, M., S. Saxena, Neetu, and S. S. Ray. 2019. Assessment of Rice Biomass Production and Yield Using Semi-Physical Approach and Remotely Sensed Data. *The International Archives of the Photogrammetry, Remote Sensing and Spatial Information Sciences, Volume XLII-3/W6, 2019 ISPRS-GEOGLAM-ISRS Joint Int. Workshop on "Earth Observations for Agricultural Monitoring"*, 18–20 February 2019, New Delhi, India.

Everingham, Y. L., J. Sexton, D. Skocaj, and N. G. Inman-Bamber. 2016. Accurate prediction of sugarcane yield using a random forest algorithm. *Agronomy for Sustainable Development*, 36(2): 27. https://doi.org/10.1007/s13593-016-0364-z

Friedman, J. H. 2001. Greedy function approximation: A gradient boosting machine. *The Annals of Statistics* 29(5): 1189–1232. https://doi.org/10.1214/aos/1013203451

Garima, S., S. S. Prabhjyot-Kaur, and K. K. Gill. 2021. Relationship between sunshine duration during different growth periods and rice productivity in central Punjab of India. Virtual National Conference on "Strategic Reorientation for Climate Smart Agriculture" (V-AGMET 2021) March 17-19th, 2021, Punjab Agricultural University, Ludhiana.

Gitelson, A. A. 2004. Wide dynamic range vegetation index for remote quantification of biophysical characteristics of vegetation. *Journal of Plant Physiology* 161: 165–173.

Gitelson, A. A., Y. Kaufman, M. N. Merzlyak. 1996. Use of a green channel in remote sensing of global vegetation from EOS-MODIS. *Remote Sensing Environ* 1996(58): 289–298.

Gitelson, A.A., A. Vina, V. Ciganda, D. C. Rundquist, and T. J. Arkebauer. 2005. Remote estimation of canopy chlorophyll content in crops. *Geophysical Research Letters* 32, L08403

Gitelson, Anatoly A., Y. Gritz, and M. N. Merzlyak. 2003. Relationships between leaf chlorophyll content and spectral reflectance and algorithms for non-destructive chlorophyll assessment in higher plant leaves. *Journal of Plant Physiology* 160(3): 271–282, ISSN 0176-1617, https://doi.org/10.1078/0176-1617-00887

Glotter, M. J., E. J. Moyer, A. C. Ruane, and J. W. Elliott. 2016. Evaluating the sensitivity of agricultural model performance to different climate inputs. *The Journal of Applied Meteorology and Climatology* 2016(55): 579–594.

Gouri, V., D. R. Reddy, S. B. S. N. Rao, and A. Y. Rao. 2005. Thermal requirement of Rabi groundnut in the southern Telangana zone of Andhra Pradesh. *The Journal of Agrometeorology* 7(1): 90–94.

Hastie, T., R. Tibshirani, and J. Friedman. 2009. *The Elements of Statistical Learning: Data Mining, Inference and Prediction* (2nd Ed). New York: Springer.

Hicks, S. C. and R. A. Irizarry Quantro. 2015. A data-driven approach to guide the choice of an appropriate normalization method. *Genome Biology* 16, 117. https://doi.org/10.1186/s13059-015-0679-0

Hoaglin, D. C., B. Iglewicz, and J. W. Tukey. 1986. Performance of some resistant rules for outlier labeling. *Journal of the American Statistical Association*, 81(396): 991–999. https://doi.org/10.2307/2289073

Hodge, V. J., and J. A. Austin. 2004. Survey of outlier detection methodologies. *Artificial Intelligence Review* 22, 85–126 (2004). https://doi.org/10.1007/s10462-004-4304-y

Horie, T., M. Ohnishi, J. F. Angu, L. G. Lewin, L. T. Tsukaguki, and T. Matano. 1997. Physiological characteristics of high-yielding rice inferred from cross-location experiments. *Field Crops Research* 52: 55–67.

Huang, J., S. Tang, and A.-I. Ousama. 2002. Rice yield estimation using remote sensing and a simulation model. *Journal of Zhejiang University - Science A: Applied Physics & Engineering* 3(4): 461–466, September 2002. https://doi.org/10.1007/BF02839491

Huete, A. R. 1988. A soil-adjusted vegetation index (SAVI). *Remote Sensing of Environment* 1988(25): 295–309.

India Meteorological Department (IMD). 2023. Office of Climate Research and Services. (2023, September 2). *Sub: Climate Summary for the Month of August 2023 (Press Release)*. New Delhi/Pune: Ministry of Earth Sciences. Retrieved from https://internal.imd.gov.in/press_release/20230902_pr_2510.pdf

Jain, V., S. Saxena, S. Dubey, K. Choudhary, S. Sehgal, Neetu, and S.S. Ray. 2019. Rice (Kharif) production estimation using SAR data of different satellites and yield models: A comparative analysis of the estimates generated under FASAL Project. *ISPRS-The International Archives of the Photogrammetry, Remote Sensing and Spatial Information Sciences, XLII-3/W6*, 99–107.

James, G., D. Witten, T. Hastie, and R. Tibshirani. 2021. *An Introduction to Statistical Learning: With Applications in R*. United States: Springer US.

Jeong, J. H., J. P. Resop, N. D. Mueller, D. H. Fleisher, K. Yun, E. E. Butler, … and S. H. Kim. 2016. Random forests for global and regional crop yield predictions. *PLoS ONE*, 11(6): e0156571. https://doi.org/10.1371/journal.pone.0156571

Jhajharia, K., P. Mathur, S. Jain, and S. Nijhawan. 2023. Crop yield prediction using machine learning and deep learning techniques. *Procedia Computer Science* 218: 406–417 (2023).

Joaquin, Q.-C., Masasi, S., Anton, S., and N. D. Lawrence. 2022. *Dataset Shift in Machine Learning*. MIT Press. ISBN: 9780262545877.

Jones, D. R. 1982. A statistical inquiry into crop–weather dependence. *Agricultural and Forest Meteorology*, 26: 91–104.

Jones, H. G. 1992. *Plants and Microclimate*, 2nd Edition, Cambridge: Cambridge University Press, p. 428.

Kalubarme, M.H., and S. C. Ahuja. 1996. Agrometeorological rice yield modelling in Karnal. *Journal of the Indian Society of Remote Sensing* 24: 125–132.

Karamat, A., A. Rehman, M. Ayyaz et al. 2019. Estimation of net rice production for the fiscal year 2019 using multi-source datasets. *International Journal of Agriculture and Sustainable Development* 1: 47–65.

Ketring, D.L. and T. G. Wheless. 1989. Thermal requirement for phenological development of peanut. *Journal of Agronomy* 8: 910–917.

Kogan, F. N. 1990. Remote sensing of weather impacts on vegetation in non-homogeneous areas. *International Journal of Remote Sensing* 1990(11): 1405–1419.

Kothawale, D. R. and K. Rupa Kumar. 2005. On the recent changes in surface temperature trends over India. *Geophysical Research Letters*, 32.

Kouw, W. M. and M. Loog. 2021. A review of domain adaptation without target labels. *IEEE Transactions on Pattern Analysis and Machine Intelligence*, 43(3): 766–785, https://doi.org/10.1109/TPAMI.2019.2945942

Kuhn, M. 2008. Building predictive models in R using the caret package. *Journal of Statistical Software* 28: 1–26.

Kuhn, M., and K. Johnson. 2013. *Applied Predictive Modeling*. New York, NY: Springer. ISBN-13: 978-1461468486.

Kumar, Y. J. N., V. Spandana, V. S. Vaishnavi, K. Neha, and V. G. R. R. Devi. 2020. Supervised machine learning approach for crop yield prediction in the agriculture sector. *5th International Conference on Communication and Electronics Systems (ICCES)*, 736–741. IEEE.

Li, T., O. Angeles, M. Marcaida, E. Manalo, M. P. Manalili, A. Radanielson, and S. Mohanty. 2017. From ORYZA2000 to ORYZA (V3): An improved simulation model for rice in drought and nitrogen-deficient environments. *Agricultural and Forest Meteorology* 157: 246–256. https://doi.org/10.1016/j.agrformet.2017.02.025

Liaw, A. and M. Wiener. 2002. Classification and regression by random forest. *R News*, 2: 18–22. https://CRAN.R-project.org/doc/Rnews/

Lobell, D. B., and M. B. Burke. 2010. On the use of statistical models to predict crop yield responses to climate change. *Agricultural and Forest Meteorology* 150(11): 1443–1452. https://doi.org/10.1016/j.agrformet.2010.07.008

Maas, S. J. 1992. GRAMI: A Crop Growth Model That Can Use Remotely Sensed Information. U.S. Department of Agriculture, National Agricultural Library. https://archive.org/details/IND93024598/

MacDonald, R.B., and F. G. Hall. 1980. Global crop forecasting. *Science*, 208: 670–679. https://doi.org/10.1126/science.208.4445.670

Mathieu, J.A. and F. Aires. 2018. Assessment of the agro-climatic indices to improve crop yield forecasting. *Agricultural and Forest Meteorology* 253: 15–30

Mia, M. S., R. Tanabe, L. N. Habibi, N. Hashimoto, K. Homma, M. Maki, T. Matsui, and T. S. T. Tanaka. 2023. Multimodal deep learning for rice yield prediction using UAV-based multispectral imagery and weather data. *Remote Sensing* 2023(15): 2511. https://doi.org/10.3390/rs15102511

Montgomery, D. C., E. A. Peck, and G. G. Vining. 2012. *Introduction to Linear Regression Analysis* (5th ed.). Wiley. https://statanaly.com/wp-content/uploads/2023/05/IntroductiontoLinearRegressionAnalysisbyDouglasC.MontgomeryElizabethA.PeckG_.GeoffreyViningz-lib.org_.pdf

Moreau, L. and Z. Li. 1996. A new approach for remote sensing of canopy-absorbed photosynthetically active radiation. II Proportion of canopy absorption. *Remote Sensing of Environment* 55: 175–191. https://doi.org/10.1016/S0034-4257(95)00098-4

Moreno-Torres, J. G., T. Raeder, R. Alaiz-Rodríguez, N. V. Chawla, and F. Herrera. 2012. A unifying view on dataset shift in classification. *Pattern Recognition*, 45(1): 521–530, ISSN 0031-3203. https://doi.org/10.1016/j.patcog.2011.06.019

Moulin, S., A. Bondeau, and R. Delecolle. 1998. Combining agricultural crop models and satellite observations: From field to regional scales. *International Journal of Remote Sensing* 1998(19): 1021–1036.

Murthy, C. S., S. Thiruvengadachari, P. V. Raji, and S. Jonna. 1996. Improved ground sampling and crop yield estimation using satellite data. *International Journal of Remote Sensing* 17: 945–956.

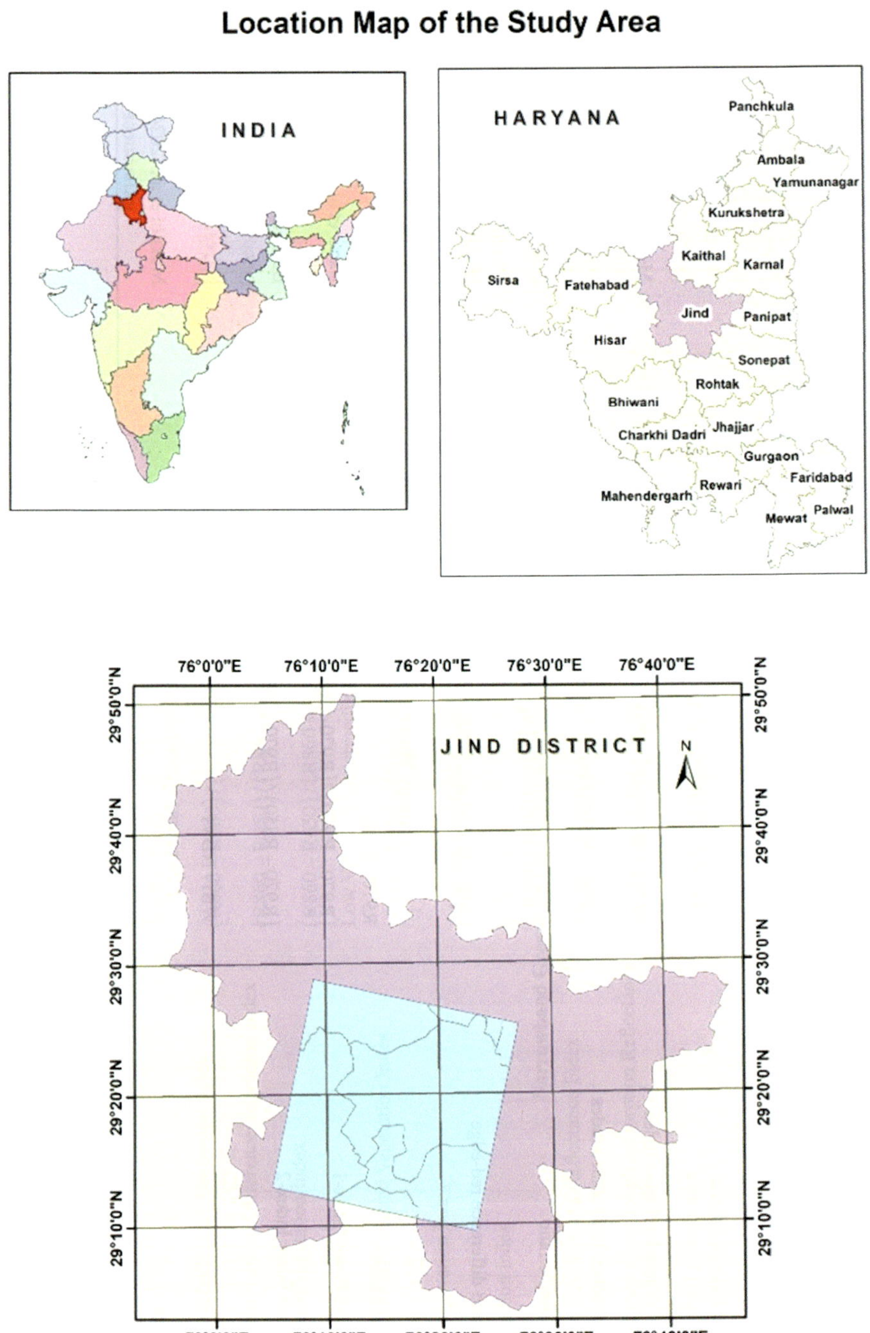

FIGURE 6.1 Location map of the Jind district in Haryana State with coverage of PRISMA Hyperspectral Data.

to loam in texture, supporting the cultivation of staple crops such as pearl millet (bajra), sorghum (jowar), and chickpea (gram). The location map of the Jind district in Haryana State is given in Figure 6.1.

6.4.2 Cultivation of Basmati and High-yielding Rice Varieties

In the Jind district, different varieties of rice are normally transplanted from the second fortnight of June to the second fortnight of July. The high-yielding varieties (HYV) are transplanted during the second fortnight of June and harvested during the first fortnight of October, almost 25 to 30 days before the harvesting of medium and long-duration Basmati varieties. The major high-yielding varieties grown in the Jind district are Pusa Rice (PR 114, PR 117, PR 118), and, two groups of aromatic Basmati varieties cultivated in the Jind district are, long duration (Pusa Basmati-1121, Pusa Basmati-1718, Pusa Basmati- 1885, Pusa Basmati -1, Pusa Basmati-6 and CSR-30) and Short duration (Pusa Basmati-1509, Pusa Basmati-1692, Pusa Basmati-1847, and Pusa Basmati-1886). The historical rice statistics collected from the Department of Agriculture, Haryana State, are given in Table 6.3. The area under different varieties of aromatic Basmati rice is continuously increasing, and during the *kharif* season of 2024, it was 158.3 thousand hectares, which is almost 68 percent of the net sown area of the Jind district.

TABLE 6.3
Historical rice statistics in Jind District

S. No.	Year	Total Rice ('000 ha)	% NSA*	Basmati ('000 ha)	% NSA*
1	2006	84.3	36.9	26.9	11.5
2	2007	98.3	42.0	63.0	26.9
3	2008	99.9	42.7	70.4	30.1
4	2009	93.4	39.9	76.5	32.7
5	2010	111.0	47.4	91.8	39.2
6	2011	105.0	44.9	89.2	38.1
7	2012	90.0	38.5	57.0	24.4
8	2013	114.0	48.7	79.5	34.0
9	2014	98.6	42.1	84.9	36.3
10	2015	114.3	48.8	95.2	40.7
11	2016	118.8	50.8	81.2	34.7
12	2023	165.3	70.6	111.3	47.6
13	2024	178.9	76.5	158.3	67.6

Geographical area of Jind District: 2736 sq km.
* Net Sown area (NSA): 2340 sq km (86 % of the Geographical area of Jind District)

6.4.3 Field Data Collection

The field data was collected in the Jind district, coinciding with the PRISMA Hyperspectral data availability periods during the second fortnight of September 2024. The large patches of paddy fields of two different varieties, namely High-Yielding Varieties (HYV) and aromatic Basmati varieties, were identified in the field, and GPS measurements of these plots were recorded for precise location on the satellite images. The field photographs of these two varieties, are given in Figure 6.2. Some of the field details of these rice varieties, like the date of sowing, irrigation availability, soil fertility status, and expected date of harvesting, were recorded for all the field sites. During the field survey, it was observed that Basmati varieties were at the flag leaf to ear head emergence stages, whereas HYV rice varieties were at the grain filling to soft dough stages. The other major crops grown along with rice during the *kharif*, as well as forest patches, were also recorded.

6.4.4 PRISMA Hyperspectral Imagery Acquisition

PRISMA generates Level-1 and Level-2 hyperspectral data products on request, which are provided in the HDF5 (Hierarchical Data Format version 5) structure. The request for PRISMA data acquisition over the Karnal, Kaithal, and Jind districts was submitted using the Mission Selection form with New Acquisition on the ASI website: http://prisma.asi.it/. The request for PRISMA hyperspectral data acquisition (A strip of 3 scenes with a 30 km swath was identified on Google image) with 20% cloud cover covering these districts during August and September 2024 was submitted. However, only one PRISMA L1 data product acquired on **11-Sep-2024**, covering parts of the Jind and Hisar districts, was acquired and downloaded for analysis.

FIGURE 6.2 Field photos of Basmati and High-yielding Rice varieties in Jind District.

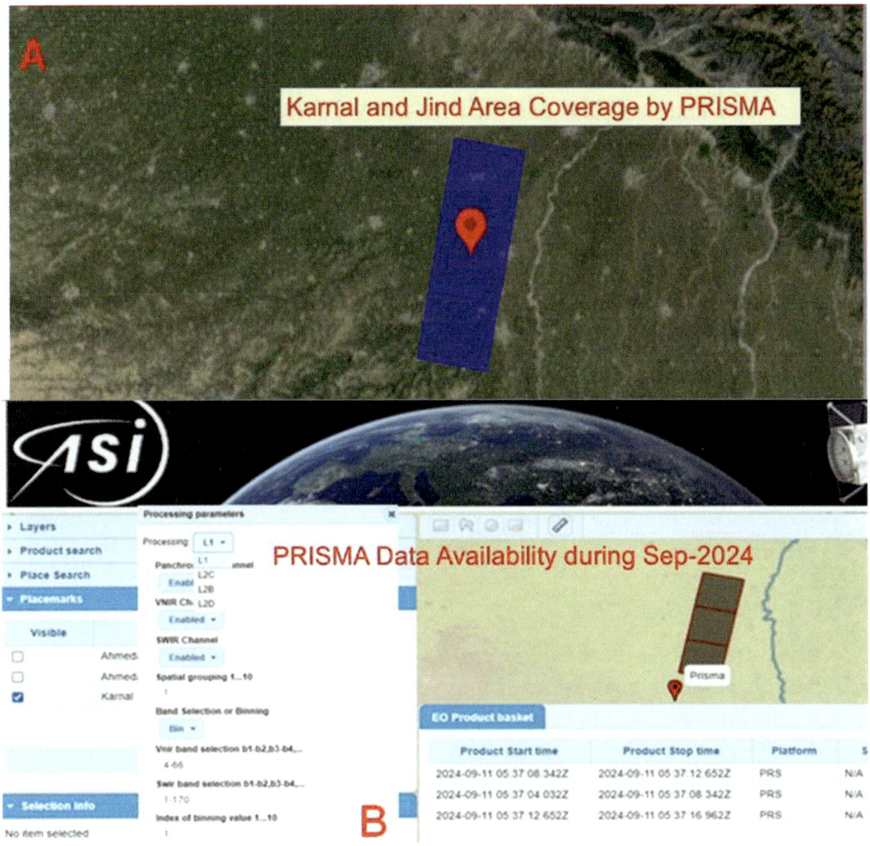

FIGURE 6.3 (A) Map of area coverage of Karnal, Jind, and Hisar districts by PRISMA Data and (B) available PRISMA data during September 2024.

The PRISMA Level 1 product contains radiance data corrected for radiometric distortions and calibrated to represent top-of-atmosphere conditions. It also includes ancillary masks for cloud cover, sun glint, and general scene classification (Loizzo et al. 2016). The map of the area coverage of 3 scenes covering parts of Karnal, Jind, and Hisar districts, and available PRISMA data are given in Figure 6.3. The cloud-free image of 11 September 2024, covering parts of the Jind and Hisar districts, was the only data available that was downloaded for further analysis.

6.4.5 PRISMA Hyperspectral Sensor Details

The PRISMA (PRecursore IperSpettrale della Missione Applicativa) satellite, developed by the Italian Space Agency (ASI), was launched on 22 March 2019 into a sun-synchronous, low-Earth orbit at an altitude of 615 km. It operates with a 29-day repeat cycle and offers revisit capabilities of under one week for specific targets through off-nadir pointing. PRISMA is equipped with a push-broom hyperspectral sensor that captures data across 240 contiguous spectral bands spanning the visible (VIS), near-infrared (NIR), and shortwave infrared (SWIR) regions (400–2500 nm), including 66 bands in the VNIR and 173 in the SWIR domain. The instrument achieves a spectral resolution finer than 12 nm and provides a spatial resolution of 30 m. Additionally, there are nine overlapping bands bridging the VNIR and SWIR ranges. A separate panchromatic sensor, operating in the 400–700 nm range, acquires imagery at a finer spatial resolution of 5 m (Chiavetta et al., 2016). The integration of hyperspectral and panchromatic imagery enhances the capability to distinguish crop types and capture their spatial structure with greater precision. PRISMA's hyperspectral sensor utilises prisms as dispersive elements to direct incoming light onto a two-dimensional detector array, employing a pushbroom scanning configuration (ASI, 2020). Image acquisition is task-driven and occurs over user-defined locations in two modes: the "standard" mode captures a 30 × 30 km scene, while the "strip" mode acquires continuous swaths 30 km in width, extending up to 1800 km in length. The combined use of hyperspectral and panchromatic data offers valuable applications in areas such as precision agriculture, forest monitoring, aquatic ecosystem evaluation, and climate-related studies.

6.4.6 Preprocessing of PRISMA Hyperspectral Data

One cloud-free PRISMA scene covering parts of the Jind and Hisar districts was acquired on 11 September 2024. The hyperspectral dataset was acquired in HDF5 format, corresponding to the Level-1 (L1) processing stage. This L1 product contains top-of-atmosphere (TOA) spectral radiance data along with associated metadata. The Level-1 product consists of radiance measurements that have undergone radiometric correction and calibration, expressed in physical units. To facilitate further analysis, the PRISMA hyperspectral L1 data, originally delivered in HDF5 format, were converted into GeoTIFF format using the EnMAP-Box plugin within QGIS. EnMAP-Box is specifically designed for the visualisation and processing of imaging spectrometer data and is optimised for handling datasets from sensors such as EnMAP (Environmental Mapping and Analysis Program, www.enmap.org).

6.5 PRINCIPAL COMPONENT ANALYSIS (PCA)

Principal Component Analysis (PCA) is a widely employed technique in digital image processing, serving as an orthogonal transformation that reprojects high-dimensional hyperspectral data into a new coordinate space. In this transformed space, the first principal component captures the maximum variance present in the dataset, followed by subsequent components representing decreasing levels of variability (Harsanyi and Chang, 1994). Dimensionality reduction is a critical preprocessing stage in hyperspectral remote sensing, as it enhances computational efficiency and mitigates redundancy without compromising data integrity. Hyperspectral imagery, which spans the 400–2500 nm range with hundreds of closely spaced spectral bands, often contains overlapping spectral information, particularly since most natural materials such as vegetation and soil exhibit minimal reflectance variation within narrow spectral intervals of 5–10 nm. Consequently, many bands convey redundant information, which must be effectively minimised to preserve the essential spectral content for land use and land cover (LULC) analysis.

Principal Component Analysis (PCA) is employed to reduce the dimensionality of high-resolution datasets by transforming the original pixel-based matrix into a set of uncorrelated components, derived through either covariance or correlation structures (Rodarmel and Shan, 2002). The technique relies on the eigenvalue decomposition of the covariance matrix associated with the hyperspectral bands under analysis. The PCA process involves mean-centering the spectral data, assessing inter-band similarity through covariance or correlation, and performing matrix decomposition using eigenvalue decomposition (EVD) or singular value decomposition (SVD) methods (Chuvieco, 2016). Dimensionality reduction techniques such as PCA are also widely applied in the preprocessing stages of machine learning workflows, including classification and clustering tasks. The number of input bands plays a crucial role in determining classification performance and contributes significantly to the development of computationally efficient and accurate predictive models. In the present study, Principal Component Analysis (PCA) was performed on PRISMA hyperspectral data using ENVI software (version 5.3). The resulting eigenvectors and associated eigenvalues, representing the principal components, were ranked in descending order based on the magnitude of variance they explain. These components are subsequently labeled sequentially, starting with Principal Component 1, corresponding to the highest explained variance.

6.5.1 Selective Principal Component Analysis (SPCA)

Although hyperspectral imagery offers rich spectral detail, it often includes high inter-band correlation and redundant information, with some bands predominantly containing noise. Therefore, feature selection becomes essential to reduce dimensionality and to identify the most informative spectral regions for tasks such as vegetation classification. Unlike traditional PCA, which considers the full set of spectral bands, Selective PCA (SPCA) applies the transformation only to a chosen subset of bands, enhancing computational efficiency and interpretability. The key distinction between PCA and SPCA lies in the targeted selection of input bands in SPCA, based on specific analytical objectives or application needs (Siljestrom et al., 1997). Shwetank et al. (2010) introduced a novel data compression technique utilising SPCA applied to the 155-band EO-1 Hyperion hyperspectral dataset. Their findings demonstrated that classification using SAM on the PCANIR segment (671.02–925.41 nm) achieved the highest accuracy in distinguishing among rice crop varieties.

In the present study, a new concept of selecting hyperspectral bands using SPCA that are specifically useful for crop classification was selected from the 240 hyperspectral bands of PRISMA Hyperspectral data, with 66 in the visible near infrared (VNIR) and 173 in the short wave infrared (SWIR) spectrum. The selected nine hyperspectral bands correspond to the Sentinel-2 MSI spectral bands. Sentinel-2 is equipped with a multispectral optical sensor that captures data across 13 distinct spectral bands. These bands are strategically selected to support a range of terrestrial applications, including agricultural monitoring, vegetation analysis, forest assessment, and land cover change detection. The selected nine PRISMA hyperspectral bands corresponding to Sentinel-2 MSI spectral bands are given in Table 6.4.

TABLE 6.4
Selected 9 PRISMA Hyperspectral Bands Corresponding to Sentinel-2 MSI Spectral Bands

Sr. No.	PRISMA Hyperspectral Bands		Sentinel-2 MSI Bands	
	Band Number	Central Wavelength (nm)	Band Number	Central Wavelength (nm)
1	3 (Blue)	446.923	1	443
2	8 (Blue)	493.846	2	490
3	15 (Green)	559.540	3	560
4	26 (Red)	662.769	4	665
5	35 (Veg. Red Edge)	747.231	6	740
6	39 (Veg. Red Edge)	784.769	7	783
7	44 (NIR)	831.692	8	842
8	129 (SWIR1)	1618.14	12	1610
9	193 (SWIR2)	2206.047	13	2190

These selected nine PRISMA channels included two vegetation red edge bands, which are sensitive to differences in leaf structure, chlorophyll content, and LAI of the vegetation. The Environment for Visualising Images (ENVI) version 5.3 has been used for hyperspectral data pre-processing operations such as spatial or spectral subsets, etc. The eigenvalue matrix of the hyperspectral data was calculated to identify which principal component (PC) was involved in the rice crop and variety information. The eigenvalue plot of the Jind district using nine PRISMA hyperspectral bands is given in Figure 6.4, which indicates that the top three eigenvalues contain the most information, and after that, the rest are mostly noise. The covariance matrix, decomposed into the principal component using the **Eigen Vector Decomposition (EVD)** method, was attempted for PCA.

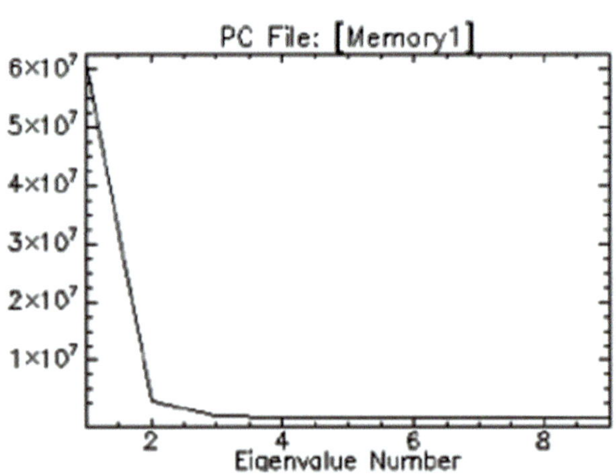

FIGURE 6.4 Eigenvalue plot of nine PRISMA hyperspectral bands for Jind districts.

The PCA was used to capture the most important information while reducing the dimensionality of the dataset and minimising information loss. This dimensionality reduction provided a concise representation of the data while retaining a significant portion of the variance. The results, as presented in Figure 6.5, show that the first two components are sufficient for capturing most of the variability present in the dataset. PC-1 accounts for 94.55% of the variance, PC-2 accounts for 4.74% of the variance, and PC-3 minimally contributes with a variance of 0.61%. The cumulative effect of the first three principal components (PC-1), second principal component (PC-2), and third principal component (PC-3) explains 99.90% of the variability observed among different high-yielding and aromatic Basmati rice varieties in the Jind district (Table 6.5).

The RGB colour composites of three combinations of PC-1, PC-2, PC-3, and PC-4 were attempted. The PRISMA hyperspectral RGB colour composites of different PC combinations along the original false colour composite (FCC) covering Jind District (Haryana state) are given in Figure 6.5. It was observed that out of three RGB colour composites of three combinations of Principal Components, the RGB colour composite using R: PC-3, G: PC-2, and B: PC-1 (Figure 6.5 B) resulted

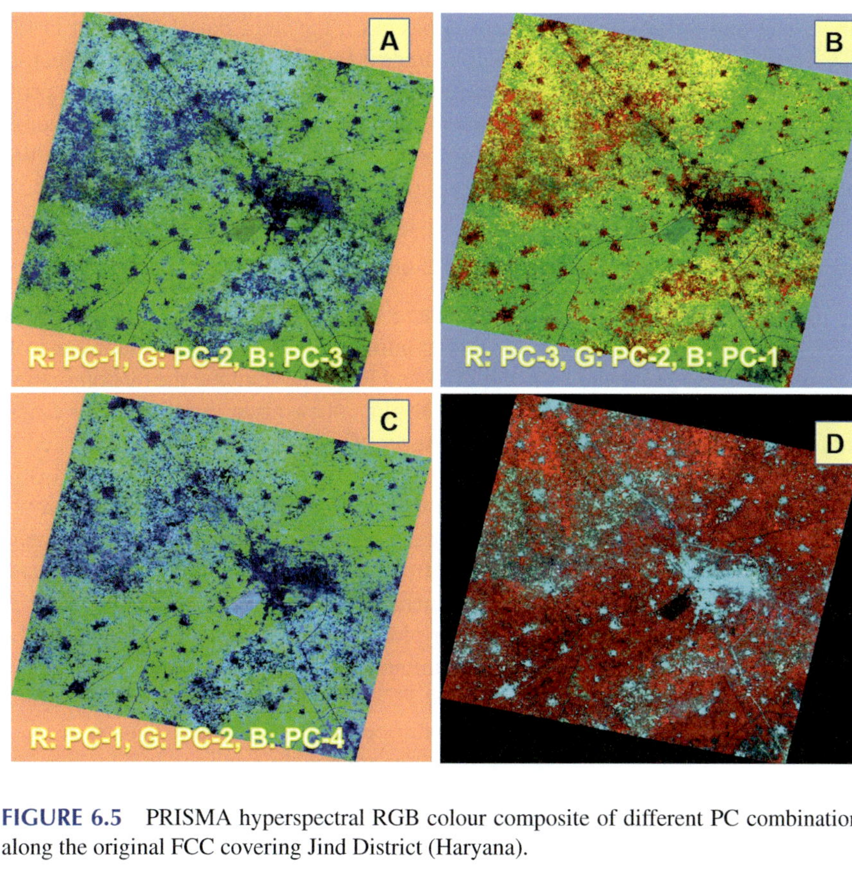

FIGURE 6.5 PRISMA hyperspectral RGB colour composite of different PC combinations along the original FCC covering Jind District (Haryana).

TABLE 6.5
The Variability Explained by the First Four Principal Components

PC No.	Min.	Max.	Mean	Std.Dev.	% Contri.
PC-1	−23761.88	11902.37	0.000035	7824.16	94.55
PC-2	−11758.15	7442.73	−000002	1751.16	4.74
PC-3	−4743.29	4943.72	0.00000	627.31	0.61
PC-4	−2203.18	2581.16	0.00000	263.53	0.10

in very good contrast and discrimination for two rice varieties as well as forest and other vegetation classes in Jind district. The visual observation of the RGB colour composite (R: PC-3, G: PC-2, and B: PC-1) image (B) indicates that the high-yielding varieties (HYV) observed in the northwestern portion and the south-east of the image are observed in prominent yellow colour whereas the aromatic Basmati rice varieties are observed in the central portion extending from northeast to southwest of the image in dark green. The rice varieties, other crops, forest, and habitation classes identified based on the field data collection in the FCC image (Figure 6.5 D). The results suggest that SPCA serves as an effective preprocessing method for hyperspectral image classification. The first three principal components collectively retain approximately 99.893% of the total spectral information, whereas the remaining components primarily capture noise or redundant content.

6.5.2 Rice Varieties Discrimination Using Principal Components

The results presented in Figure 6.4 indicated that the PC-1 and PC-2 account for 99.3% of the variance; therefore, the digital numbers of HYV and aromatic Basmati rice varieties were generated from the PC-1 and PC-2 for studying the discrimination of rice varieties and other crops. The scatter plots of two rice varieties, other crops, and forest in the two-dimensional spectral space of PC-1 and PC-2 were generated to analyse their spectral separability (Figure 6.6). The two-dimensional scatter plots in Figure 6.6 indicate that the clusters of rice varieties and other crops, and forest cover are quite distinct and well separated from each other. The two rice varieties, namely, HYV and Basmati, are also quite distinct from each other, and there are significant differences in rice varieties in the two-dimensional scatter plots of PC-1 and PC-2. The results of the two-dimensional scatter plots of PC-1 and PC-2 indicate that the two rice varieties can effectively be discriminated against using the PC-1 and PC-2 of the PRISMA hyperspectral data. In the PC-1 and PC-2 Scatter plots, there is a clear distinction between HYV and Basmati rice varieties. The distribution of PC-1 and PC-2 values indicates that the PC-1 values are negative and the PC-2 values are positive. The distribution of these values indicates that there is an inverse relationship between these two principal components.

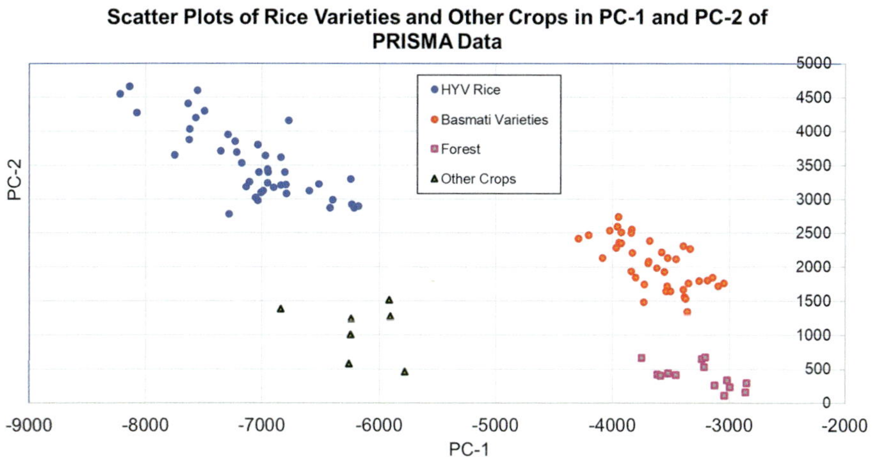

FIGURE 6.6 Scatter plots of PC-1 and PC-2 for rice varieties and other crops.

6.6 HYPERSPECTRAL VEGETATION INDICES FOR RICE VARIETY DISCRIMINATION

Hyperspectral remote sensing, characterised by narrow and contiguous spectral bands, is instrumental in assessing crop biophysical traits, forecasting yields, and determining the most informative spectral regions and acquisition times for crop discrimination. In classification tasks, spectral signatures of distinct surface features serve as end-members, enhancing the performance of supervised classification algorithms, particularly when integrated with machine learning techniques. Thenkabail et al. (2004) conducted a comprehensive evaluation of hyperspectral sensor data across the 400–2500 nm range for crop classification, applying data reduction and pattern recognition techniques such as principal component analysis, stepwise discriminant analysis, and greenness-based derivative vegetation indices. More recent findings by Sellami et al. (2022) highlighted the effectiveness of red-edge vegetation indices, such as the Chlorophyll Absorption Reflectance Index (CARI), Double Difference Index (DD), Red-Edge Inflection Point (REIP), and Chlorophyll Red-Edge Index (Clred-edge), in predicting crop yield and physiological responses. In the analysis of PRISMA hyperspectral imagery, a dominant methodological approach involves calculating VIs, which are widely recognised for their utility in estimating crop conditions from reflectance spectra. Given the fine spectral resolution of PRISMA data, a range of VIs has been explored for identifying rice crops and distinguishing between rice varieties.

6.7 PRISMA HYPERSPECTRAL INDICES GENERATION FOR RICE VARIETY DISCRIMINATION

Vegetation indices, which incorporate information from multiple spectral bands, can enhance the discrimination of various features (Carrasco et al., 2022). Vegetation indices, like enhanced vegetation index (EVI) and the land surface water index

TABLE 6.6
PRISMA Hyperspectral Indices Generated Using Visible and Near-infrared Wavelengths and Their Mathematical Formulae

Index	Formula	Reference
RVI: Ratio Vegetation Index NIR/RED	$\text{NIR}(860\,\text{nm})/\text{RED}(670\,\text{nm})$	Mukiibi et al. (2024)
Ratio: SWIR/Green	$\text{SWIR}\,1646\,\text{nm})/\text{Green}(569\,\text{nm})$	Thenkabail et al. (2013)
GNDVI: Green Normalised Difference Vegetation Index	$(\text{R}860-\text{R}550)/(\text{R}860+\text{R}550)$	Gitelson et. al. (1998)
Clgreen: Chlorophyll green Index	$(\text{R}730/\text{R}530)-1$	Gitelson et al. (2003)
PSRI: Plant Senescence Reflectance Index	$(\text{R}680-\text{R}500)/\text{R}752$	Merzlyak et al. (1999)
NDRE: Normalised Difference Red Edge Index	$(\text{RE}2-\text{RE}1)/(\text{RE}2+\text{RE}1)$ $(\text{RE}1=701\text{nm}),(\text{RE}2=752\text{nm})$	Lin et al. (2019)
RERI: Red Edge Rice Index	$2^{*}\text{NIR}(860\,\text{nm})+\text{Rededge}2(749\text{nm})$ $-3^{*}\text{Red}(670\,\text{nm})/1+\text{Red}(670\text{nm})$	Yiliang, et al. (2024)

(LSWI), were generated for extracting paddy rice crop area (Wei et al., 2022). Hyperspectral remote sensing provides notable improvements over traditional multi-spectral approaches, particularly in the accurate identification and differentiation of crop types and species (Thenkabail et al., 2013). In this study, various hyperspectral vegetation indices using high spectral resolution of PRISMA hyperspectral data were generated to discriminate between two types of rice varieties, aromatic Basmati rice and HYV in Jind District. The list of different indices attempted in this study, their formulae, and references are given in Table 6.6.

6.7.1 Rice Variety Discrimination Using Red and Near-infrared PRISMA Hyperspectral Bands

Thenkabail et al. (2000) identified a strong association between specific crop traits and distinct narrow spectral bands, particularly within the longer wavelengths of the red region (650–700 nm). Additional spectral groupings linked to crop characteristics were observed in the shorter green wavelengths (500–550 nm) and within a defined segment of the near-infrared spectrum (900–940 nm). To understand the relationship between PRISMA hyperspectral red and near infrared (NIR) bands for the discrimination of rice varieties and other crops, the spectral values in these two hyperspectral bands were generated. The spectral values of two rice varieties, namely HYV and aromatic Basmati rice, along with other crops and deciduous forests, were generated using the PRISMA hyperspectral data of 11 September 2024 covering parts of Jind district to study the rice variety discrimination. The aromatic Basmati rice varieties were at the flag leaf initiation to earhead emergence stage, whereas HYV was at the grain-filling to dough stage. There was a significant noticeable difference in the phenological stages of these two rice varieties.

The scatter plots of spectral signatures of two rice varieties along with other crops and forests in the PRISMA hyperspectral red (670 nm) and NIR (860 nm) bands are given in Figure 6.7. The scatter plots of spectral signatures of two rice varieties using PRISMA hyperspectral red (670 nm) and NIR (860 nm) indicate that these two varieties are quite distinct as well, and these rice varieties are also separable from the other crops and deciduous forests in the study area. Compared to other crop types, rice exhibits relatively lower reflectance in the SWIR region, likely due to the presence of standing water in paddy fields. It also shows reduced absorption in the red spectral range (630–680 nm) and increased reflectance between 1300 and 2400 nm when contrasted with perennial vegetation.

6.7.2 Relationship of NIR/RED and SWIR/Green Ratios

Vegetation indices derived from spectral reflectance in the red and near-infrared (NIR) regions, including ratio-based indices such as NIR/Red, are highly responsive to variations in crop development and physiological status. More recently, hyperspectral indices that utilise narrow spectral bands—particularly in the red-edge and NIR regions—have proven effective for distinguishing between crop species. A comprehensive evaluation of hyperspectral data spanning 400–2500 nm for agricultural classification was conducted using advanced analytical methods, including principal component analysis, λ–λ plots, stepwise discriminant analysis,

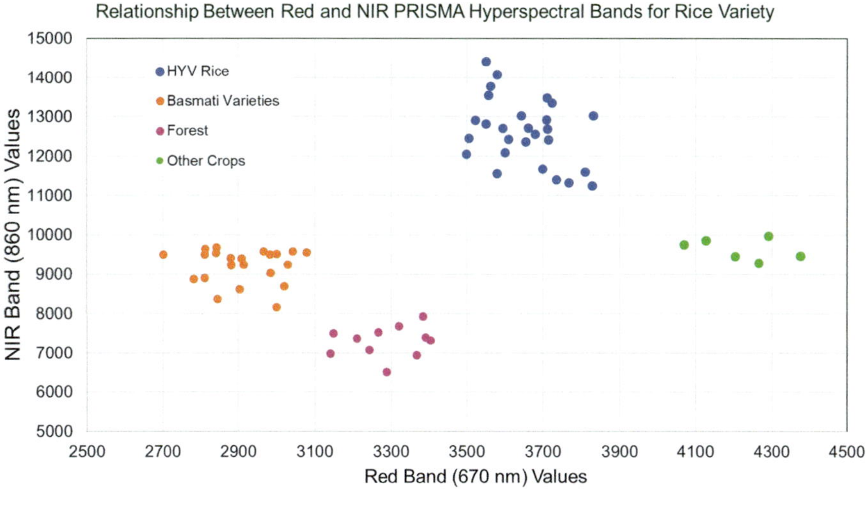

FIGURE 6.7 Relationship between red and near infrared (NIR) PRISMA hyperspectral bands for rice variety discrimination.

and derivative greenness indices (Thenkabail et al., 2004). The ratios of spectral reflectance in red (670 nm) and near infrared (NIR: 860 nm) wavelengths (NIR/RED) and green (569 nm) and SWIR (1646 nm) (SWIR/Green) PRISMA hyperspectral data were generated for two rice varieties namely HYV and Basmati and other crops and forest in the study area. The scatter plots of ratios of NIR/RED and SWIR/Green of two rice varieties, along with other crops and forests, generated using PRISMA hyperspectral data are given in Figure 6.8. The scatter plots of ratios of NIR/RED and SWIR/Green indicate that these two rice varieties are quite distinct from each other as well, and these rice varieties are separable from the other crops and deciduous forests.

6.7.3 Green Normalised Difference Vegetation (GNDVI)

Under favourable growth conditions, crops typically exhibit strong reflectance in the near-infrared range (760–900 nm), elevated reflectance in the green region (520–600 nm), and relatively low reflectance in the blue (450–520 nm) and red (630–690 nm) portions of the spectrum (Jain et al., 2007) where they absorb almost all of the incident light. This spectral reflectance data is used to compute various spectral vegetation indices, which are helpful in crop variety discrimination and crop condition assessment. The Green Normalised Difference Vegetation Index (GNDVI), introduced by Gitelson and Merzlyak (1998), utilises reflectance values from the NIR and green spectral regions to assess plant chlorophyll activity and photosynthetic efficiency. Compared to NDVI, GNDVI exhibits greater sensitivity to variations in chlorophyll content and demonstrates a higher threshold before signal saturation.

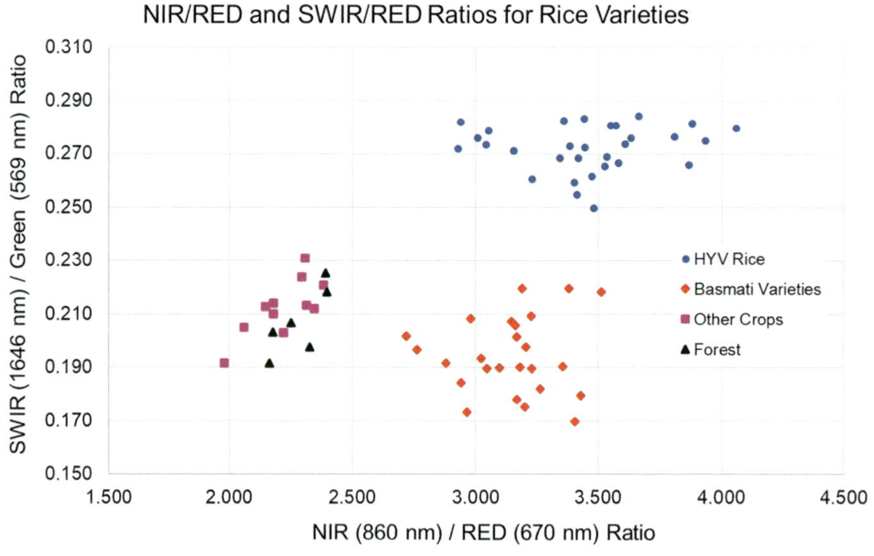

FIGURE 6.8 Relationship of NIR/RED and SWIR/green ratios of PRISMA hyperspectral bands for rice variety discrimination.

This makes it particularly effective for monitoring crops with dense foliage or during later stages of growth, whereas NDVI is generally more suitable for assessing crop vigor during early development.

6.7.4 Chlorophyll Green Index (ClGreen)

Spectral regions near the chlorophyll absorption peak around 700 nm and the green region near 550 nm have been found to be particularly responsive to variations in chlorophyll concentration. Recent research has shown that vegetation indices derived from narrowband reflectance measurements in these regions are strongly associated with leaf chlorophyll content (Gitelson and Merzlyak, 1994). According to Blackmer and Schepers (1995), reflectance in the green spectral region around 550 nm demonstrates a strong association with both leaf nitrogen levels and chlorophyll concentration. In the present study, GNDVI and Clgreen indices were generated using the PRISMA hyperspectral data for two rice varieties, namely Basmati and HYV, along with other crops and deciduous forest in the study area covering the parts of Jind district. The scatter plots of GNDVI and Clgreen indices given in Figure 6.9 were generated to study the discrimination of two rice varieties, other crops, and forest. The results indicated that Basmati and high-yielding rice varieties are well separable from each other. It also indicates that these two rice varieties are quite distinct from the other crops and deciduous forests.

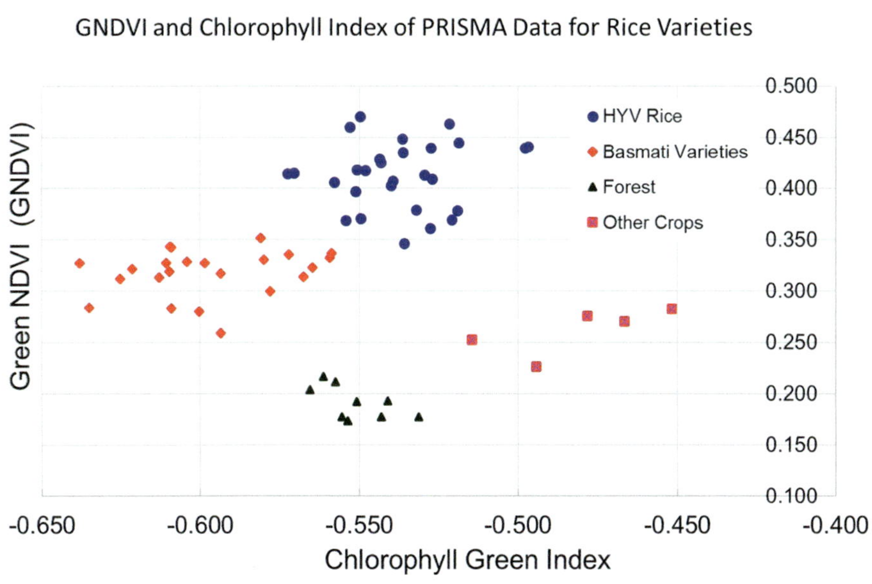

FIGURE 6.9 Scatter plots of GNDVI and chlorophyll green indices for rice varieties discrimination.

6.7.5 Normalised Difference Red Edge (NDRE) Index and Plant Senescence Reflectance Index (PSRI)

The red-edge–based vegetation indices NDRE705 and PSRI effectively capture temporal changes in vegetation development and senescence. These indices have demonstrated strong correlations with crop biophysical parameters throughout different growth stages (Jiang et al., 2021). Zhao et al. (2025) proposed a dual-index method that combines the Normalised Difference Red Edge Index (NDRE705) and the Plant Senescence Reflectance Index (PSRI), using time-series data from Sentinel-2 imagery. This integrated approach outperformed conventional single-index models by providing a more complete representation of crop development. NDRE705 demonstrated strong sensitivity to chlorophyll concentration, while PSRI effectively captured changes related to phenological stages and plant stress. Together, the indices enabled detailed monitoring of vegetation dynamics across the growing season. NDRE, an advancement over NDVI, offers improved responsiveness to chlorophyll variation, canopy condition, and early signs of senescence, making it particularly valuable for precision agriculture and crop stress detection. $NDRE_{705}$ (Lin et al., 2019) is calculated using the red-edge bands RE1 (B5) and RE2 (B6) of Sentinel-2. The red-edge-based vegetation indices NDRE705 and PSRI effectively represent the temporal progression of vegetation development and senescence. Their time-series patterns exhibit strong associations with key crop biophysical parameters, such as absorbed photosynthetically active radiation (APAR), canopy coverage, and LAI (Lin et al., 2019; Jiang et al., 2021). NDRE705 is particularly responsive to changes in chlorophyll concentration, minor fluctuations in canopy structure, and early signs of senescence. In contrast, PSRI is more sensitive to the ratio of carotenoids to chlorophyll, with elevated values reflecting heightened physiological stress and the onset of senescence in vegetation. The scatter plot of PSRI and NDRE (Figure 6.10) for two rice varieties along with other crops and forests indicates that the high-yielding varieties of rice and other crops show higher values of PSRI index, indicating that they are at the physiological maturity stages as compared to Basmati rice varieties. Based on the phenological stage differences between Basmati and high-yielding rice varieties, these two varieties are quite distinct, showing high separability in the PSRI and NDRI scatter plots.

6.8 HYPERSPECTRAL DATA CLASSIFICATION USING SPECTRAL ANGLE MAPPER (SAM)

Hyperspectral imagery is represented as a three-dimensional data cube, where the two spatial dimensions capture the spatial structure of the scene, and the third dimension records reflectance values across a dense set of contiguous spectral bands. The Spectral Angle Mapper (SAM) algorithm is a supervised classification technique widely used for hyperspectral and multispectral data. It evaluates the spectral similarity between image pixels and reference spectra by computing the angle between their corresponding spectral vectors in an n-dimensional feature space, where n equals the number of spectral bands. A smaller angle indicates a higher degree of spectral similarity. Mathematically, for small angles, the SAM metric approximates the

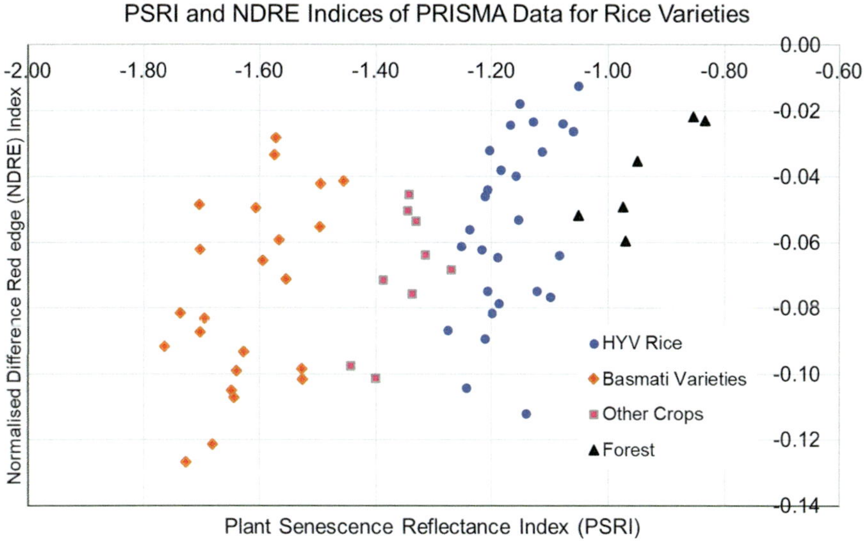

FIGURE 6.10 Scatter plots of PSRI and NDRE indices for rice varieties discrimination.

Euclidean distance, making it particularly effective in distinguishing materials with subtle spectral differences. Due to its robustness to illumination variations, SAM is especially useful in material identification and crop classification tasks (Kruse et al., 1993). The spectral angle θ is defined as:

$$\theta(x,y) = \cos^{-1}\left(\frac{\sum_{i=1}^{n} x_i y_i}{\left(\left(\sum_{i=1}^{n} x_i^2\right) * \left(\sum_{i=1}^{n} y_i^2\right)\right)^{\frac{1}{2}}}\right)$$

where $\boldsymbol{x}$ = spectral signature vector of an image pixel; $\boldsymbol{y}$ = spectral signature vector of a training area; $\boldsymbol{n}$ = number of image bands.

The SAM algorithm evaluates the similarity between spectral signatures by calculating the angle between a reference class spectrum and the spectral vector of each image pixel within an n-dimensional spectral space. The SAM calculates the angle between two spectra to identify their spectral similarity, and smaller angles between the two spectra indicate high similarity and vice versa. The solar illumination factors do not affect this classifier. Nigam et al. (2017) conducted a study across various regions in India to classify crops and soils at the field level and assess crop health using AVIRIS-NG hyperspectral imagery. The classification was carried out using a supervised approach based on the SAM algorithm. The findings demonstrated that AVIRIS-NG data provided high classification accuracy for distinguishing crop types, with SAM achieving approximately 86% accuracy in homogeneous agricultural areas and around 81% in more heterogeneous landscapes. In comparison, LISS-IV imagery of similar spatial resolution yielded significantly lower accuracy, below 40%, across both site types. Additionally, empirical models developed using hyperspectral vegetation indices enabled effective estimation of crop parameters, including plant nitrogen content. PRISMA hyperspectral imagery was utilised for crop type discrimination in selected regions of Punjab, employing the Spectral Hourglass Wizard (SHW) tool within the ENVI software environment (Lata et al., 2024). The SHW workflow involves a sequence of steps, including Minimum Noise Fraction (MNF) transformation, assessment of data dimensionality, computation of the Pixel Purity Index (PPI), extraction of spectral endmembers, and classification using the SAM algorithm.The spectral reflectance spectra of different crops like maize, sunflower, moong, sugarcane, and chili were derived for crop type separability analysis. The reflectance spectra in visible and near-infrared bands of each crop were generated, and the sensitive wavelength region for crop type discrimination was identified.

6.8.1 Rice Crop Classification Using Spectral Angle Mapper (SAM)

Spectral reflectance in the VNIR, red-edge, and NIR regions plays a critical role in differentiating rice varieties. Reflectance in the VNIR region is typically elevated during the initial stages of crop establishment and tends to decline after approximately 12 weeks. Patterns of increasing or decreasing reflectance across the VIS to red-edge (447–701 nm), red-edge to NIR (701–1027 nm), and NIR to SWIR (1027–1104 nm) spectral intervals can be indicative of different phenological phases in rice growth. A study was carried out to identify and classify paddy fields on Landsat-8 OLI data during the paddy growing season using the SAM method in Indonesia (Fathoni et al., 2021). The spectral reflectance patterns of paddy and other vegetation collected during field surveys were used as training signatures for classification. Pixels that have the closest angle to the threshold are classified as the same class as the target spectral, while angles outside the threshold are not classified as the same class as the reference spectral signature. The results indicated that paddy fields were accurately classified from non-paddy field areas with an overall classification accuracy of 70%.

6.8.2 Hyperspectral Data Classification for Rice Variety Discrimination Using SAM

While the SAM is a commonly applied supervised classification technique in hyperspectral remote sensing, its application in crop-type discrimination has been relatively limited. In vegetation mapping, the spectral angle derived by SAM can serve as an indirect indicator of plant health and density. The PRISMA hyperspectral narrow band data of 11 September 2024, during the *kharif* season, was acquired, covering parts of the Jind district in Haryana State. In this study, nine spectral bands corresponding to the Sentinel-2 MSI bands that are specifically useful for crop classification were selected using SPCA from the 240 hyperspectral bands of PRISMA. The spectral profiles of two rice varieties, other crops, and forest classes were generated to understand the spectral separability of these classes in the 240 hyperspectral bands of PRISMA data covering the area in Jind District (Figure 6.11).

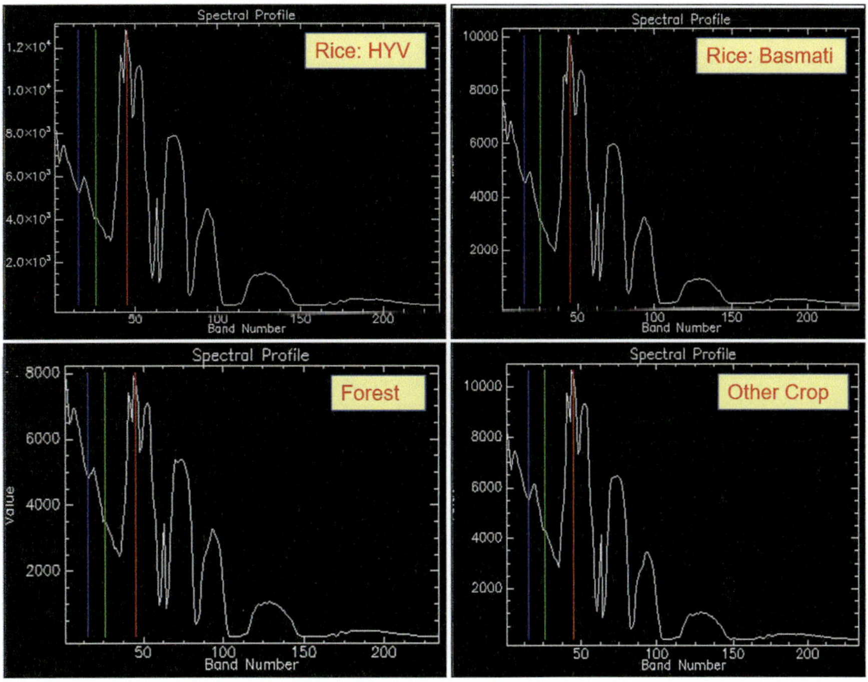

FIGURE 6.11 Spectral profiles of different classes in 234 hyperspectral channels of PRISMA data.

The SAM algorithm was used for rice classification and discrimination of two rice varieties and other crops. The false colour composite of the nine-channel PRISMA hyperspectral image is displayed to identify rice varieties and other classes. Based on the field information, seven reference spectra of different classes, namely, two rice varieties, HYV and Basmati, other crops, forest, fallow land, wasteland, and habitation, were generated from the nine-channel PRISMA hyperspectral image covering the Jind district area. These are also called spectral endmembers, which are used as a reference for classifying a hyperspectral image. After selecting spectral signatures of various classes, image classification was performed using the SAM algorithm, which is a supervised classification technique. One of the key strengths of the SAM algorithm is its reliance on the angular relationship between spectral vectors rather than their magnitude. This makes the method robust against variations in illumination and sensor gain, ensuring consistent classification regardless of changes in lighting conditions. The process followed for endmember selection of different classes for classification using SAM in ENVI software is given in Figure 6.12.

The image was classified using the SAM, a supervised classification approach that assigns each pixel to the most spectrally similar class by computing the angle between the pixel's spectral signature and those of predefined reference spectra (endmembers). During the supervised classification process, SAM identifies distinct classes within the image by evaluating spectral similarity, enabling differentiation based on

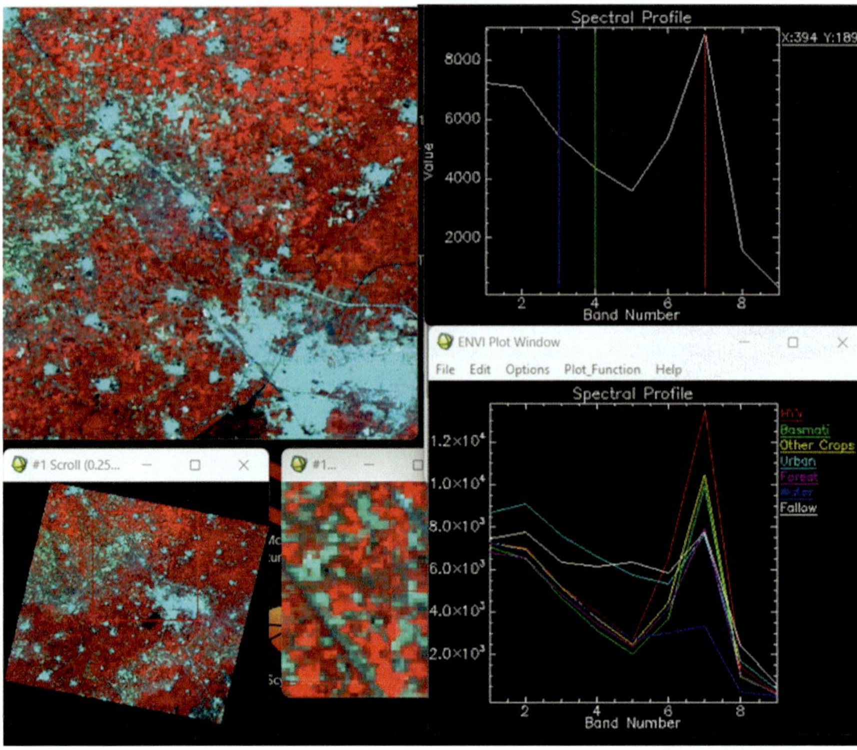

FIGURE 6.12 End-member selection for supervised classification using SAM for rice varieties and other classes.

angular measurements in multi-dimensional spectral space. The final classified image is a colour composite image with different colours as assigned during the spectral signature generation. The hyperspectral FCC image is shown along with the supervised classified image using the SAM classifier, with different classes for comparison given in Figure 6.13.

6.8.3 Validation of Hyperspectral Data Classification for Rice Variety Discrimination Using SAM Classifier

Fortnightly composite imagery from Sentinel-1 SAR and Sentinel-2 MSI sensors, covering the Jind district, was processed and classified using the RF algorithm within the Google Earth Engine (GEE) environment to delineate rice varieties like HYV and Basmati. The supervised classification workflow and associated processing steps are outlined in Chapter 4. From the classified output, the spatial subset corresponding to the PRISMA hyperspectral data footprint was extracted to enable direct comparison with the PRISMA–derived classification, which was performed using the SAM algorithm.

To evaluate the effectiveness of rice variety discrimination, the single-date PRISMA hyperspectral image classified via SAM was validated with the classified

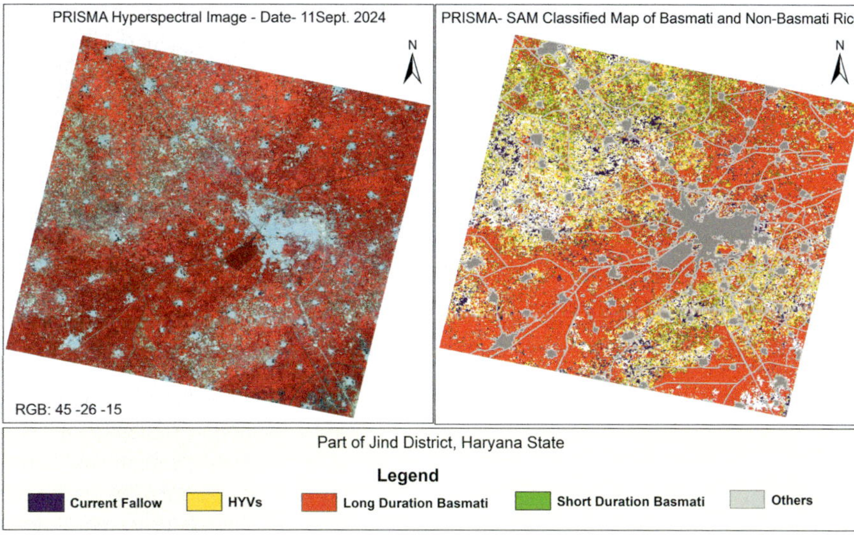

FIGURE 6.13 PRISMA hyperspectral FCC (a) and supervised classified image (b) using Spectral Angle Mapper for rice varieties and other classes.

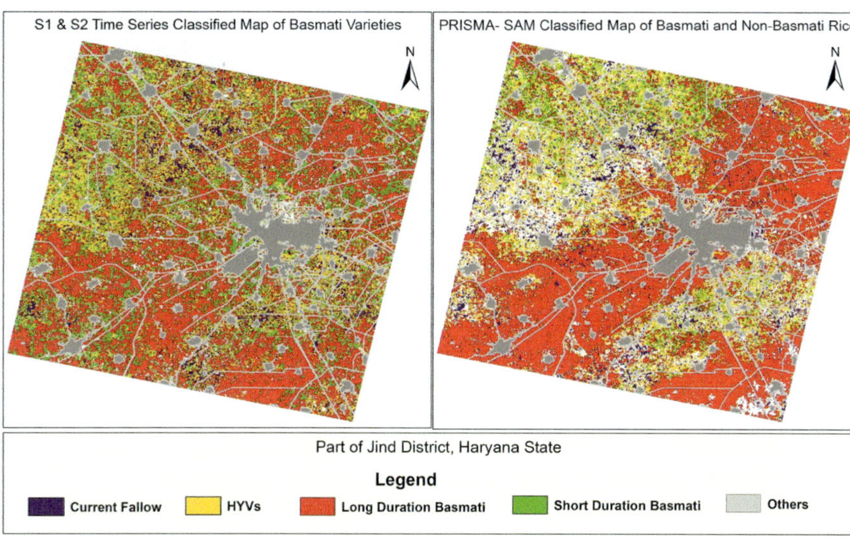

FIGURE 6.14 Comparison of PRISMA hyperspectral supervised classified image with time series S1 and S2 data using RF classifier and Spectral Angle Mapper for rice varieties discrimination.

output of multi-temporal Sentinel-1 and Sentinel-2 composite images, as shown in Figure 6.14. The Sentinel-1 and Sentinel-2 time-series data, with a finer spatial resolution of 10 m, captured crop-specific phenological dynamics, which are markedly different between HYV and Basmati due to variations in growth duration and physiological development. This temporal dimension enhanced the model's ability to distinguish not only between HYV and Basmati but also among three Basmati varietal segments, traditional, long-duration, and short-duration cultivars. In contrast, the classification based on the single-date PRISMA hyperspectral data (30 m resolution) was effective in separating HYV and Basmati rice only. However, it lacked sufficient spectral and temporal detail to resolve within Basmati varieties, limiting its ability to discriminate between traditional and evolved Basmati varieties. This separability could be improved with the availability of one or two more dates of PRISMA data during different phenological growth stages of rice varieties.

6.9 CONCLUSIONS

Hyperspectral data from the PRISMA sensor, acquired on 11 September 2024 over parts of the Jind district, Haryana, was used to assess its capability in identifying different rice varieties, including Basmati. To reduce dimensionality and enhance class separability, Selective Principal Component Analysis (SPCA) was applied to the full 240-band dataset. The SPCA selected a set of nine spectral bands corresponding to Sentinel-2 MSI wavelengths for their relevance for rice discrimination and classification. Principal Component Analysis (PCA) of the chosen bands indicated that the first three components accounted for 99.90% of the total spectral variability. Visualisation through PC-1 vs. PC-2 scatterplots highlighted distinct spectral signatures between high-yielding rice types and aromatic Basmati varieties.

To further characterise rice types, several narrow-band red-edge vegetation indices were computed, namely, the Ratio Vegetation Index (RVI), Green NDVI (GNDVI), Chlorophyll Index (Clgreen), Normalised Difference Red Edge Index (NDRE), and the Plant Senescence Reflectance Index (PSRI). These indices proved valuable in distinguishing between HYV and Basmati rice. Classification of the selected band subset was conducted using the Spectral Angle Mapper (SAM), a supervised method that utilises spectral similarity for categorisation. The classification separated HYV from Basmati. Accuracy assessment using multi-temporal Sentinel-1 and Sentinel-2 derived classification maps validated the findings. The analysis confirms that PRISMA's single-date hyperspectral imagery at 30 m spatial resolution enables effective discrimination between HYV and Basmati rice. However, additional spectral or temporal information may be needed to accurately distinguish between traditional and improved Basmati cultivars (evolved), which could not be separated with a single date of hyperspectral data acquisition.

REFERENCES

ASI. 2020. Prisma products specification document issue 2.3. Available online: https://prisma.asi.it/missionselect/docs/PRISMA%20

Atzberger, C., M. Guérif, F. Baret, W. Werner. 2010. Comparative analysis of three chemometric techniques for the spectroradiometric assessment of canopy chlorophyll content in winter wheat. *Computers and Electronics in Agriculture*, 73 (2010): 165–173.

Babar, M.A., M. P. Reynolds, M. van Ginkel, A. R. Klatt, W. R. Raun, and M. L. Stone. 2006. Spectral reflectance to estimate genetic variation for in-season biomass, leaf chlorophyll, and canopy temperature in wheat. *Crop Science* 46: 1046.

Bannari, A., K. Staenz, C. Champagne, and K. Khurshid. 2015. Spatial variability mapping of crop residue using Hyperion (EO-1) hyperspectral data. *Remote Sensing* 7: 8107–8127.

Bhojaraja, B., A. Shetty, M. K. Nagaraj, and P. Manju. 2015. Age-based classification of areca nut crops: A case study of Channagiri, Karnataka, India. *Geocarto International* 31(9): 995–1005.

Blackmer, T.M. and J.S. Schepers. 1995. Use of a chlorophyll meter to monitor nitrogen status and schedule fertigation for corn. *Journal of Production Agriculture*, 8(1), January-March 1995: 56–60. https://doi.org/10.2134/jpa1995.0056

Buschkamp, P., J. Hofmann, D. Rio-Fernandes, P. Haberler, M. Gerstmeier, C. Bartscher, S. Bianchi, P. Delpet, H. Weber, H. Strese, et al. 2023. CHIME's hyperspectral imager (HSI): Status of instrument design and performance at *PDR. International Conference on Space Optical Systems and Applications* 2023(12777): 1379–1398.

Carlson, T. N. and D. A. Ripley. 1997. On the relation between NDVI, fractional vegetation cover, and leaf area index. *Remote Sensing of Environment* 62: 241–252.

Carrasco, L., G. Fujita, K. Kito, and T. Miyashita. 2022. Historical mapping of rice fields in Japan using phenology and temporally aggregated landsat images in google earth engine. *ISPRS Journal of Photogrammetry and Remote Sensing* 191:277–289. https://doi.org/10.1016/j.isprsjprs.2022.07.018.

Carter, G.A. and A. K. Knapp. 2001. Leaf optical properties in higher plants: Linking spectral characteristics to stress and chlorophyll concentration. *American Journal of Botany* 88: 677–684.

Chen, J. M. 1996. Evaluation of vegetation indices and a modified simple ratio for boreal applications. *Canadian Journal of Remote Sensing* 22(3): 229–242. https://doi.org/10.1080/07038992.1996.10855178

Chiavetta, U., N. Camarretta, V. Garfì, M. Ottaviano, G. Chirici, M. Vizzarri, and M. Marchetti. 2016. Harmonized forest categories in central Italy. *Journal of Maps* 12: 98–100.

Chuvieco, E. 2016. *Fundamentals of Satellite Remote Sensing: An Environmental Approach* (2nd ed.). CRC Press, Taylor & Francis Group.

Cristiano, P., M. G. Posse, C. M. Di Bella, and F. R. Jaimes. 2010. Uncertainties in fPAR estimation of grass canopies under different stress situations and differences in architecture. *International Journal of Remote Sensing* 31: 4095–4109.

Din, M., W. Zheng, M. Rashid, S. Wang, and Z. Shi. 2017. Evaluating hyperspectral vegetation indices for leaf area index estimation of Oryza sativa L. at diverse phenological stages. *Frontiers in Plant Science* 8: 820.

EnMAP. 2022. Welcome to EnMAP: The German spaceborne imaging spectrometer mission. Available: https://www.enmap.org/

Fathoni, M. N., G. K. Anintika, D. Firda, P. Kricella, P. Widayani, S. A. Widayani, and M. Kamal. 2021. Identification of paddy fields based on temporal phase using Spectral Angle Mapper (SAM) on Landsat-8 OLI. *Proceedings of the SPIE 12082, Seventh Geoinformation Science Symposium*. 120820B. https://doi.org/10.1117/12.2619435

German Aerospace Center (formerly DLR) and Teledyne Brown. 2022. TCloud: Teledyne technologies. Available: http://tcloudhost.com/

Gitelson, A. A., Y. Gritz, and M. N. Merzlyak. 2003. Relationships between leaf chlorophyll content and spectral reflectance and algorithms for non-destructive chlorophyll assessment in higher plant leaves. *Journal of Plant Physiology* 160: 271–282.

Gitelson, A. A., G. P. Keydan, and M. N. Merzlyak. 2006. Three-band model for noninvasive estimation of chlorophyll, carotenoids, and anthocyanin contents in higher plant leaves. *Geophysical Research Letters*. 33(11). Doi:10.1029/2006GL026457

Gitelson, A. A. and M. N. Merzlyak. 1994. Spectral reflectance changes associated with autumn senescence of Aesculus hippocastanum L. and Acer platanoides L. leaves. Spectral features and relation to chlorophyll estimation. *Journal of Plant Physiology* 143: 286–292.

Gitelson, A. A., and M. N. Merzlyak. 1998. Remote sensing of chlorophyll concentration in higher plant leaves. *Advances in Space Research* 22: 689–692.

Gopinath, G., U. Surendran, J. Vishak, N. Sasidharan, and C. T. M. Fasil. 2024. Hyperspectral data and vegetative indices for paddy: A case study in Kerala, India. *Remote Sensing Applications: Society and Environment* 33: 101109.

Green, R. O., N. Mahowald, C. Ung, D. R. Thompson, L. Bator, M. Bennet, M. Bernas, N. Blackway, C. Bradley, and J. Cha. 2020. The Earth surface mineral dust source investigation: An Earth science imaging spectroscopy mission. In *Proceedings of the 2020 IEEE Aerospace Conference*, Big Sky, MT, USA. 1–15.

Haboudane, D. 2004. Hyperspectral vegetation indices and novel algorithms for predicting green LAI of crop canopies: Modeling and validation in the context of precision agriculture. *Remote Sensing of Environment (RSE)* 90: 337–352.

Haboudane, D.; J. R. Miller, N. Tremblay, P. J. Zarco-Tejada, L. Dextraze. 2002. Integrated narrow-band vegetation indices for prediction of crop chlorophyll content for application to precision agriculture. *Remote Sensing of Environment (RSE)* 81: 416–426

Harsanyi, J. C., and C. Chang. 1994. Hyperspectral image classification and dimensionality reduction: An orthogonal subspace projection approach. *IEEE Trans Geoscience and Remote Sensing* 32(4): 779–785.

Hayes, A. 2021. Multiple Linear Regression (MLR), 2021. Available online: https://www.investopedia.com/terms/m/mlr.asp

Heiden, U., M. Bachmann, K. Alonso, E. Carmona, D. Cerra, D. Dietrich, M. Langheinrich, R. de los Reyes, R. Mueller, N. Pinnel, and V. Ziel. 2019. DESIS imaging spectrometer data access and synergistic use with other ISS Earth observing instruments. In *Proceedings Workshop on International Cooperation in Spaceborne Imaging Spectroscopy*, held in Frascati, Italy, 9–11 July 2019, https://elib.dlr.de/129175/

Horler, D. N. H., M. Dockray, and J. Barber. 1983. The red edge of plant leaf reflectance. *International Journal of Remote Sensing* 1983(4): 273–288.

Huete, A., K. Didan, T. Miura, E. Rodriguez, X. Gao, and L. Ferreira. 2002. Overview of the radiometric and biophysical performance of the MODIS vegetation indices. *Remote Sensing of Environment (RSE)* 2002(83): 195–213.

Huete, A. A. 1988. Soil-adjusted vegetation index (SAVI). *Remote Sensing of Environment (RSE)* 1988(25): 295–309.

ISRO. 2017. ISRO develops optical imaging detector array for hyperspectral imaging applications. https://web.archive.org/web/20170809134705/https://www.isro.gov.in/isro-develops-optical-imaging-detector-array-hyperspectral-imaging-applications

Italian Space Agency(ASI). 2022. ASI data portal, 2022. Available: https://www.asi.it/en/earth-science/prisma/

Itiya, A., D. Foley, P. Thenkabail, A. Oliphant, and P. Teluguntla. 2022. New Generation Hyperspectral Data From DESIS compared to high spatial resolution planetscope data for crop type classification. *IEEE Journal of Selected Topics In Applied Earth Observations And Remote Sensing* 15: 2022.

Itiya, A. and P. S. Thenkabail. 2022. New Generation Hyperspectral Sensors DESIS and PRISMA Provide Improved Agricultural Crop Classifications. *Photogrammetric Engineering & Remote Sensing*, 88(11): 715–729, November 2022.

Jain, N., S. S. Ray, J. P. Singh, and S. Panigrahy. 2007. Use of hyperspectral data to assess the effects of different nitrogen applications on a potato crop. *Precision Agriculture* 2007(8): 225–239.

Jiang, X. Q., S. H. Fang, X. Huang, Y. H. Liu, and L. L. Guo. 2021. Rice mapping and growth monitoring based on time series GF-6 images and red-edge bands. *Remote Sensing* 2021(13): 579.

Kruse, F. A., A. B. Lefkoff, Y. J. Boardman, K. B. Heidebrecht, A. T. Shapiro, P. J. Barloon, and A. F.H. Goetz. 1993. The spectral image processing system (SIPS): Interactive visualization and analysis of imaging spectrometer data. *Remote Sensing of Environment* 44(2–3): 145–163.

Krutz, D. et al. 2019. The instrument design of the DLR earth sensing imaging spectrometer (DESIS)," *Sensors* 19, 2019, Art. no. 1622.

Lata, K., M. Arora, and N. Kaur. 2024. Spectral discrimination of crop types based on hyperspectral sensors. *Engineering Proceedings* 82(1): 107.

Lin, S. R., J. Li, Q. H. Liu, L. H. Li, J. Zhao, and W. T. Yu. 2019. Evaluating the effectiveness of using vegetation indices based on red-edge reflectance from Sentinel-2 to estimate gross primary productivity. *Remote Sensing* 2019(11): 1303.

Loizzo, R., C. R. Ananasso, E. Guarini, L. Lopinto, L. Candela, and A. Pisani. 2016. The PRISMA hyperspectral mission. In *Proceedings Workshop on Living Planet Symposium*, held in Prague, Czech Republic, 9–13 May 2019. https://lps16.esa.int/

Lu, B., P. D. Dao, J. Liu, Y. He, and J. Shang. 2020. Recent advances of hyperspectral imaging technology and applications in agriculture. *Remote Sensing*, 12(16), 2020, Art. no. 2659.

Main, R., M. A. Cho, R. Mathieu, M. M. O'Kennedy, A. Ramoelo, and S. Koch. 2011. An investigation into robust spectral indices for leaf chlorophyll estimation. *The ISPRS Journal of Photogrammetry and Remote Sensing* 2011(66): 751–761.

Merton, R. N. 1998. Monitoring community hysteresis using spectral shift analysis and the red-edge vegetation stress index. In *Proceedings of the Seventh Annual JPL Airborne Earth Science Workshop*, Pasadena, CA, USA, 12–16 January 1998; pp. 12–16.

Merzlyak, M. N., A. A. Gitelson, O. B. Chivkunova, and V. Y. Rakitin. 1999. Non-destructive optical detection of pigment changes during leaf senescence and fruit ripening. *Physiologia Plantarum* 1999(106): 135–141.

Mirzaei, S., S. Pascucci, M. F. Carfora, R. Casa, F. Rossi, F. Santini, A. Palombo, G. Laneve, and S. Pignatti. 2024. Early-season crop mapping by PRISMA images using machine/deep learning approaches: Italy and Iran test cases. *Remote Sensing* 16(13): 2431. https://doi.org/10.3390/rs16132431

Mishra, S. and D. R. Mishra. 2012. Normalized difference chlorophyll index: A novel model for remote estimation of chlorophyll-a concentration in turbid productive waters. *Remote Sensing of Environment*, 117: 394–406.

Morcillo-Pallarés, P., J. P. Rivera-Caicedo, S. Belda, C. De Grave, H. Burriel, J. Moreno, and J. Verrelst. 2019. Quantifying the robustness of vegetation indices through global sensitivity analysis of homogeneous and forest leaf-canopy radiative transfer models. *Remote Sensing* 2019(11): 2418

Mukiibi, A., A. T. B. Machakaire, A. C. Franke, and J. M. Steyn. 2024. A systematic review of vegetation indices for potato growth monitoring and tuber yield prediction from remote sensing. *Potato Research*, 2024, https://doi.org/10.1007/s11540-024-09748-7

Muller, R. et al. 2021. The spaceborne imaging spectrometer DESIS: Data access and scientific applications," in *Proceedings of the IEEE International Geoscience and Remote Sensing Symposium*, Brussels, Belgium, pp. 1–4.

Navalgund, R. R. and S. S. Ray. 2011. Hyperspectral data, analysis techniques and applications, *Indian Society of Remote Sensing* 2011: 156.

Nigam, R., K. Rajsi, and B. K. Bhattacharya. 2017. Crop discrimination using AVIRIS-NG data over homogeneous and heterogeneous agriculture area. Science Results from Phase – 1 Airborne Hyperspectral Campaign with AVIRIS-NG over India, Scientific Report, ISRO-JPL/AVIRIS-NG/SR01, July 2017.

Nizom, F., K. Amankulova, J. Szatmári, A. Sharifi, D. Abbasi-Moghadam, S. M. Mirhoseini Nejad, and L. Mucsi. 2023. Crop Type Classification by DESIS Hyperspectral Imagery and Machine Learning Algorithms.

Peñuelas, J., J. Pinol. R. Ogaya, and I. Filella. 1997. Estimation of plant water concentration by the reflectance Water Index WI (R900/R970). *International Journal of Remote Sensing* 1997(18): 2869–2875

Prabhakar, M., K. A. Gopinath, N. Ravi Kumar, M. Thirupathi, U. Sai Sravan, G. Srasvan Kumar, G. Samba Siva, P. Chandana, and V. K. Singh. 2024. Mapping leaf area index at various rice growth stages in Southern India using airborne hyperspectral remote sensing. *Remote Sensing* 2024, 16, 954. https://doi.org/10.3390/rs16060954

Prasad, B., B.F. Carver, M.L. Stone, M.A. Babar, W.R. Raun, and A.R. Klatt. 2007. Genetic analysis of indirect selection for winter wheat grain yield using spectral reflectance indices. *Crop Science* 2007(47): 1416.

Qian, S.-E. 2021. Hyperspectral Satellites, Evolution, and Development History. *IEEE Journal of Selected Topics In Applied Earth Observations and Remote Sensing*, 14, 2021

Rodarmel, C. and J. Shan. 2002. Principal component analysis for hyperspectral image classification. *Surveying and Land Information Systems* 62(2): 115–122.

Rondeaux, G., M. Steven, and F. Baret. 1996. Optimization of soil-adjusted vegetation indices. *Remote Sensing of Environment* 1996(55): 95–107.

Rouse, J.W., R. H. Hass, J. A. Schell, D. W. Deering, and J. C. Harlan. 1974. Monitoring the Vernal Advancement and Retrogradation (Green Wave Effect) of Natural Vegetation; Final Report, RSC 1978-4; Texas A&M University: College Station, TX, USA, 1974; pp. 1–120.

Sahoo, R. N., S. S. Ray, and K. R. Manjunath. 2015. Hyperspectral remote sensing of agriculture. *Current Science*, 108(5), 10 March 2015.

Sellami, M.H., R. Albrizio, M. Colovic, M. Hamze, V. Cantore, M. Todorovic, L. Piscitelli, and A.M. Stellacci. 2022. Selection of hyperspectral vegetation indices for monitoring yield and physiological response in sweet maize under different water and nitrogen availability. *Agronomy* 2022(12), 489. https://doi.org/10.3390/agronomy12020489.

Shwetank, A., K. Jain, and K. Bhatia. 2010. Hyperspectral Data Compression Model Using SPCA (Segmented Principal Component Analysis) and Classification of Rice Crop Varieties. In *Contemporary Computing. IC3 2010. Communications in Computer and Information Science* ed. Ranka, S., et al., vol 94. Berlin, Heidelberg: Springer, pp. 360–372. https://doi.org/10.1007/978-3-642-14834-7_34

Siljestrom, P. A., A. Moreno, K. Vikgren, and L. M. Caceres. 1997. Technical note: the application of selective principal components analysis (SPCA) to a thematic mapper (TM) image for the recognition of geomorphologic features configuration. *International Journal of Remote Sensing*, 18(18): 3843–3852. https://doi.org/10.1080/014311697216658

Spiller, D., L. Ansalone, F. Carotenuto, and P. P. Mathieu. 2021. Crop-type mapping using PRISMA hyperspectral images and one-dimensional convolutional neural network. In *Proceedings of the IEEE International Geoscience and Remote Sensing Symposium IGARSS*, Brussels, Belgium, 11–16 July 2021; pp. 8166–8169.

Thenkabail, P., I. Aneece, P. Teluguntla, and A. Oliphant. 2021. Hyperspectral narrowband data propel gigantic leap in the Earth remote sensing, highlight article. *Photogrammetric Engineering & Remote Sensing* 87(7): 461–467. [Online]. Available:https://www.asprs.org/a/publications/pers/2021journals/0721_July_Flipping_Public.pdf

Thenkabail, P. S., E. A. Enclona, M. S. Ashton, and Van Der Meer, B. 2004. Accuracy assessments of hyperspectral waveband performance for vegetation analysis applications. *Remote Sensing of Environment*, 91(3–4): 354–376.

Thenkabail, P.S., I. Mariotto, M. K. Gumma, E. M. Middleton, A. D. R. Landis and F. K. Huemmrich. 2013. Selection of hyperspectral narrowbands (HNBs) and composition of hyperspectral two-band vegetation indices (HVIs) for biophysical characterization and discrimination of crop types using field reflectance and Hyperion/EO-1 data, *IEEE Journal of Selected Topics in Applied Earth Observations and Remote Sensing*, 6(2): 427–438.

Thenkabail, P. S., R. B. Smith, and E. De Pauw. 2000. Hyperspectral vegetation indices and their relationships with agricultural crop characteristics. *Remote Sensing of Environment* 71(2), February 2000: 158–182.

Vogelmann, J. E., B. N. Rock, and D. M. Moss. 1993. Red edge spectral measurements from sugar maple leaves. *International Journal of Remote Sensing* 1993(14): 1563–1575. https://doi.org/10.1080/01431169308953986

Wang, L. and J. J. Qu. 2007. NMDI: A normalized multi-band drought index for monitoring soil and vegetation moisture with satellite remote sensing. *Geophysical Research Letters* 2007: 34.

Wei, J., Y. L. Cui, S. H. Zhou, and Y. F. Luo. 2022. Regional water-saving potential calculation method for paddy rice based on remote sensing. *Agricultural Water Management* 267: 107610. https://doi.org/10.1016/j.agwat. 2022.107610

Wu, C., Z. Niu, Q. Tang, and W. Huang. 2008. Estimating chlorophyll content from hyperspectral vegetation indices: Modeling and validation. *Agricultural and Forest Meteorology* 2008(148): 1230–1241.

Zhao, H., C. Chang, Z. Wang, and G. Zhao. 2025. A large-scale agricultural land classification method based on synergistic integration of time series red-edge vegetation index and phenological features. *Sensors* 2025(25): 503. https://doi.org/10.3390/s25020503

7 Traditional Methods of Crop Area and Yield Estimation in India

7.1 HISTORICAL EVOLUTION OF AGRICULTURAL STATISTICS

India has a deep-rooted tradition of maintaining agricultural records, with the earliest references found in *Arthashastra* by Kautilya. During the Mughal period, agricultural statistics were compiled primarily for revenue administration. A notable advancement occurred under Emperor Akbar (1556–1605), when Raja Todarmal introduced a systematic method of crop data collection. These developments were comprehensively documented in *Ain-i-Akbari*, authored by Abul Fazl in 1596–97. The formalisation of modern agricultural statistics began under British rule. In 1866, land revenue administration initiated a structured data collection. Crop forecasting commenced in 1884, initially focusing on wheat to support British trade interests. This marked the beginning of systematic record-keeping for agricultural production. By 1900, other major crops such as rice, oilseeds, cotton, jute, and sugarcane were incorporated into crop forecast surveys.

Recognising the need for institutional reform, the Royal Commission on Agriculture (RCA) was established in India during 1926. Among its key recommendations were the formation of the Imperial Council of Agricultural Research, later renamed as the Indian Council of Agricultural Research (ICAR) after independence (Nath, 2008). Post-independence, further modernisation of the statistical system was pursued. In 1949, the Technical Committee on Coordination of Agricultural Statistics (TCCAS) was formed to identify and rectify deficiencies in the system. One of the most significant innovations came from Prof. P.C. Mahalanobis, who introduced the grid sampling method for crop estimation—a statistically rigorous approach where sample units of defined area are selected randomly for data collection (Mahalanobis, 1944). Subsequently, several national institutions, including the Indian Agricultural Statistics Research Institute (IASRI), Central Statistical Organisation (CSO), National Sample Survey Organisation (NSSO), and Indian Statistical Institute (ISI) have played instrumental roles in enhancing the quality and methodology of agricultural statistics in India.

7.2 CONTEMPORARY FRAMEWORK OF CROP STATISTICS COLLECTION IN INDIA

Accurate and timely agricultural statistics are essential for informed policymaking related to procurement, food distribution, and international trade. However, existing

DOI: 10.1201/9781032659398-7

conventional methods, although useful for estimating aggregate crop production, are often inadequate for rapid, real-time decision-making. In India, state governments play a primary role in gathering data related to crop area and production. The estimation of crop yield is conducted through Crop Estimation Surveys (CES) implemented by the states. The NSSO of the Government of India supports the process by offering training, supervision, and document verification at various stages.

Based on their land revenue systems, states and union territories in India are grouped under the category of temporarily settled or Land Record States, permanently settled states (Non-Land Record States), and other States. Temporarily settled states, which comprise around 86% of India's geographical area, have well-established land record systems. Village-level revenue officials, known as *Patwaris* in northern states, are responsible for inspecting fields each season and recording crop area data. These records are based on cadastral maps and updated land registers. Crop area data is compiled at the village level and aggregated at the circle, *tehsil/taluka*, and district levels. The State Agricultural Statistics Authority (SASA) compiles district-level data and forwards it to the Directorate of Economics and Statistics (DES) under the Ministry of Agriculture and Farmers' Welfare, Government of India. The DES is responsible for generating the national-level crop estimates (Parihar and Oza, 2006).

Permanently settled states, such as Kerala, West Bengal, and Odisha, lack comprehensive land records. Here, area statistics are collected through a 20% sample of villages each year under the *Establishment of an Agency for Reporting of Agricultural Statistics (EARAS)* scheme. Sampling rotation is done in such a way that, in 5 years, all the villages are covered. This system has also been extended to the northeastern states, including Arunachal Pradesh, Nagaland, Sikkim, and Tripura, which collectively cover nearly 9% of the area reported. For crop area, a system of complete enumeration at the village level is employed in land record states, while non-land record states rely on rotational sampling. Crop area estimates are then aggregated at multiple administrative levels, village, *taluka/tehsil*, district, and state, which can result in considerable delays in the release of national crop forecasts (Parihar and Oza, 2006). For crop yield estimation, crop cutting experiments (CCEs) are employed. These are scientifically designed and statistically valid field experiments, forming the core of the country's crop production estimation system (Anon, 2005). Currently, crop statistics on area and production are compiled for 51 food crops and 15 non-food crops.

7.3 SCIENTIFIC ADVANCES IN CROP YIELD ESTIMATION

Historically, crop yield estimates in India were computed using the *Annawari* method, which involved multiplying the normal yield by a "condition factor" based on subjective visual assessment. While widely used, this approach lacked statistical rigor, was costly, time consuming, and prone to significant bias. To overcome these limitations, CCEs were developed as a scientifically grounded alternative. These involve harvesting and measuring yield from small, randomly selected plots within farmers' fields. The methodology was pioneered by Dr. P.V. Sukhatme of the Indian Agricultural Statistics Research Institute (IASRI) under ICAR and marked a critical innovation in yield estimation based on statistical sampling (Sukhatme et al.,

1984). Today, crop yield estimation in India is conducted under the General Crop Estimation Surveys (GCES) by the NSSO. These surveys use CCEs and cover 68 crops—including 52 food crops and 16 non-food crops—across different states and union territories. Conducted twice annually, GCES ensures seasonal coverage of both *kharif* and *rabi* crops. To improve the timeliness and reliability of agricultural production statistics, particularly for major crops, the Timely Reporting Scheme (TRS) was introduced during 1968–70. Under this scheme, approximately 20% of villages are surveyed using scientifically designed sampling frames. TRS operates across most Indian states and enables more responsive and data-driven agricultural policy decisions (Ghiya, 2013). This scientific evolution, from subjective estimations to statistically designed sampling, has greatly enhanced the reliability, timeliness, and credibility of India's crop statistics system.

7.3.1 Sampling Design with Quality Assurance in Yield Estimation

In India, crop yield assessment relies on CCEs, a systematic approach central to the GCES. The GCES employs a multistage stratified random sampling methodology to ensure statistical reliability and national representativeness (Aditya et al., 2020). The design structure comprises the following levels:

a. Strata: Districts, tehsils, or blocks
b. First-stage sampling units: Villages within strata
c. Second-stage sampling units: Agricultural fields within selected villages
d. Ultimate sampling units: Plots of a specified size within selected fields

This scientifically robust sampling design aims to estimate crop yields with a 2% sampling error at the national level. The number of CCEs per crop per district is determined based on the area proportion of the crop in different districts (Ahmed et al., 2024). Within each district, experiments are proportionally distributed among strata to ensure coverage across regions with varying cultivation intensity. The selection of villages for conducting CCEs is done randomly from the TRS. In each selected village, two CCEs are carried out on randomly chosen survey/sub-survey numbers. TRS is designed to facilitate priority enumeration of approximately 20% of villages, producing advanced estimates of crop area to enable timely government decision-making regarding procurement, distribution, and market interventions. To ensure data quality, the improvement in the Crop Statistics (ICS) scheme is implemented alongside GCES. The ICS provides sample checks at three levels:

a. Area enumeration
b. Area aggregation
c. Yield per unit area estimation

The scheme not only verifies data accuracy but also provides feedback to state agencies for capacity building and methodological improvement (Parihar and Oza, 2006). This layered framework ensures that India's agricultural statistics maintain both scientific integrity and policy relevance.

7.3.2 Design for Crop Cutting Experiments

Crop yield estimation in India follows a multistage stratified random sampling design, ensuring representative and statistically sound results. In this sampling framework, tehsils, taluks, or blocks function as strata; villages act as the primary sampling units; fields represent the second stage; and designated plots within fields serve as the ultimate sampling units. The number of crop-cutting experiments conducted in each district is determined in proportion to the crop area. Similarly, within each stratum, the experiment count is distributed based on the crop's cultivation area. From the TRS sample, random villages are chosen, and two crop cutting experiments are carried out in each selected village on separate, randomly selected survey plots (Nath, 2008). This methodology enhances the representativeness, objectivity, and precision of crop yield estimates at various administrative levels and is termed as the gold standard yield estimation method by FAO (Sud et al., 2017).

7.4 BASMATI RICE FIELD SURVEY

As Basmati rice is a key export commodity, periodic information on changing cropping patterns, farmers' varietal preferences, pest and disease management practices, and harvesting methods are essential. However, variety-specific data on Basmati is not captured in village revenue records or through satellite-based remote sensing. In this context, a well-designed crop survey using an appropriate sampling methodology emerges as the most viable approach to obtain such detailed and disaggregated information. Moreover, the exporter community requires advanced forecasts of Basmati rice production to effectively plan their procurement strategies.

The Basmati Export Development Foundation (BEDF), under the Agricultural and Processed Food Products Export Development Authority (APEDA), has been conducting Basmati crop surveys through third-party agencies in India since 2004. Between 2013 and 2016, a robust field survey methodology was employed for Basmati estimation in the states growing Basmati varieties. Similarly, in Pakistan, the Crop Reporting Service (CRS) of the Agriculture Department of Punjab, along with the Trade Development Authority of Pakistan (TDAP), conducts comprehensive field surveys on Basmati rice cultivation (Crop Reporting Services (CRS), 2021). These surveys focus on production practices, export potential, and value chain dynamics, providing valuable insights into the Basmati rice sector in their respective countries and offering a clear understanding of farmers' cultivation practices, market trends, and the overall industry outlook. A comprehensive field survey was conducted across the key Basmati rice-producing regions in India, encompassing the states of Punjab, Haryana, Western Uttar Pradesh, Uttarakhand, and Jammu and Kashmir. The purpose of the survey was to gather ground-level data on Basmati cultivation practices, varietal distribution, and crop condition during the season (Agri Net Solutions, 2013). During the field survey, a hand-held Garmin GPS device was employed for systematic field traversing, with two primary objectives: (i) to collect variety-specific acreage information directly from the farmer's survey, and (ii) to record geospatially referenced field-level data corresponding to the crop varieties being cultivated. In many cases, this process involved the acquisition of over 500 to 1,000 georeferenced data points. These location-specific records were subsequently utilised to train satellite imagery within image processing software, enabling the development of supervised classification models for varietal mapping. In subsequent years, the advancement of GPS–enabled smartphones and the integration of high-resolution cameras facilitated the collection of geotagged photographs. These photographs provided accurate latitude and longitude coordinates for specific crop fields, thereby enhancing the precision and efficiency of location-based data collection and varietal identification.

7.4.1 Crop Field Survey Report

To ensure consistency and data quality, a standardised Crop Survey Proforma was developed for filling crop details. This structured tool was used by survey teams to capture a wide range of information from the field, including sowing dates, irrigation methods, pest/disease incidence, and preferred Basmati varieties.

7.4.2 District-level Sample Survey

For sampling, 4–5 villages were selected from each prominent block within the Basmati-dominant districts of each state. The selection of villages was based on both geographical representation and known intensity of Basmati cultivation. The summary of surveyed villages and their distribution across states and districts is presented in **Table 7.1**.

7.4.3 Farmer-level Data Collection and Area Estimation Methodology

In each selected village, a few available farmers (Ghiya, 2013; Kaustav et al., 2020) were interviewed to gather detailed crop-level information. The survey teams recorded the total area transplanted under rice and other crops, along with the variety-wise area of Basmati rice cultivated by each individual farmer on a questionnaire form. A tactfully drafted questionnaire by a knowledgeable expert is a capable channel

TABLE 7.1
Number of Villages and Blocks in the District-level Basmati Survey

State	No. of Districts	Total No. of Blocks	No. of Villages Surveyed
Punjab	22	141	900
Haryana	21	115	800
Uttar Pradesh	30	260	600
Uttarakhand	04	27	70
Jammu & Kashmir	03	30	70
Himachal	02	20	30
Delhi	01	01	30

for eliciting the needed information (Singh and Mangat, 1996). This granular data collection helped in accurately mapping the varietal spread and cropping patterns at the village level. The collected data from individual farmers was subsequently aggregated at the block level. Using this, the proportion of Basmati rice area was calculated as a percentage of the total rice area in each block. In addition to Basmati rice, the area under non-Basmati rice was also estimated to provide a comprehensive view of rice cultivation trends in the region.

Further, the block-wise, variety-wise Basmati area percentages were converted into absolute area figures. These were then aggregated at the district level to derive district-wise estimates of area under each Basmati variety. This step ensured spatial consolidation of data, useful for both policy and trade-level planning. The surveyed villages included both villages covered in the previous year and newly selected villages, ensuring continuity in data trends while also expanding the coverage. Additionally, a Questionnaire-Based Farmer Survey was conducted to capture socio-economic and varietal preference data, along with insights into input usage and crop management practices. The sample size for this survey was determined based on regional representation and statistical reliability, covering:

- Punjab: 4,000 farmers
- Haryana: 3,000 farmers
- Western Uttar Pradesh: 3,000 farmers
- Uttarakhand: 300 farmers
- Jammu & Kashmir: 500 farmers

This large-scale farmer-level survey provided robust, representative data for varietal area estimation and crop characterisation across the Basmati-growing states.

BASMATI SURVEY – 2025

Date .... Latitude .... Longitude ....

Village .... Farmer's Name .... Phone No ....

Block .... Tehsil .... Dist. ....

Total Land Area........................ Acres Own: Leased:

Variety	Area (Acre)		Variety (Yield)	Area (Acre)	
	Current Year	Last Year		Current Year	Last Year
PR – 111/112			PB-1637		
PR-113/114			PB-1		
Pusa - 44			PB-1886		
PB-1509			PB-1401		
Pusa 1692			C S R – 30		
Pusa 1847			Sugarcane		
Pusa-1121			Fodder or any other millet crop		

Seed treatment: Yes/No
Source of seed: 1. Own 2. Public sector 3. Private

Crop Stage:
1) Booting Stage 2) Tillering 3) Flowering 4) Milking 5) Grain Filling 6) Maturing

Was the crop damaged due to flooding? If yes, which variety?

Is there any disease in the paddy? If yes, which?...

How much loss in percentage?...............

Specific...

Name & Signature of the Surveyor
Date:

REFERENCES

Aditya, Kaustav, Hukum Chandra, Pradip Basak, Vandita Kumari, and Shrila Das. 2020. District-level crop yield estimation with reduced number of crop cutting experiments. *The Indian Journal of Agricultural Sciences* 90(6): 1185–1194. https://doi.org/10.56093/ijas.v90i6.104798

Agri Net Solutions. 2013. *Report on Basmati Crop Survey conducted by Agrinet Solutions for BEDF/ APEDA, Govt. of India.*

Ahmed, T., S. P. Misra, M. Singh, and A. Biswas. 2024. Monograph on crop cutting experiment: Techniques for determination of yield rates of field crops. Monograph. ICAR-Indian Agricultural Research Institute. Monograph No. I.A.S.R.I./M-01/2024.

Anon, 2005. *Directorate of Economics and Statistics*, Department of Agriculture, Cooperation and Farmers Welfare, Govt. of India. https://eands.da.gov.in/

Crop Reporting Services (CRS). 2021. Dept. of Ag., Punjab, Pakistan 2021. https://crs-agripunjab.punjab.gov.pk/system/files/Bhatti-3-2-0026-jgss-31122021-main.pdf

Ghiya, A. 2013. The methodology used to provide advanced estimates of crop production for major domains in India. *International Journal of Agricultural and Statistical Sciences*, 9: 501–510.

Kaustav, A., H. Chandra, P. Basak, V. Kumari, and S. Das. 2020. District Level Crop Yield Estimation with Reduced number of crop cutting experiments. *Indian Journal of Agriculture and Allied Sciences* 90(6): 1185–1189.

Mahalanobis, P. C. 1944. On large-scale sample surveys. *Philosophical Transactions of the Royal Society of London. Series B, Biological Sciences*, 231(584): 329–451, (Oct. 31, 1944).

Nath, S.K. 2008. *Manual on Area and Crop Production Statistics*. New Delhi: Central Statistical Organisation (CSO), *CSO-M-AG-01*. https://www.mospi.gov.in

Parihar, J. S. and M. P. Oza. 2006. FASAL: An integrated approach for crop assessment and production forecasting. Agriculture and Hydrology Applications of Remote Sensing, In *Proc. of SPIE*, eds, R. J. Kuligowski, J. S. Parihar, and G. Saito Vol. 6411, 641101. 0277-786X/06/$15 https://doi.org/10.1117/12.713157

Singh, R. and N. P. S. Mangat. 1996. *Elements of Survey Sampling*. New York: Springer-Verlag.

Sud, U. C., T. Ahmad, V. K. Gupta, H. Chandra, P. M. Sahoo, K. Aditya, M. Singh, and A. Biswas. 2017. Crop1&2: Measuring Crop Area and Yield under Pure Stand, Mixed and Continuous Cropping. Working Paper No. 16. Global Strategy, FAO Publication, FAO, Rome. https://gsars.org/wp-content/uploads/2017/01/WP-16.01.2017-Findings-from-the-Field-Tests-Conducted-inThree-Countries.pdf

Sukhatme P. V., B. V. Sukhatme, S. Sukhatme, and C. Asok. 1984. *Sampling Theory of Surveys with Applications*, 3rd Edition, USA: Iowa State University Press.

8 Global Trade in Aromatic Basmati Rice

8.1 INTRODUCTION

Aromatic rice varieties such as Basmati and Jasmine are globally prized for their unique sensory attributes, including a delicate, nutty aroma and distinctively fluffy texture upon cooking. These qualities command a premium price in international agricultural markets, positioning aromatic rice as a high-value commodity. Among these, Basmati rice, grown in a distinct geographical region in India and Pakistan, holds a special place due to its long-standing cultural heritage, genetic distinctiveness, and its specific agro-climatic requirements (APEDA, 2025; Islam et al., 2024). Basmati rice exhibits distinct physicochemical and sensory traits, including kernel elongation upon cooking (typically more than double its original length), a soft and non-sticky texture, and the presence of aromatic compounds such as 2-acetyl-1-pyrroline, which contribute to its characteristic fragrance and flavour profile. These characteristics make Basmati unique among the consumers and traders of long-grained rice, popularly known as *the scented pearl*.

India leads globally in both the cultivation and export of Basmati rice. Its annual output typically ranges between 8 to 12 million tonnes, with approximately 65–70% of the total production going to international markets, while the rest is utilised for domestic consumption (J M Baxi Group, 2021). Historically, Basmati was cultivated mainly on small-scale farms using traditional methods. However, the post-independence period in India and Pakistan witnessed significant expansion and modernisation of irrigation infrastructure during the 1950s and 1960s, enabling Basmati rice to transition into large-scale commercial cultivation. The green revolution, which started during the late 1960s to early 1970s, further enhanced productivity and expanded Basmati acreage in both countries. In parallel, trade liberalisation policies in the 1990s opened international markets, significantly boosting Basmati exports and establishing it as a globally recognised commodity. Currently, India exports Basmati rice to over 100 countries, with the Middle East, Europe, and the United States being the key markets (Singh and Singh 2009; Adhikari et al. 2016).

Pakistan's Basmati production is similarly concentrated, with approximately 95% grown in the Punjab province (which together with the present Indian states of Punjab, Haryana, and Himachal Pradesh was a part of the combined Punjab Province during the British period), particularly in the fertile the conventional growing region located between the Ravi and Chenab river basins. Due to this eco-geographical affinity, this region of Pakistan shares its agro-climatic characteristics with the Indian Basmati belt, contributing to comparable grain quality and aroma (Ashfaq et al., 2015). In Pakistan, Basmati rice contributes approximately 3.1% to the overall value

DOI: 10.1201/9781032659398-8

addition in the agriculture sector and represents around 0.6% of the country's Gross Domestic Product (GDP) (TDAP, 2022; FAOSTAT, 2025). Pakistan's Basmati rice commands strong demand internationally, notably in the Middle East, the United States, and Europe, where it competes closely with the Indian export.

Economically, Basmati rice fetches prices nearly three times higher than non-aromatic high-quality rice varieties (Mahajan et al., 2018), reflecting its premium position in both domestic and international markets. This price premium incentivises farmers and exporters in both countries to prioritise Basmati cultivation, despite challenges such as water scarcity, climate variability, and evolving trade trends. The growing global demand for aromatic rice is underpinned by changing consumer preferences towards premium and specialty rice products, along with increasing diaspora populations who value traditional rice varieties (OECD, 2021).

However, sustaining and expanding Basmati rice production requires addressing critical issues related to water resource management, yield stability, and quality maintenance under changing environmental and economic conditions. The unique terroir-based qualities of Basmati rice also underscore the importance of geographic indications in protecting the economic value and authenticity of this prized commodity in an increasingly competitive global market.

Basmati and non-Basmati rice represent the largest agricultural export commodities from India, accounting for 21.4% of the country's total agricultural exports during 2020–21 (Ministry of Commerce and Industry, 2020–21). In the 2021–22 agricultural year, India produced an estimated 130.29 million tonnes of rice, including Basmati, from approximately 46 million ha, ranking it as the world's second-largest producer and contributing over 11% of global rice output. Rice exports accounted for more than 19% of India's total agricultural export value during the same period. In the year 2022–23, out of total 22.1 million metric tonnes (MT) rice export from India, 23% (4.6 million MT) was Basmati rice, which increased to 6.07 million MT of Basmati (30.06%) in 2024–25, out of total rice export of 20.19 million MT rice.

8.2 BASMATI RICE PRODUCTION IN INDIA

In India, Basmati rice cultivation is concentrated in the northwestern states of Jammu and Kashmir, Himachal Pradesh, Punjab, Haryana, Delhi, and Western Uttar Pradesh. The unique quality traits of Basmati rice, long grains, superior aroma, and elongation on cooking are inherently linked to the agro-ecological conditions of the northwestern Himalayan foothills and the Indo-Gangetic Plains. This geographic specificity has led to Basmati rice being granted Geographical Indication (GI) status in 2016, which legally protects its distinct identity and origin within the Indian subcontinent (Murai et al. 2017).

Udhayakumar et al. (2021) studied the Basmati rice area, production, and productivity in the growing states of Punjab, Haryana, Uttar Pradesh, and others (Jammu & Kashmir, Himachal Pradesh, and Uttarakhand) from the triennium ending (TE) 1998 to TE 2018. They observed that the area under Basmati during the study period had doubled, from 0.73 million ha in TE 1998 to 1.56 million ha in TE 2018. The production increased four times, from 1.22 million MT in TE 1998 to 5.61 million MT in

TE 2018. The productivity also doubled, from 1.69 MT/ha in TE 1998 to 3.58 MT/ha in TE 2018. The rate of growth during the 20 years was found to be 6.01% in Basmati rice area, 10.55% in production, and 4.28% in productivity (Udhayakumar et al., 2021). When the study period was split into two timeframes, a higher growth rate in production and productivity was observed during recent years (2010–11 to 2018–19), compared to earlier years (1995–96 to 2009–10).

The long-term rate of growth in Basmati rice acreage, production, and productivity has also been studied, which also showed a similar trend. The compound growth rate (CGR) of Basmati rice area, production, and productivity was calculated from the year 1988–89 to 2017–18 (Tiwari et al., 2024). CGR in the Basmati rice area during the 30 years was found to be 5.45%, while it was 9.40% in production, and 3.75% in productivity, indicating the productivity of Basmati has outpaced the area increase in India (Kamboj et al., 2021).

8.3 DOMESTIC CONSUMPTION OF BASMATI IN INDIA

Basmati rice was initially considered a rich person's rice in India. Chandrasekaran (2020) has stated that Basmati rice had moved to the wealthiest kitchen from the royal kitchen even prior to the independence of India. However, as the Basmati production in the country increased over the years, domestic consumption also increased. The increase in disposable income of India's middle class and branding by traders in recent years have increased the consumption of Basmati rice in the domestic market significantly.

Reliable published data on the annual domestic consumption of Basmati rice in India is not available. Estimates from different trade-related survey organisations have shown the domestic consumption to vary between 25% to 33%. A recent trade report (J M Baxi Group, 2021) details the Indian domestic consumption to be one-third (33%) of total Basmati production, and about two-thirds being exported. Similar views have also been expressed by Joshi and Kumar (2024b). However, another report from the rating agency ICRA (2024) assumes the domestic Basmati consumption in India to be about 25%.

8.4 AROMATIC BASMATI RICE EXPORT FROM INDIA

8.4.1 Basmati Rice Trade During the Early Years

During the early post-independence years, India exported Basmati rice only in limited quantities, as most of the Indian Basmati rice production was consumed in the domestic market. It has been reported that in 1955, the Basmati rice export merely exceeded 100,000 tonnes annually, mainly for the consumption of Indians living abroad (Parker, 1970, quoted in Chandrasekaran, 2020). In Indian Punjab (which then included the present states of Haryana and Himachal Pradesh also), Basmati rice formed a quite substantial proportion of the total rice production in the year 1957. Farmers' interests were also kept protected, and in case of price sagging, the Central Government was prepared to buy Basmati paddy (unhusked grain) from the farmers in order to protect the supply of Basmati rice to the wealthy consumers living in the commercial metropolis of Calcutta (now Kolkata), Bombay (now Mumbai), and other cities (Chandrasekaran,

2020). In order to control admixture of other varieties, a Basmati rice standard specification was also prepared by the Ministry of Finance, in consultation with technical experts in 1958. However, the government never considered exporting Basmati rice as an option to earn foreign exchange in those days, probably considering the adverse psychological effect of export on the internal market (Chandrasekaran, 2020).

In the Green Revolution phase, when India became self-sufficient in rice during the early 1970s, the government took initiative in increasing area under Basmati cultivation, and its export through the State Trading Corporation (an Indian Government Enterprise). Basmati rice export increased from 14,879 metric tonnes (MT) in 1972–73 to 68,000 MT in 1974–75 (Chandrasekaran, 2020). However, the quantum of export continued wavering, as is evident from the reply of the Indian Minister of State for Commerce in the Parliament on 11 August 1978, stating that Basmati rice production of the country was 250,000 MT, and export does not exceed 42,000 MT, which does not disturb internal prices and food security (quoted in Chandrasekaran, 2020).

During the 1980s, the Indian government introduced several measures to enhance agricultural trade, one of which was the formation of the Agricultural and Processed Food Products Export Development Authority (APEDA) in 1985. APEDA actively promotes rice exports through coordinated collaborations with various stakeholders along the value chain, facilitating India's position as a leading exporter. To promote Basmati export, a specialised agency, the Basmati Export Development Foundation (BEDF) has been created as a registered society by APEDA, under the Ministry of Commerce and Industry. BEDF plays a pivotal role in enhancing the productivity and quality of Basmati rice by overseeing research initiatives, promotional activities, dissemination of advanced technologies, and infrastructure development related to Basmati cultivation (Siddiq et al., 2012; APEDA, 2017). The Directorate General of Foreign Trade (DGFT) has accredited the BEDF as a designated laboratory for the molecular characterisation and varietal authentication of Basmati rice. In line with this, samples collected by customs authorities for verifying varietal purity through DNA fingerprinting are routinely submitted to BEDF for analytical testing. BEDF also has facilities for measurements of pesticide residue and heavy metals in the rice grain, and draws out samples from farmers' fields periodically for pesticide residue analysis. It also produces breeder and foundation seeds of Basmati rice varieties and provides demonstrations of Basmati varieties on the farmers' fields.

Trade promotion initiatives of the Indian government during the late 1970s and early 1980s increased the Basmati rice production and export significantly, and Basmati export reported as a mere 42,000 MT in 1977–78, exceeded 0.247 million MT during 1985–86, and reached the level of 0.442 million MT during 1994–95. Likewise, the unit export value of Basmati rice export registered a positive growth rate of 13.74% per annum. The higher positive growth rate was mainly due to an increase in export unit value driven by higher demand (Udhayakumar and Karunakaran, 2020).

8.4.2 Export During the Post-GATT/WTO Period

Post–GATT, and the creation of WTO in 1995, Basmati exports from India increased further, mainly because of the liberalisation policy of the government in 1991. During

1987–88, India's aromatic Basmati rice exports were modest at 0.36 million MT, valued at USD 224.65 million. However, since 2007–08, Basmati rice exports from India have witnessed consistent growth. Despite fluctuating growth in global rice trade over time, the dynamic and complex trade networks evolved, resulting in significantly improved network efficiency (Chen and Zhao, 2023). With its strong trade network, India has emerged as the largest and core among the rice exporting countries, not only in Basmati rice, but also in total rice export, including non-Basmati rice. In the year 2022–23, out of a total of 22.1 million MT of rice export from India, 23% (4.6 million MT) was Basmati rice, which increased to 5.94 million MT of Basmati (29.7%) in 2024–25, out of a total rice export of 19.9 million MT. Statistics on Indian Basmati rice export (quantity and export value) from the year 1990 to 2024 have been presented in Table 8.1. A graphical illustration of annual Basmati export (quantity and its value) from India between 1987–88 to 2024–25 has been presented in Figure 8.1. India's Basmati rice production has shown steady growth, driven by increasing buffer stock availability and rising global demand. The export market for Basmati rice from India is projected to expand from INR 495 billion in 2024 to INR 537 billion by 2033 at a compound Annual Growth rate (CAGR) of 0.9% (IMARC Group, 2025).

TABLE 8.1

Export of Basmati Rice from India (quantity in million MT) and Export Value (million US$) from 1990 to 2024

S. No.	Year	Qty (in Million MT)	Value (in Million USD)
1	1990–91	0.22	154.05
2	1991–92	0.24	177.79
3	1992–93	0.29	228.34
4	1993–94	0.53	338.36
5	1994–95	0.44	275.61
6	1995–96	0.37	254.31
7	1996–97	0.52	351.45
8	1997–98	0.59	453.45
9	1998–99	0.60	446.14
10	1999–2000	0.64	410.88
11	2000–01	0.85	474.16
12	2001–02	0.67	391.87
13	2002–03	0.71	425.53
14	2003–04	0.77	442.8
15	2004–05	1.16	629.35
16	2005–06	1.17	687.39
17	2006–07	1.05	619.52
18	2007–08	1.18	1073.8
19	2008–09	1.56	2060.67
20	2009–10	2.02	2297.29

(*Continued*)

TABLE 8.1 (Continued)

S. No.	Year	Qty (in Million MT)	Value (in Million USD)
21	2010–11	2.37	2491.14
22	2011–12	3.18	3222.31
23	2012–13	3.46	3564.04
24	2013–14	3.76	4866.3
25	2014–15	3.70	4518.11
26	2015–16	4.05	3477.98
27	2016–17	3.99	3216.59
28	2017–18	4.06	4169.48
29	2018–19	4.41	4722.52
30	2019–20	4.45	4330.69
31	2020–21	4.63	4018.71
32	2021–22	3.95	3540.4
33	2022–23	4.56	4787.5
34	2023–24	5.24	5837.12
35	2024–25	6.07	5944.48

Source: APEDA

8.4.3 Aromatic Basmati Rice Export Destinations from India

India exports Basmati rice to approximately 132 countries worldwide, underscoring its global reach (APEDA, 2025). India's Basmati rice exports are predominantly directed to Middle Eastern countries such as Iran, Saudi Arabia, Iraq, Kuwait, and the United Arab Emirates. Notably, during 2023–2024, India exported Basmati rice to 106 countries, with Iran and Saudi Arabia alone accounting for over 50% of these exports. A detailed list of 106 countries importing Basmati rice from India, with its quantity, value, and percentage share by country, is provided in Table 8.2.

India is the world's leading rice exporter, driven by robust international demand for its rice varieties. Globally, around 15 to 18 million tonnes of rice are traded annually, with Basmati rice comprising less than 10% of this volume. Despite its smaller share, Basmati rice commands significantly higher returns, approximately three times those of non-Basmati rice, in both domestic and international markets. The export composition of rice from India has fluctuated, with Basmati rice contributing between 30% and 40% of the total rice exports, while non-Basmati rice constitutes the remaining 60–70%.

8.4.4 Growth Trend in Basmati Export from India

In 2019, the international market for Basmati rice was estimated at around USD 10.29 billion. It is expected to expand to approximately USD 18.32 billion by 2026, reflecting a projected compound annual growth rate (CAGR) of 8.5% during the period from 2021 to 2026 (Joshi and Kumar, 2024a).

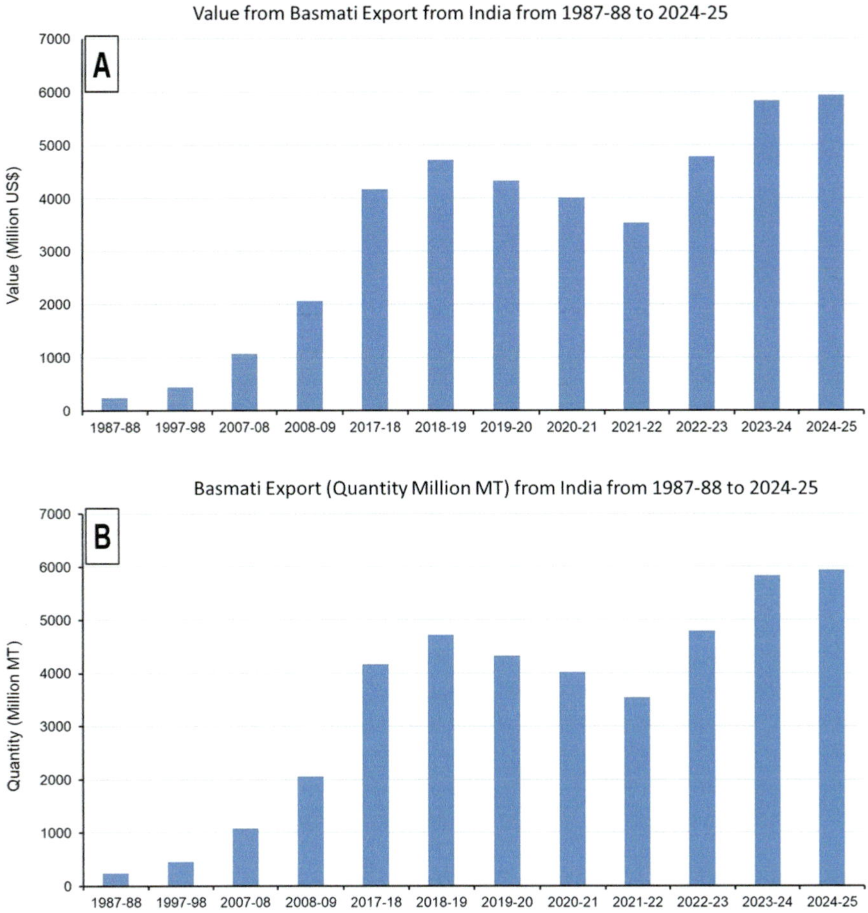

FIGURE 8.1 Value of Basmati export (A) and quantity of Basmati export from India from 1987–88 to 2024–25.

Basmati rice export data from India, spanning 2010–2024 and sourced from APEDA, Agri Exchange, and trade reports, reveal a strong upward trajectory. This translates to an average export price of around USD 978.5 per metric ton. A detailed growth trend analysis shows a consistent rise in both export value and volume. Specifically, between April 2024 and March 2025, Basmati rice exports increased by 1.73% in value, reaching USD 5.94 billion compared to the previous year. Correspondingly, the export volume grew from 4.68 million MT in financial year 2022–2023 to 6.07 million MT by the end of financial year 2024–2025, reflecting an average annual increase of about 15%. This sustained growth underscores India's dominant position in the global Basmati rice market.

TABLE 8.2
Basmati Rice Export to Different Countries from India (Quantity, Value, and % Share) 2024–2025

Sr. No.	Country	Quantity (MT)	Value (Million USD)	Share (%)
1	Saudi Arabia	11,73,833.31	1,203.67	19.35
2	Iraq	9,05,598.90	850.08	14.93
3	Iran	8,55,132.87	753.2	14.10
4	Yemen	3,92,180.74	358.34	6.47
5	United Arab Emirates	3,89,148.26	364.55	6.42
6	United States	2,74,213.14	337.1	4.52
7	United Kingdom	1,80,951.39	190.93	2.98
8	Kuwait	1,75,000.19	180.39	2.89
9	Oman	1,49,148.11	144.73	2.46
10	Qatar	1,24,557.34	122.93	2.05
11	Israel	99,665.12	100.53	1.64
12	Jordan	1,01,853.82	100.52	1.68
13	Canada	83,496.54	94.92	1.38
14	Netherlands	76,710.21	78.19	1.26
15	Australia	59,544.63	72.47	0.98
16	Egypt	76,940.38	72.04	1.27
17	Malaysia	78,939.63	71.42	1.30
18	Nepal	68,967.73	58.1	1.14
19	Kenya	68,365.99	56.53	1.13
20	Turkey	55,700.89	50.18	0.92
21	Baharain	52,395.20	50.16	0.86
22	Algeria	48,934.90	45.14	0.81
23	Italy	36,934.48	38.59	0.61
24	Mauritius	31,507.42	34.4	0.52
25	South Africa	28,712.47	31.24	0.47
26	Somalia	24,458.73	22.83	0.40
27	Syria	23,583.13	22.12	0.39
28	Belgium	21,826.59	21.32	0.36
29	Singapore	18,598.44	20.4	0.31
30	Germany	16,355.66	20.2	0.27
31	Russia	21,313.52	19.94	0.35
32	Spain	20,961.21	19.55	0.35
33	Lebanon	17,909.49	17.63	0.30
34	Djibouti	18,937.22	17.48	0.31
35	Nigeria	14,557.58	16.23	0.24
36	Bhutan	19,102.14	15.82	0.31
37	Libya	15,790.11	15.71	0.26
38	Portugal	13,345.26	12.69	0.22

(Continued)

TABLE 8.2 (Continued)

Sr. No.	Country	Quantity (MT)	Value (Million USD)	Share (%)
39	Mauritania	12,248.26	12.08	0.20
40	New Zealand	11,783.10	11.91	0.19
41	Tanzania	12,325.08	11.78	0.20
42	Reunion	9,256.91	11.39	0.15
43	Indonesia	11,174.89	11.26	0.18
44	Sweden	10,036.51	10.94	0.17
45	France	10,168.87	10.87	0.17
46	Sri Lanka	9,159.37	8.7	0.15
47	Comoros	7,739.17	8.52	0.13
48	Azerbaijan	6,491.89	7.36	0.11
49	Japan	5,038.36	7.05	0.08
50	Georgia	7,008.44	6.62	0.12
51	China	5,833.87	6.39	0.10
52	Morocco	6,068.33	6.32	0.10
53	Seychelles	5,901.94	6.15	0.10
54	Ghana	6,311.00	5.87	0.10
55	Maldives	4,998.49	5.79	0.08
56	Togo	5,429.04	5.4	0.09
57	Benin	5,816.97	4.96	0.10
58	Poland	5,227.56	4.91	0.09
59	Thailand	4,251.20	4.75	0.07
60	Romania	4,280.83	4.24	0.07
61	Angola	3,280.72	3.69	0.05
62	Ireland	2,710.38	3.43	0.04
63	Sudan	3,836.32	3.42	0.06
64	Greece	3,303.33	3.31	0.05
65	Philippines	2,656.10	2.94	0.04
66	Ethiopia	3,116.41	2.67	0.05
67	Cyprus	2,159.58	2.64	0.04
68	Tunisia	2,765.64	2.64	0.05
69	Mozambique	2,285.02	2.5	0.04
70	Norway	1,867.08	2.41	0.03
71	Jamaica	1,636.54	2.17	0.03
72	Bangladesh	1,715.04	2.05	0.03
73	Denmark	1,621.20	1.91	0.03
74	Ukraine	1,847.79	1.81	0.03
75	Cote D Ivoire	1,892.18	1.75	0.03
76	Hong Kong (HK)	1,246.57	1.63	0.02
77	Cameroon	1,452.49	1.62	0.02
78	Senegal	1,343.51	1.48	0.02
79	Pakistan	1,625.10	1.39	0.03

(Continued)

TABLE 8.2 (Continued)

Sr. No.	Country	Quantity (MT)	Value (Million USD)	Share (%)
80	Malta	1,104.23	1.36	0.02
81	Fiji	1,131.77	1.28	0.02
82	Mexico	924.22	1.09	0.02
83	Vietnam	1,083.72	1.07	0.02
84	Brunei	886.3	1.02	0.01
85	Korea, South	951.84	0.98	0.02
86	Slovenia	599.98	0.92	0.01
87	Cambodia	737.85	0.81	0.01
88	Switzerland	573.83	0.78	0.01
89	Kazakhstan	851.87	0.76	0.01
90	Trinidad	494.58	0.64	0.01
91	Guinea	604.55	0.6	0.01
92	Uzbekistan	664.57	0.58	0.01
93	Armenia	513.01	0.57	0.01
94	Liberia	648.71	0.56	0.01
95	Finland	429.61	0.51	0.01
96	Gambia	474.27	0.51	0.01
97	Madagascar	412.56	0.48	0.01
98	Botswana	438.94	0.47	0.01
99	Chile	385.6	0.47	0.01
100	Haiti	424.67	0.42	0.01
101	Lithuania	364.75	0.38	0.01
102	Guadeloupe	362.25	0.37	0.01
103	Congo Democratic Republic (Congo)	318.08	0.36	0.01
104	Congo P Rep	333.36	0.35	0.01
105	Tajikistan	355.67	0.33	0.01
106	Mali	305.5	0.32	0.01

Source: APEDA, Ministry of Commerce and Industry, Government of India

8.4.5 Major Aromatic Basmati Rice Exporting States of India

Among the Basmati-producing states, Punjab, often referred to as the "Rice Bowl of India," stands as the major contributor to the country's rice production, contributing close to 27% of the total Basmati production. In the fiscal year 2024–2025, India retained its status as the world's leading exporter of Basmati rice, exporting 6.07 million MT, valued at USD 5.94 billion globally. Punjab alone contributed around

40% of the country's total Basmati rice exports. Other key Basmati-exporting states include Haryana, Uttar Pradesh, Uttarakhand, and Himachal Pradesh.

8.4.6 Determinants of Basmati Exports/Factors Influencing Export from India

Key factors influencing the export potential of food commodities include variations in comparative advantage, spatial and political proximity between trading partners, and the extent of existing trade restrictions (Kumar et al., 2008). Adhikari et al. (2016) conducted an empirical analysis to identify the determinants of rice exports from India by applying multiple regression techniques to a time series dataset spanning 33 years (1980/81 to 2012/13). The export volume was modelled as a function of variables such as export price, exchange rate, global market price, domestic consumption levels, and lagged production figures. Export price and international price were found to be significantly influencing the export (at 1% level of significance), while lagged production and domestic consumption were significant at 5% level. The exchange rate was found to be non-significant. Apart from these, government policies also sometimes affect exports. For example, in August 2023, Govt. of India fixed a Minimum Export Price (MEP) of USD 1200 per ton on Basmati rice. This acted as a trade barrier and subsequently, in October 2023, the government had to reduce the MEP to USD 900 per ton (ICRA, 2024).

A comprehensive study analysing long-term data from 1980 to 2022 (sourced from the Economic Survey of India and the Ministry of Commerce) analysed the impact of policy interventions and macroeconomic variables on the export performance of India's Basmati rice sector (Joshi and Kumar, 2024b). The findings revealed that production volumes, export prices, and exchange rates significantly impact the total value of Basmati exports. A 1% increase in export volume or in export price leads to an 0.82% increase in export value in the following year, whereas a 1% appreciation in the domestic currency leads to a 0.64% increase in export value (Joshi and Kumar, 2024b). Domestic inflation raises the price of Basmati rice, reduces domestic consumption, and makes more volume available for export. According to their study, a 1% increase in inflation increases export volume by 0.22%. Increased trade openness has shown an adverse effect on Basmati rice exports. A 1% increase in the terms of trade index has been shown to reduce the value of exports by 1.0%.

8.5 BASMATI EXPORT FROM PAKISTAN

8.5.1 Basmati Area, Production, and Yield in Punjab Province of Pakistan

In Pakistan, the majority of Basmati rice cultivation is concentrated in the traditional 'kallar tract' region of Punjab province, renowned for its unique agro-climatic conditions ideal for aromatic rice production. The principal Basmati varieties grown in this region include Basmati-385, Basmati-198, Basmati Pak, Punjab Basmati, and Super Basmati. The area under Basmati rice cultivation is about 2.03 million ha. The production in 2024–25 hovered around 4.3 million tonnes (Crop Reporting

Services—Punjab, Pakistan, 2025). The annual domestic consumption is 2.4 million tonnes. During recent years, the Basmati rice area and production have been increasing. This increase is driven by rising international demand for Pakistani Basmati rice, which in turn generates higher income for both farmers and exporters. The change in cultivated area, production, and productivity of aromatic Basmati rice in Pakistan from 2009–10 to 2024–25 is detailed in Table 8.3.

8.5.2 Basmati Export from Pakistan

Basmati rice represents a vital component of Pakistan's agricultural export portfolio, with its contribution estimated at around 3.1% of the agricultural value addition and approximately 0.6% of the national GDP. As the world's second-largest exporter of Basmati rice, Pakistan primarily supplies key markets in the Middle East, North America, and Europe. Traditionally, Basmati rice accounts for around 6% of Pakistan's total annual export earnings, making it the country's second most important agricultural export commodity after cotton. The top importers of Pakistani Basmati rice include UAE, Kazakhstan, Saudi Arabia, Oman, Kenya, and Somalia (TDAP, 2022).

The present level of Basmati rice export from Pakistan should be seen from a historical perspective since the early post-independence years. In the year 1955, the

TABLE 8.3
Area, Production, and Productivity of Basmati Rice in Pakistan (2009–10 to 2024–25)

Sr. No.	Year	Area (000′ ha)	Yield (tonnes/ha)	Production (000′tonnes)
1	2009–10	1414.57	1.75	2475
2	2010–11	1334.41	1.77	2365
3	2011–12	1121.46	1.68	1889
4	2012–13	995.55	1.77	1758
5	2013–14	1193.12	1.72	2057
6	2014–15	1320.65	1.77	2337
7	2015–16	1254.66	1.82	2279
8	2016–17	1353.44	1.86	2524
9	2017–18	1417.00	1.99	2817
10	2018–19	1473.68	2.00	2949
11	2019–20	1662.75	1.95	3244
12	2020–21	1872.47	2.14	4014
13	2021–22	1772.47	2.06	3650
14	2022–23	1679.35	1.89	3168
15	2023–24	2114.9	2.24	4736
16	2024–25	2029.1	2.12	4296

Source: Crop Reporting Services, Punjab, Pakistan (2025).

export of Basmati from Pakistan was a mere 629 tonnes (Chandrasekaran, 2020). In response to the sharp rise in global oil prices during the early 1970s and the consequent escalation in rice demand across Middle Eastern nations, Pakistan significantly expanded its Basmati rice exports to the region. By 1972–73, export earnings from Basmati rice reached Rs. 112.9 crores, marking a threefold increase compared to the export revenues recorded in 1958–59.

In 1980, the export value of Pakistani rice (Basmati and non-Basmati together) was modest, at just USD 385 million, but by 2009–10 this figure had surged to USD 2.2 billion. The share of Basmati in this export value was USD 866 million, slightly less than half, with the quantity crossing 1 million MT (Akhter and Haider, 2020). The significant expansion in Basmati rice production and trade can be linked to the release of high-yielding cultivars such as Basmati 385, the liberalisation and growing involvement of the private sector beginning in 1988, and the accelerated growth of export activities post-1992. However, the export growth could not keep the earlier pace, and from 2012–13 onwards, the export started declining (Figure 8.2).

In 2024–25, Pakistan's Basmati rice exports reached a record high of 1 million metric tonnes, reflecting an 18% increase. This surge has been fuelled by strong demand from key markets such as Saudi Arabia, Iraq, and the UAE. Other significant export destinations include Oman, Yemen, China, the United Kingdom, Kazakhstan, the United States, and Kenya.

8.6 COMPARATIVE ANALYSIS OF AROMATIC BASMATI RICE EXPORTS FROM INDIA AND PAKISTAN

India and Pakistan are the foremost producers of Basmati rice, collectively accounting for approximately 90% of the global exports of this premium aromatic variety. In the fiscal year 2024–25, Pakistan exported about 1.0 million metric tonnes of

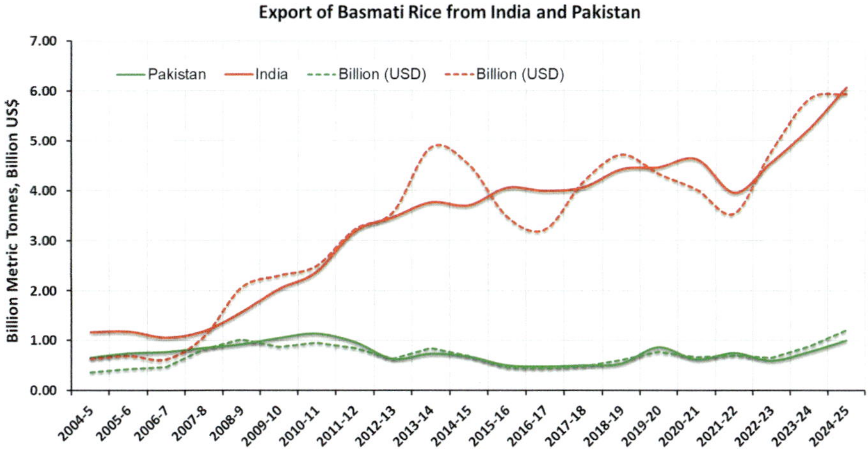

FIGURE 8.2 Aromatic Basmati Rice Exports from India and Pakistan from 2001–02 to 2024–25.

TABLE 8.4
Comparison of Basmati Rice Export from India and Pakistan from 2019–20 to 2024–25

	Quantity (million MT)		Value (millions USD)	
Year	**India**	**Pakistan**	**India**	**Pakistan**
2019–20	4.45	0.87	4330.69	783.28
2020–21	4.63	0.62	4018.71	569.49
2021–22	3.95	0.75	3540.40	693.32
2022–23	4.56	0.60	4787.50	650.42
2023–24	5.24	0.77	5837.12	876.94
2024–25	5.94	1.0	5944.48	1200.00

Source: Pakistan: Rice Exporters Association of Pakistan (www.reap.com.pk/exporthistory), India: APEDA (AgriXChange, 2025)

Basmati rice to key markets, including the UAE, Saudi Arabia, Oman, Yemen, China, the UK, Kazakhstan, the USA, and Kenya, at an average export price of around USD 900 per ton. Both countries dominate the export of this highly prized long-grain rice, renowned for its distinctive aroma and quality, to major destinations such as Iran, Iraq, Yemen, Saudi Arabia, the United Arab Emirates, and the United States. A detailed comparative analysis of aromatic Basmati rice exports from India and Pakistan over the last five years (2019–20 to 2024–25) is provided in Table 8.4.

Both India and Pakistan have experienced substantial growth in their Basmati rice exports from 2000 to 2024, with India consistently leading in terms of export volume. In 2024, India exported approximately 6.07 million MT of Basmati rice, marking a 16% increase compared to the previous year. Pakistan's Basmati rice exports also surged during the same period, stood at around 1.0 million MT. As illustrated in Figure 8.2, there is a significant disparity between the export volumes of India and Pakistan, with India maintaining a clear dominance. Indian Basmati exports have followed a steady upward trajectory since the early 2000s, culminating in a peak export volume of 6.07 million MT in 2024–25. Meanwhile, Pakistan's Basmati rice exports have also shown notable growth in recent years, driven largely by competitive pricing strategies and increased production capacity.

8.6.1 Major Factors Influencing Basmati Rice Export from India and Pakistan

Basmati rice exporters from India and Pakistan face multiple challenges that affect export stability and growth. Key issues include volatile global market prices influenced by changing demand, currency fluctuations, and competition from other rice-exporting nations (FAO, 2022b). Regulatory factors, including export tariffs, quality standards, and trade policies, further complicate market access and competitiveness (APEDA, 2023; Pakistan Ministry of Commerce, 2024). Government interventions such as export promotion schemes, infrastructure improvements, and research support play a crucial role in sustaining export growth (World Bank, 2021).

The major determinants of Basmati exports include domestic production levels, export prices, exchange rates, and global demand trends. A recent study covering 1980–2022 found these factors significantly influence export values, while trade openness poses both opportunities and risks (Joshi and Kumar, 2023). Over the past 16–17 years, India's export growth has also been facilitated by proactive marketing efforts, enhancements in logistics and storage systems, and the wider adoption of grain drying technologies. In Pakistan, elevated inflation levels and currency depreciation have been found to exert a significant negative impact on the volume of Basmati rice exports (Javed et al., 2020). There is a need for targeted monetary and fiscal measures in the country to curb inflationary pressures and to improve exports.

***Export restrictions by India in 2024*:** In response to India's decision on the removal of minimum export price (MEP) in mid-October 2024, Pakistan also announced the removal of its MEP for all rice varieties, a measure that had been in place since 2023. Pakistan's MEP had set prices at USD 1,300 per metric ton for Basmati rice and USD 550 for non-Basmati rice. This policy was originally implemented to capitalise on the higher demand and limited supply caused by India's previous export restrictions. Rice trade later intensified as India and Pakistan removed export restrictions (Trade diggie, 2024). Despite challenges, the rising demand for premium aromatic rice in markets like the Middle East, North America, and Europe continues to drive growth. Coordinated efforts are needed to enhance price stability, climate resilience, regulatory harmonisation, and innovation in production and marketing. Such strategies will ensure consistent supply, high quality, and competitiveness in the evolving international marketplace.

***Influence of export price in competition between India and Pakistan*:** There being only two market players in Basmati export, a price comparison between the two countries during 2017 to 2021 has been made by the Trade Development Authority of Pakistan (TDAP, 2022). In 2018, the per unit export price of both countries was comparable, with India at USD 1080/ton and Pakistan at USD 1093/ton. In 2019, Pakistan decreased its price to USD 911/ton, while India's price was USD 1050/ton. Pakistan's export value increased to USD 771 million in 2019, in comparison to its USD 613 million in 2018. In 2020, India reduced its per unit export price to USD 870 per ton, which significantly undermined Pakistan's competitiveness in the Basmati rice export market (TDAP, 2022). Pakistani exporters, in order to compete with India's Basmati price, reduced their unit price to USD 898/ton. In 2021, India further decreased the per unit price to USD 864/ton, while Pakistan's price was 874 USD/ton.

***Common importing markets for both India and Pakistan*:** Market competition between India and Pakistan also exists because of common importing countries, like the UAE and Saudi Arabia (TDAP, 2022). India exported Basmati to Saudi Arabia worth USD 1 billion in 2020, whereas Pakistan exported Basmati worth USD 74 million. India's exports to the UAE in

2020 were worth USD 234 million, whereas Pakistan's exports were worth USD 126 million. Pakistan's share in the markets of Kazakhstan, Kenya, Somalia, and Italy is increasing, whereas India seems to have capitalised on the markets of Oman, Yemen, the UK, Netherlands, Australia, and Canada (TDAP, 2022).

8.7 FUTURE SCENARIO IN BASMATI TRADE FROM BOTH INDIA AND PAKISTAN

Basmati rice cultivation in both India and Pakistan is increasingly under stress from environmentally unsustainable practices. In Pakistan's Punjab province, the core of its Basmati-growing region, the Indus Basin Irrigation System (IBIS) is nearing its operational limits. A detailed water balance study covering the period 1981–2012 revealed a decline in canal water deliveries across many commands, while actual crop water consumption (evapotranspiration) continued to rise (Ahmad et al., 2021). This divergence underscores a growing dependence on groundwater, leading to falling water tables, particularly alarming during the dry rabi season presently, which may also affect the water availability during the Basmati season (*kharif*) as well in the future.

Likewise, India's Punjab, Haryana, and western Uttar Pradesh states, which are central to the Basmati rice production, also face severe groundwater depletion. The causes include high water demand from rice cultivation, unregulated extraction facilitated by free or subsidised electricity, and limited enforcement of sustainable water use policies. As in Pakistan, groundwater over-extraction not only threatens the environmental viability of Basmati production but also endangers the economic stability derived from export earnings (Husain et al., 2023; API, 2020).

The major Basmati rice-growing regions in India and Pakistan are situated within the Ganges and Indus River basins, both of which are heavily reliant on consistent water availability to sustain rice cultivation. Climate models project an overall increase in precipitation across these basins; however, the implications vary significantly by region. In the Ganges basin, there is reported scope for expanding reservoir capacity, which could partially buffer against seasonal variability in water supply. Furthermore, the basin shows only a limited dependence on glacier melt. Nevertheless, growing population pressure and increasing water demand, coupled with uncertainty in future precipitation patterns, place the Ganges basin at an intermediate level of water scarcity risk (Immerzeel and Bierkens, 2012). In contrast, the Indus basin, home to Pakistan's Basmati heartland, faces much higher vulnerability. This stems from an overreliance on unpredictable glacier melt, severe groundwater depletion, rapid population growth, and an uncertain precipitation regime.

Given the centrality of water to Basmati production, the future of this iconic export commodity is inextricably linked to transboundary water governance, climate variability, and local resource management. Both countries must prioritise sustainable agricultural practices, invest in water-efficient irrigation technologies, and fortify regional cooperation mechanisms to safeguard the long-term viability and global competitiveness of Basmati rice.

The export potential in the future will depend on the production in the two major exporting countries and international trade relations, and networks. Production is also affected by weather factors during the crop season; for example, during the year 2022–23, the production in Pakistan dropped significantly due to floods, and it worked as an opportunity for India to export a higher quantity.

REFERENCES

Adhikari, A., M. K. Sekhon, and M. Kaur. 2016. Export of rice from India: Performance and determinants. *Agricultural Economics Research Review* 29: 135–150.

Ahmad, M., J. L. Peña-Arancibia, J. P. Stewart, and J. M. Kirby. 2021. Water balance trends in irrigated canal commands and its implications for sustainable water management in Pakistan: Evidence from 1981 to 2012. *Agricultural Water Management*, 245(C). https://doi.org/10.1016/j.agwat.2020.106648

Akhter, M. and Z. Haider. 2020. Basmati rice production and research in Pakistan. In *Sustainable Agriculture Reviews 39*, ed., E. Lichtfouse. Springer Nature Switzerland AG. https://doi.org/10.1007/978-3-030-38881-2_5

APEDA. 2017. *Agricultural and Processed Food Products Export Development Authority (APEDA), Ministry of Commerce and Industry, Government of India*, New Delhi. Annual Report 2017–18.

APEDA. 2023. *Annual Export Report 2022–23*. New Delhi, India: Agricultural and Processed Food Products Export Development Authority.

APEDA. 2025. Basmati rice. (Accessed on 17 April 2025 from www.apeda.gov.in)

API. 2020. *Rice Paddy Policy Analysis for 2020-21 Crops*. API Series No. 276. Islamabad: Agricultural Policy Institute, Ministry of National Food Security and Research, Government of Pakistan.

Ashfaq, M., M. S. Haider, I. Saleem, M. Ali, A. Ali, S. A. Chohan. 2015. Basmati rice- a class apart (A review). *Journal of Rice Research* 3: 156. https://doi.org/10.4172/2375-4338.1000156

Chandrasekaran, S. 2020. *Basmati Rice- The Natural History, Geographical Indication*. New Delhi: Satyam Law International.

Chen W. and X. Zhao. 2023. Understanding global rice trade flows: Network evolution and implications. *Foods*. Sep 2;12(17):3298. https://doi.org/10.3390/foods12173298. PMID: 37685236; PMCID: PMC10486664.

FAO. 2022b. *Rice Market Monitor*. Rome: Food and Agriculture Organization.

FAOSTAT. 2025. Production/yield quantities of rice in the World (2016–2023). [2025-06-01] (2023) https://www.fao.org/faostat/en/#data/QCL/visualize

Hussain, S., M. Mubeen, W. Nasim, S. Fahad, M. Ali, M. A. Ehsan, and A. Raza. 2023. Investigation of irrigation water requirement and evapotranspiration for water resource management in southern Punjab, *Pakistan. Sustainability* 15(3): 1768. https://doi.org/10.3390/su15031768

ICRA 2024. Export of rice by India. *Report* March 2024.

IMARC Group 2025. *India Basmati Rice Market Size, Share, Trends, and Forecast By Type, Size, Distribution Channel, And Region 2025–2033*. 100 p.

Immerzeel, W. W. and M. F. P. Bierkens. 2012. Asia's water balance. *Nature Geoscience.* 5: 841–842.

Islam, S., S. Rehman, M. Munir. 2024. The Basmati rice war between India and Pakistan. *Trends in Intellectual Property Research* 1: 1–6. https://doi.org/10.69971/tipr.2.1.2024.21

J M Baxi Group. 2021. Tidings. *Newsletter* (35), December 2021. 24 p. (www.jmbaxi.com).

Javed, I., A. Rehman, I. Nabi et al. 2020. Performance and macroeconomic determinants of basmati rice exports from Pakistan. *Sarhad Journal of Agriculture* 36: 617–624.

Joshi, N and A. Kumar. 2023. Basmati rice exports from India: Towards global food security. *International Journal of Environment and Climate Change* 13(8): 996–1000. https://doi.org/10.9734/ijecc/2023/v13i82037

Joshi, N. and A. Kumar. 2024a. Investigating the long-term dynamics of basmati rice agricultural exports in India: A var approach. *International Journal of Research in Agronomy* SP-7(6): 546–549. https://doi.org/10.33545/2618060X.2024.v7.i6Sh.945

Joshi, N. and A. Kumar. 2024b. Navigating growth and instability: Analyzing the dynamics of India's Basmati rice exports. *International Journal of Advanced Biochemistry Research* 8: 546–551. https://doi.org/10.33545/26174693.2024.v8.i6g.1361

Kamboj et al. 2021. Trend analysis of APY of Basmati rice in India and Haryana. *Pharma Innovation Journal* 10: 488–493.

Kumar, N. R., A. B. Rai, and M. Rai. 2008. Export of cucumber and Gherkin from India: performance, destinations, competitiveness and determinants. *Agri. Economics Research Review.* 21: 130–138.

Mahajan. G., A. Matloob, R. Singh, V. P. Singh, and B. S. Chauhan. 2018. Basmati rice in the Indian subcontinent: Strategies to boost production and quality traits. In *Advances in Agronomy*, ed., D. L. Sparks, Academic Press, 151: 159–213, https://doi.org/10.1016/bs.agron.2018.04.002

Murai, A. S., P. Mangai, P. Bhadauriya, and A. Kumar. 2017. Tale of geographical indication tag. *Indian Farming* 67 (6): 28–29.

OECD. 2021. https://www.oecd.org/content/dam/oecd/en.publications/reports/2021/04/reviewing-indica-and-japonica-rice-market-developments_4d78ea81/0c500e05-en.pdf

Pakistan Ministry of Commerce. 2024. *Year-Book-2023–24, Ministry of Commerce*, Government of Pakistan, 152 p. www.commerce.gov.pk

Siddiq, E.A., L. R. Vemireddy, and J. Nagaraju. 2012. Basmati rices: Genetics, breeding, and trade. *Agricultural Research* 1: 25–36. https://doi.org/10.1007/S40003-011-0011-5)

Singh, V.P., and A. K. Singh. 2009. History of Basmati rice research and development in India. *Indian Farming* 59 (4): 4–8.

TDAP. 2022. *Exploring potential of rice exports from Pakistan*. Report: TDAP-Agro-Food Division, Research Wing. Series 02/2022. Trade Development Authority of Pakistan. 46p.

Tiwari, A., V. K. Choudhary, N. Pandey, and S. Choubey. 2024. Analysis on area, production, and productivity of Basmati rice in India. *International Journal of Research in Agronomy.* 7:333–336. https://doi.org/10.33545/2618060X.2024.v7.i8e.1275

Trade diggie. 2024. Rice trade intensifies as India and Pakistan remove export restrictions. 17 October 2024. (www.tradologie.com)

Udhayakumar, M. and K. Karunakaran. 2020. Trade dynamics of Basmati and non-Basmati rice exports from India. *Journal of Economics, Management and Trade* 26(10): 68–76. https://doi.org/10.9734/JEMT/2020/v26i1030299

Udhayakumar, M., K. Karunakaran, M. Thilagarathi and K. R. Ashok. 2021. State-wise production performance of Basmati and non-Basmati rice in India. *Asian Journal of Agricultural Extension, Economics and Sociology.* 39(4):17–31. https://doi.org/10.9734/AJAEES/2021/v39i430559

World Bank. 2021. *India Country Economic Memorandum: Sowing the Seeds of Equity & Prosperity*. Agricultural Export Competitiveness Report. The World Bank Group.

9 Influence of Climate Change on Aromatic Basmati Rice

9.1 INTRODUCTION

The United Nations Framework Convention on Climate Change (UNFCCC) characterises climate change as a significant alteration in the Earth's climate system attributable to anthropogenic influences that modify the atmospheric composition, occurring in addition to the background variability arising from natural processes over comparable temporal scales. The Intergovernmental Panel on Climate Change (IPCC) in its most recent Sixth Assessment report (AR6) in 2022 has firmly established the unequivocal evidence that anthropogenic activities are the primary driver of recent global warming, which has triggered extensive and measurable alterations across the atmosphere, oceans, cryosphere, and biosphere. This warming is evident in rising global temperature, melting glaciers and ice sheets, and rising sea levels (IPCC, 2022). Human-driven climate change has led to a documented rise in both the frequency and severity of extreme weather phenomena, including prolonged heatwaves, intense precipitation events, drought conditions, and flooding. The primary contributors to global warming from anthropogenic sources include carbon dioxide emissions from fossil fuel combustion and industrial activities (CO_2-FFI), carbon fluxes resulting from land use changes and forestry practices (CO_2-LULUCF), as well as emissions of methane (CH_4), nitrous oxide (N_2O), and various fluorinated greenhouse gases (F-gases). To make valid comparisons, all these GHG emissions are converted in terms of CO_2–equivalent (IPCC, 2022). In order to combat climate change, the UNFCCC in its Conference of the Parties (COP)–21 at Paris in 2015, a global consensus was reached to enhance the pace and scale of efforts and financial commitments necessary to transition toward a low-carbon development pathway. The Paris Agreement reiterated the objective of restricting the rise in global average temperature by 2100 to well below 2°C above pre-industrial levels, with an additional aspiration to constrain warming to no more than 1.5°C. However, to limit warming to 1.5°C or 2°C above the pre-industrial level (1850 AD) by the year 2100, there is a necessity for a deep, rapid, and sustained reduction in GHG emissions, particularly CO_2 (IPCC, 2022).

Recent decades have exhibited a markedly accelerated pace of global warming compared to earlier periods. In its Fourth Assessment Report (AR4) published in 2007, the Intergovernmental Panel on Climate Change (IPCC) noted that the Earth's mean surface temperature rose by approximately 0.6°C in the 20th century (1901–2000), with the highest temperatures of that century recorded during its final decade. Climate model projections indicate that during the 21st century, global surface

DOI: 10.1201/9781032659398-9

temperature is likely to rise 0.3°C to 1.7°C for their lowest emission scenario using stringent mitigation, and 2.6–4.8°C for the business-as-usual carbon-intensive scenario (IPCC, 2014; Chaudhry, 2017). In the current century, the period from 2010 to 2019 recorded the highest average annual greenhouse gas (GHG) emissions compared to all previous decades. In 2019, global net anthropogenic GHG emissions were estimated at 59 ± 6.6 $GtCO_2$–equivalent, representing an increase of approximately 12% (6.5 $GtCO_2$–eq) relative to 2010 levels and 54% (21 $GtCO_2$–eq) compared to 1990 (IPCC, 2022). Despite this rise, the emission growth rate during 2010–2019 was slower than the rate observed between 2000 and 2009.

Despite the lower rate of growth observed during 2010–2019, high GHG emissions have brought the global temperature increase to 1.5°C earlier than projected. The Copernicus Climate Change Service (C3S) has recently confirmed 2024 to be the warmest year on record globally, and the first calendar year when the average global temperature exceeded 1.5°C above its pre-industrial level (Copernicus Global Climate Highlights, 2025). The global CO_2 concentration has increased from a pre-Industrial Revolution value of 280 ppm in 1780 to 383 ppm in 2007, according to the Fourth Assessment Report (AR 4), and is projected to increase to 550 ppm by 2050 (IPCC, 2007). The global increase in CO_2 concentration is due primarily to the use of fossil fuels and industry, and land use, land use change, and forestry, while those of methane and nitrous oxide are primarily due to agriculture (Aggarwal, 2007). During the year 2019, the share of CO_2 from fossil fuel and industry (CO_2-FFI) in the global net anthropogenic GHG emissions was 61%, while from land use, land use change, and forestry (CO_2-LULUCF), it was 11%. The share of other gases measured in terms of CO_2–equivalent was estimated to be 18% from methane, 4% from nitrous oxide (N_2O), and 2% from fluorinated gases (IPCC, 2022).

9.2 NATIONAL ASSESSMENTS OF GHG EMISSIONS FROM AGRICULTURE

9.2.1 Assessment of Greenhouse Gas (GHG) Emissions from Agriculture in India

The inventory of GHG emissions in India during the year 2007 showed that methane emissions from rice cultivation, N_2O emissions from soils, and crop residue burning were the main sources from agriculture in India. Methane emissions from rice cultivation in India were estimated at 3.37 million tonnes (Mt), while nitrous oxide emissions from agricultural soils accounted for 0.22 Mt. Additionally, open field burning of crop residues released approximately 0.25 Mt of methane and 0.01 Mt of nitrous oxide. In terms of global warming potential, methane is approximately 22 times and nitrous oxide around 298 times more potent than carbon dioxide over a 100-year time horizon. Hence, when calculated in terms of CO_2–equivalent, the total emission from these three sources in India was estimated to be 157.15 Mt CO_2–eq (million tons of CO_2 equivalent) (ICAR Network Project, 2012). The Ministry of Environment, Forest and Climate Change (MoEFCC), Government of India, recently submitted its Third National Communication to the UNFCCC, providing annual estimates of the country's total greenhouse gas emissions for the period 2011–2019 (MoEFCC,

2023). It showed that the agriculture sector contributed 13.44% of the total GHG emissions in the country in 2019, of which the major contributor was the enteric fermentation from livestock, followed by emissions from agricultural soils due to fertiliser use, methane emissions from rice cultivation, and other sources. The contribution of different sources in the agriculture sector in GHG emissions during 2011 and 2019 has been summarised in Table 9.1.

It is apparent from Table 9.1 that the emission from all sources except crop residue burning has been increasing over the years. Methane emissions from rice cultivation have only slightly increased in 2019 compared to 2011, whereas N_2O emissions from agricultural soils have increased due to increasing fertiliser use in India. The amount of GHG emissions from agriculture in different states of India varies. Some states like Punjab, Haryana, Uttar Pradesh, and Andhra Pradesh emit higher amounts of N_2O-N because of higher amounts of N fertiliser use in these states. States like Andhra Pradesh, Bihar, Jharkhand, Odisha, West Bengal, and northeastern states emitted higher amounts of methane per ha of rice cultivation (ICAR Network Project, 2012).

9.2.2 Assessment of GHG Emissions from Agriculture in Pakistan

Assessment of GHG emissions from different sources and sectors in Pakistan has been carried out since 1994, which was followed by the assessments in 2008 and 2012. According to the national GHG inventory of Pakistan for 2011–12, prepared by the Global Change Impact Study Centre (GCISC), Islamabad, the total GHG emission from all sources was estimated at 369 Mt CO_2–eq (million tons CO_2–equivalent), of which the share of agriculture and livestock was 44.8% (165 Mt CO_2–eq). The emission from agriculture and livestock is projected to increase to 1395 Mt CO_2–eq in 2050, out of a total emission of 4200 Mt CO_2–eq in Pakistan (Mir and Ijaz, 2015). Estimated GHG emissions from the agriculture sector (including livestock) in Pakistan during different years, and projections made for the 2020s and 2050s, using the IPCC scenario business as usual, have been summarised in Table 9.2. According

TABLE 9.1
Estimated Total GHG Emission from the Agriculture Sector in India in 2011 and 2019

Emission source	2011 (MtCO_2–eq)	2019 (MtCO_2–eq)
Enteric fermentation from cattle (CH_4)	219.24	223.25
Manure management (CH_4 and N_2O)	27.22	27.51
Rice cultivation (CH_4)	72.67	73.43
Agricultural soils (N_2O)	81.70	88.41
Agricultural residue burning (CH_4 and N_2O)	8.53	8.35
Total	409.37	420.96

Source: Extracted from MOEFCC, 2023.

TABLE 9.2
Estimated GHG Emissions from the Agriculture Sector in Pakistan and Projection for 2020s and 2050s Scenario

Year	GHG Emissions from the Agriculture Sector ($MtCO_2$-eq)	% of Total GHG Emissions from All Sectors
1994	72	NA
2008	120	38.8
2012	165	44.8
2020s[a]	245	—
2040s[a]	1395	—

Source: Extracted from Chaudhry, 2017.
[a] Projections for the business-as-usual climate scenario.

to the projections shown in Table 9.2, GHG emission levels for Pakistan are expected to increase many times in the coming decades.

9.3 GHG EMISSION SCENARIOS USED IN FUTURE PROJECTIONS IN SIMULATION STUDIES

The IPCC highlights that future greenhouse gas (GHG) emissions are shaped by a complex interplay of factors, including population dynamics, economic growth, and technological advancements. To explore possible trajectories of these influences, the IPCC has formulated a range of emission scenarios that represent different pathways the future might take. These scenarios serve as analytical tools to evaluate how varying socio-economic and technological trends could affect future emissions and to examine related uncertainties. The Special Report on Emissions Scenarios (SRES) outlines six scenario families (A1T, A1F1, A1B, A2, B1, and B2), each defined by assumptions about demographic trends, economic development, technological progress, and energy use patterns (IPCC, 2000).

The scenario A1 represents a world where the global population will peak in 2050, and decline thereafter till 2100, and a rapid economic growth with the use of new technology for development. This scenario has been subdivided into three based on the fuel use into A1T, A1F1, and A1B. Scenario A1T uses non-fossil energy sources, A1F1 is a fossil-fuel intensive scenario, and A1B represents a balanced scenario, which does not rely on one energy source (uses a mix of both fossil and non-fossil fuel). The scenario A2 represents a higher emission scenario in a heterogeneous world where population increases continuously till 2100, and has a slower technological growth (IPCC, 2000). Due to high population growth, energy consumption is high, and economic development is at a moderate level.

Scenario B1 represents a low-emission, ideal world scenario with the population growth same as A1, and introduces clean, efficient technology, but does not take additional climate initiatives. Scenario B2 represents a world with continuous population increase, but slower than A2; the economic development is of intermediate level, and technological change is less compared to A1 and B1 (IPCC, 2000).

Subsequently, IPCC in its Fifth Assessment Report (AR5) in 2014 defined a set of four new scenarios, which is based on radiative forcing so as to provide scientists with standardised inputs for climate modelling and research. Radiative forcing quantifies the alteration in the Earth's energy balance resulting from differences between incoming solar energy and outgoing longwave radiation at the top of the atmosphere. A positive radiative forcing indicates an energy surplus that contributes to warming, while a negative value reflects an energy deficit, leading to cooling. The emission trajectories used in climate modelling are referred to as representative concentration pathways (RCPs), labelled $RCP_{2.6}$, $RCP_{4.5}$, $RCP_{6.0}$, and $RCP_{8.5}$. The numerical values correspond to the estimated radiative forcing levels (in watts per square meter) by the year 2100, relative to pre-industrial conditions around 1750, ranging from 2.6 to 8.5 watts per square meter (W/m^2). Each RCP includes geographically detailed datasets on land use dynamics and sector-specific emissions of atmospheric pollutants, along with yearly projections of greenhouse gas concentrations from human-induced sources extending through the year 2100 (Chaudhry, 2017).

$RCP_{8.5}$ is considered a high emissions pathway, where radiative forcing does not peak by 2100 due to minimal efforts to reduce emissions, rapid population expansion, combined with limited advancements in technology and slow progress in improving energy efficiency. CO_2 concentration reaches ~ 936 ppm by 2100, and the mean temperature rise is 4.3°C (range 3.5°C– 5.4°C) in this pathway (IPCC, 2013). RCP_6 is the intermediate stabilisation pathway, where emissions peak around 2060, then begin to decline, and stabilise after 2100. This pathway requires meaningful mitigation efforts but avoids most aggressive emission cuts needed in lower scenarios. The CO_2 concentration reaches approximately 670 ppm, the mean global temperature rises by 2.8°C (range 2.0–3.7°C), and sea level rises by 0.48 m (range 0.33–0.65 m) by 2100 in this pathway (IPCC, 2013).

$RCP_{4.5}$ is a moderate mitigation pathway, and represents a scenario where emissions peak around 2040, and then decline substantially. It requires strong mitigation efforts starting relatively soon, with emission reduction becoming a global priority. GHG emissions stabilise by 2100 in this pathway (IPCC, 2013; Chaudhry, 2017). On the other hand, $RCP_{2.6}$ is the aggressive emission pathway, and represents the most ambitious pathway, where stringent climate policies lead to a peak emission around 2020, followed by a consistent decline. This is the only RCP pathway likely to keep global warming below 2°C compared to pre-industrial temperatures—the goal established in the Paris Agreement.

9.4 CLIMATE CHANGE VULNERABILITY TO AGRICULTURE AND RICE CULTIVATION

Climate change has introduced significant risks in agricultural production through irregular monsoons, rising temperatures, and water scarcity, leading to yield fluctuations of up to 15–20%, which directly impact export volumes and quality (IPCC, 2022; Ishtiaque et al., 2024). An increase in the occurrence of extreme weather events due to climate change is likely to impact farmers in the stress-prone environments even more, resulting in an additional reduction in crop productivity (GRiSP, 2013). Altering precipitation patterns may lead to frequent droughts and floods in different regions, damaging crops and yield loss (Aggarwal, 2008; Ishtiaque et al. 2024; Srinivasa Rao et al., 2024). Extreme precipitation and flooding events have affected several parts of India in recent years. A study conducted by NRSC/ISRO over the last four years shows the number of states affected by major floods has increased from eight states in 2017 to 15 states in 2021 (MOEFCC, 2023). Changes in temperature and precipitation patterns may also influence the severity of pests and diseases. Some pests or diseases may become more prevalent and widespread, requiring additional challenges for crop protection and management (Srinivasa Rao et al., 2024).

Simulations conducted with the IMPACT model, created by the International Food Policy Research Institute (IFPRI), indicate that by 2050, climate change could reduce global rice output by approximately 12–14% relative to production levels in the year 2000 (GRiSP, 2013). The model further projects South Asia to be the worst-affected region, where rice production may be reduced from 169 million tons projected in 2050 under a no-climate change scenario to slightly above 100 million tons under a climate-change scenario, a difference of more than 65 million tons (GRiSP, 2013). Rising temperatures have significantly disrupted the rice-growing season by accelerating phenological development, shortening the crop's growth duration, and diminishing both the rate and total accumulation of biomass (Saseendran et al. 2000; Guhey et al. 2009; van Oort et al. 2011; Tao et al. 2013; Figueiredo et al. 2015).

9.4.1 Observed Past Climate Trends and Future Projections in Basmati Rice Growing Areas in India

Past climate trends in Basmati growing states: The ICAR National Network Project on *Impact, adaptation, and vulnerability of Indian agriculture to climate change (2004–2012)* conducted an analysis of long-term historical climate records to find out the existing trends in Indian climate scenario (rainfall, temperature, occurrence of extreme weather events, occurrence of warm and cold days and nights). The study covered all the states of India (ICAR Network Project, 2012). Basmati rice is grown in a large area in the states of Punjab, Haryana, and western Uttar Pradesh, and in relatively smaller areas in the southern parts of Uttarakhand, Himachal Pradesh, and Jammu and Kashmir. The trend observed in climate parameters in Basmati growing states has been summarised in Table 9.3.

Table 9.3 shows that among the larger Basmati growing states of Punjab, Haryana, and western Uttar Pradesh, the rainfall in Punjab and the eastern part of Haryana has increased in the past 50 years, while temperature shows an increasing trend in Punjab and western Uttar Pradesh, with no change observed in Haryana (ICAR Network Study, 2012). The increase in the occurrence of warm nights is also indicative of warming in the Basmati growing states. A recent study conducted in Punjab, Haryana, and western Uttar Pradesh, based on the analysis of rainfall data of 1983–2022, showed an increasing trend in the rainfall in all zones, except Zone I of Punjab, with the south-west monsoon being the predominant source of increase (Muruganandham

TABLE 9.3
Trend of Climate Parameters Observed in Basmati Rice Growing States of India Based on the Analysis of Long-term Historical Records

Climate Parameter	Trend Assessment in Basmati Rice Growing States	Database (years)
Rainfall	Decrease in U.P., and Haryana (western part); Increase in remaining states	50
Temperature (Annual & Seasonal)	Increase in Punjab and western U.P.; No change in Haryana	30
Extreme weather events (max. one-day precipitation)	Decrease in Western U.P.; No change in remaining states	60
Cold days	No significant change in all states, except the northern part of Jammu & Kashmir (increase)	50
Warm days	Decrease in all states, except northern part of Jammu & Kashmir	50
Cold nights	Decrease in all states, except the northern part of Jammu & Kashmir	50
Warm nights	Increase in all states	50

Source: Summarised from ICAR Network Project on Climate change and Indian agriculture (2012).

and Singh, 2024). However, the rainfall increase was more in Punjab and Haryana, compared to western Uttar Pradesh.

9.4.2 Projection of Future Climate Trends by the End of the 21st Century in India

Projections from the India Meteorological Department and the Indian Institute of Tropical Meteorology, Pune, indicate that by the late 21st century, India could experience a 15–40% rise in precipitation levels, along with an increase in average annual temperatures ranging from 3°C to 6°C as a result of global warming (NATCOM, 2004). The temperature rise is expected to be more significant over land, with northern regions likely to witness the greatest warming. Recent analysis of annual and seasonal rainfall trends from 1983 to 2022 in the green revolution states of India (Punjab, Haryana, and western Uttar Pradesh) has also shown a visible upward trend in the southwest monsoon rainfall across all zones except Zone I of Punjab (Muruganandham and Singh, 2024). However, the increase was more in Punjab and Haryana compared to western Uttar Pradesh. The third National Communication from India to UNFCC in 2023 has also recorded the recent warming trend in India. It shows that out of the past 15 years (2008–2022), 11 years were the warmest in India on record since 1901 (MOEFCC, 2023). The past 10 years (2013–2022) were the warmest decade on record since 1901, indicating the continued warming trend in India, as in other regions of the globe.

9.4.3 Observed Past Climate Trends and Projections of Future Trends in Pakistan

Past climate trend in Basmati growing region of Pakistan: The warming trend in Pakistan's annual mean temperature during the past century (1901–2000) was recorded to be 0.57°C (Chaudhry 2017), which is less than the annual mean temperature rise for the same period in South Asia region of 0.75°C (Ahmed and Suphachalasai, 2014). The overall temperature increase is largely attributed to higher winter temperatures. Meanwhile, total annual rainfall has shown an approximate 25% rise. Global Change Impact Studies Centre (GCISC), Islamabad, has carried out studies on trends in the past climate changes in Pakistan based on meteorological data from 54 stations for the period 1951–2000. These trends for the main Basmati rice growing province of Punjab have been summarised in Table 9.4.

9.4.4 Projections for Future Climate Trends in the Basmati Rice Growing Area of Pakistan

In 2007, the Global Change Impact Study Centre (GCISC), Islamabad, utilised a General Circulation Model (GCM) to simulate future changes in temperature and

TABLE 9.4
Trend Observed in Climatic Parameters in the Basmati Growing Region (Punjab) of Pakistan During 1950–2000

Climate Parameter	Trend Observed in Punjab (Pakistan)	Data Analysis Period
Rainfall	Increase in annual rainfall (0.63%); increase in monsoon rainfall (0.57%)	1950–2000
Mean temperature	Increase in mean annual temperature. (0.11°C), decrease in monsoon season temp. (−0.25°C)	1950–2000
Mean maximum temperature	Decrease in annual (−0.14°C), and monsoon temp. (−0.20°C), increase in summer temp. (0.78°C)	1950–2000
Mean minimum Temperature	Warmer in all seasons except monsoon; an Increase in annual (0.41°C), and summer temp. (0.76°C); Decrease in monsoon (−0.35°C)	1950–2000
Extreme wet events	Increasing trend	1950–2000
Extreme monthly max. temp.	Increasing trend	1950–2000
Heat wave events (continuous 5-7 days > 40°C and >45°C temp.)[a]	Increasing trend	1961–2009

Source: Extracted from Sheikh et al., 2009.
[a] from Zahid and Rasul, 2012.

precipitation for the years 2020, 2050, and 2080 under two emission pathways: A2 (business-as-usual) and A1B (balanced growth). The projections indicated that by 2080, temperatures in Pakistan could rise by up to 4.38°C (Chaudhry, 2017). Temperature projections indicated a greater rise in both summer and winter seasons across northern Pakistan compared to the southern regions. Additionally, for both areas, the warming trend was more pronounced during the winter months than in summer. Precipitation patterns, however, were not expected to undergo substantial changes. However, precipitation increased in summer and decreased in winter in southern Pakistan (Iqbal and Zahid, 2014; Chaudhry, 2017).

Another study was conducted by the Pakistan Meteorological Department (PMD) that projected region-wise changes in temperature and precipitation from 2011 to 2050 using different scenarios. The model showed that the maximum rise in temperature will occur in the northern areas of Pakistan, in the central and south Punjab (the region growing Basmati rice), and the lower part of Khyber Pakhtunkhwa province (Chaudhry et al. 2009). Precipitation showed a mixed trend in different regions. The climate change projection for Pakistan and the Basmati rice growing Punjab region for alternate scenario A2 (business as usual), A1B (balanced scenario), and B1(ideal world scenario) has been presented in Table 9.5.

The historical change (1980–2010) in rainfall and temperature during both rice and wheat growing seasons in the northern part of Punjab province of Pakistan was used in climate change projection for the 2050 scenario (2040–2069) (Bokhari et al., 2017). The projection using $RCP_{8.5}$ showed a clear tendency towards a warmer and wetter future climate. By mid-century, average maximum temperatures are expected to rise by approximately 2.0–2.5°C during the *kharif* season, which coincides with rice cultivation, and by about 2.4–2.7°C during the rabi season, when wheat is typically grown. The mean minimum temperature was projected to increase by 2.6–3°C during *kharif*, and by 2.5°C during the *rabi* season. The rainfall during the *kharif* season was projected to increase by 25–35%, while a minimal change was expected during the *rabi*, indicating that the increase in crop water demand due to the temperature rise may be satisfied by the increased rainfall (Bokhari et al., 2017).

TABLE 9.5

Climate Projections for Pakistan and the Basmati Growing Punjab Region in the Alternate Emission Scenario (2011–2050)

Area	**Precipitation (mm/decade)**			**Temperature °C per Decade**		
Emission scenario	**A2**	**A1B**	**B1**	**A2**	**A1B**	**B1**
Pakistan	+ 1.73	+ 1.26	-0.89	+0.51	+0.41	+0.24
Central/ southern Punjab and lower NWFP	−2.98	−1.78	−3.5	+0.63	+0.71	+0.05

Source: Extracted from Chaudhry et al., 2009.

9.5 IMPACT OF CLIMATE CHANGE ON BASMATI RICE

9.5.1 Impact of Rising Temperature on Basmati Rice

In Basmati rice, for effective grain filling and aroma development, bright, clear days with temperatures ranging from 25 to 32°C, cooler night temperatures (20–25°C), and moderate humidity levels are required during the flowering and grain development stages (Singh, 2000). Elevated temperatures during the grain filling phase reduce its duration and negatively influence yield, amylose concentration in starch, and the functioning of key enzymes involved in starch synthesis in Basmati rice (Ahmed, N. et al., 2015). A study explored how elevated temperature impacts rice grain development by cultivating Basmati rice under two temperature regimes, 32 and 22°C, during the grain filling stage. Exposure to higher temperatures led to a shorter grain filling duration (from 32 to 26 days), a 6% drop in yield, and noticeable declines in both total starch (by 3.1%) and amylose content (by 22%). Heat stress has been shown to significantly affect chlorophyll content, relative water content, and relative cell injury at the seedling stage in Basmati rice varieties; however, the varietal response varies. The varieties Basmati-515, Super Basmati, and NIAB-16 exhibited superior thermal tolerance and better performance, whereas Kisan Basmati and PK-1121 Aromatic varieties constantly showed poor performance, and hence were graded as thermo-sensitive in experiments conducted in Pakistan Punjab (Baloch et al., 2024).

High temperature during the flowering and grain filling stages affects the pollen germination and spikelet sterility (ICAR Network Study, 2012). Temperatures exceeding 33°C during anthesis increased pollen sterility in Basmati as well as non-Basmati varieties. At 35.5°C, the Basmati cultivar exhibited 17% pollen sterility, along with a 26% decline in pollen germination (Chakrabarti et al., 2010). Elevated temperature during the grain-filling phase also led to a rise in spikelet sterility in Basmati, whereas the non-aromatic variety was found to be less affected (ICAR Network Study, 2012). High temperature and low light stress (3°C higher than ambient temperature and 20% reduction in light intensity) throughout the growth period has also been reported to cause spikelet sterility and resultant yield reduction in Basmati rice varieties (Johnson et al. 2011). However, some aromatic varieties like Taraori Basmati and Pusa Basmati-1 varieties showed greater tolerance to high temperature and low radiation as compared to Basmati-370, Basmati-386, and Pusa Sugangh-2. This differential growth and yield sensitivity of some aromatic varieties may be helpful in the development of heat-tolerant, climate-smart aromatic Basmati rice varieties in the future.

Over the past two decades, climate change has been associated with a more rapid rise in nighttime temperatures compared to daytime temperatures (Nagarajan et al. 2010). Nighttime temperatures exceeding 22°C during flowering and the initial grain-filling phase have been found to adversely influence both the yield and grain quality in Basmati and other rice varieties. Heat stress during these critical stages, especially at night, tends to lower yield levels and reduce the characteristic aroma in Basmati rice. Research examining temperature effects on grain aroma compounds in different rice types found that under heat stress, the mutated form of the BADH2 gene became less active, resulting in little to no production of the key aromatic compound 2-acetyl-1-pyrroline (2-AP) during grain development, which ultimately suppressed aroma expression (Mishra et al. 2024). High temperature also affects the quality of cooked Basmati rice. A reduction in test weight and grain elongation upon cooking, and a slight reduction of aroma have been reported due to temperature increase (ICAR Network Study, 2012). High temperature is also likely to affect the pest population in rice. An increase of 0.5°C in average temperature in Delhi in the year 2006 caused a slight reduction in the population of brown plant hopper (BPH), a common rice pest, while a temperature rise of 1–3°C resulted in a significant decline (ICAR Network Study, 2012).

9.5.2 Impact of Elevated Atmospheric CO_2 Concentration

An increase in ambient CO_2 concentration promotes growth and productivity of C3 crops like rice and wheat, due to the increased photosynthesis in these crops (Aggarwal, 2008). The results from multiple CO_2 enrichment studies conducted under open-field conditions indicate that an atmospheric CO_2 level of 550 ppm can enhance yields by approximately 8–10% in rice and wheat, with soybean showing increases up to 15%. However, crops like maize and sorghum exhibit minimal yield response under similar conditions (Long et al., 2005). A yield increase of 9% in Basmati rice has been reported from Pakistan due to the enhancement of CO_2 level from 375 ppm to 550 ppm (Iqbal et al., 2009b). Elevated CO_2 also causes a reduction in protein content of rice and a slight reduction of aroma in Basmati rice (ICAR Network Study, 2012). A combined effect of elevated CO_2 concentration and heat stress has also been shown to decrease the protein content in rice (Chaturvedi et al., 2017).

9.5.3 Interactive Impact of Changing Rainfall, Increased Temperature, and Elevated CO_2 Concentration on Basmati Rice Grain Yield, Quality, and Aroma Using Simulation Models

Crop growth simulation models offer a valuable approach for examining the combined effects of carbon dioxide levels, precipitation, and temperature. By incorporating daily weather data, these models replicate plant development and yield responses based on physiological functions and soil dynamics at specific locations (Aggarwal 2008). Several crop models, such as DSSAT (Decision Support System for Agrotechnology Transfer), ORYZA, WTGROWS, CROPGRO, CERES-Rice, and InfoCrop, have been used. Simulation-based assessments across key rice-producing areas in Asia indicate that a 1°C rise above the current average temperature can result in a 7–10% decline in rice yield under present atmospheric CO_2 levels. While elevated CO_2 enhances biomass accumulation, panicle formation, and grain-filling efficiency—thereby improving yield and water productivity, it simultaneously heightens the vulnerability of spikelets to heat-induced sterility. Consequently, the positive effects of increased CO_2 on rice productivity are counterbalanced by the adverse impacts of high temperature (GRiSP, 2013).

The ICAR Network Project conducted a regional study on the impact of climate change on major crops of India—rice, wheat, maize, and sorghum—using the InfoCrop model. The model used $1^0 \times 1^0$ gridded climate data for the baseline period (1969–1990) from India Meteorological Department (IMD), gridded soil data from the National Bureau of Soil Survey & Land Use Planning (NBSS&LUP) and the ISRIC database, crop management from farmers' practices, and genetic coefficients of varieties used in the area. The model was run for three time periods and four emission scenarios according to IPCC (2000) (ICAR Network Project, 2012). The three time periods used in the model for projection were 2020 (representing years 2010–2040), 2050 (2041–2070), and 2080 (2071–2100). The four emission scenarios considered were A1B (balanced scenario), A2, B1, and B2. Projection for the MIRO and PRECIS climate scenario showed a reduction in the yield of irrigated rice by ~4% in 2020, 7% in 2050, and by ~10% in the 2080 scenario. The yield of rainfed rice is, however, likely to be reduced by ~6% in 2020, but likely to only marginally decrease (less than 2.5%) in 2050 and 2080. In regions like Punjab and Haryana, where Basmati rice cultivation is extensive, irrigated rice is expected to experience a relatively higher yield decline (estimated around 6–8%) by 2020 compared to other rice-growing areas in the country. The yield loss in 2050 is likely to be 15 - 17% (ICAR Network Study, 2012).

A study conducted in Uttar Pradesh, where Basmati rice is grown on a large area in India, shows that due to the impact of climate change, temperature, rainfall, and CO_2 concentration are projected to increase in the state (Singh et al. 2024). The study employed the CERES-Rice model to simulate crop performance across three future timeframes (2030, 2050, and 2090) under two distinct climate scenarios: SSP2-4.5 and SSP5-8.5. SSP2-4.5 represents a stabilisation pathway with moderate radiative forcing of 4.5 W/m^2 by the end of the century, reflecting balanced socio-economic development and intermediate land use trends. In contrast, SSP5-8.5 assumes a trajectory of rapid economic growth driven by intensive fossil fuel usage, leading to a high radiative forcing level of 8.5 W/m^2 by 2100. Projections from the model show that while the increasing CO_2 concentration is likely to increase rice yield in both irrigated and rainfed rice, the associated increase in temperature will reduce the yield on irrigated rice lands in Uttar Pradesh (Singh et al., 2024). While the seasonal daily average maximum temperature in the state is already above optimal (above 32 °C) level, the projected further increase in temperature would negate the positive impact of CO_2 fertilisation (Singh et al., 2024). However, in the rainfed condition, CO_2 fertilisation combined with increased rainfall would compensate for the adverse effect of temperature increase. The Basmati growing area in the state lies in the irrigated zone, hence it will not benefit from the positive impact of the CO_2 increase. Climate change is also likely to adversely impact soil health in the future. Soil profiles from 117 soil series covering 60 agro-ecological sub-regions across India were incorporated into simulation models such as RothC, Century, and InfoCrop. Model outputs indicated that a temperature rise of 2.5°C led to a noticeable decline in total organic carbon (TOC), with variations in the extent of reduction across different soil depths and regions (ICAR Network Study, 2012).

Global Change Impact Studies Centre of Pakistan has used crop growth simulation model CERES to simulate the effect of climate change on the growing season length and yield of major crops of Pakistan under IPCC scenarios A2 (business as usual) and B2 (intermediate level emission). A decrease in the yield of major crops, wheat, and in the length of growing season in Pakistan has been reported (Iqbal et al. 2009c). Using the CERES-Rice model, the study projected a decrease in Basmati rice yield under the A2 scenario by 2% in 2020, 7% in 2050, and 18% in 2080. Under the B2 scenario, the yield reduction in aromatic Basmati rice projected was 4% in 2020, 10% in 2050, and 15% in 2080, if climate change is allowed to go unchecked (Iqbal et al., 2009b; Iqbal et al., 2009c; Chaudhry, 2017). The model also showed that under the A2 scenario, the growing season length of Basmati rice shortened by 5 days in the 2020s, 11 days in the 2050s, and 15 days in the 2080s. Under the B2 scenario, the growing season is shortened by 6 days in the 2020s, 10 days in the 2050s, and 13 days in the 2080s (Iqbal et al. 2009c).

Crop growth simulations using the DSSAT and APSIM models for the rice–wheat systems in Pakistan's Punjab region suggest a projected decline in rice yields by mid-century (2040–2069) due to climate change. The models estimated an average reduction in rice productivity of approximately 15–17%, with APSIM indicating a slightly greater yield loss than DSSAT under future climate scenarios (Ahmad, A. et al., 2015). Gummadi et al. (2016) addressed uncertainties associated in the assessment of climate change impacts on crop productivity through simulation modelling, highlighting that the credibility of future projections largely depends on how well baseline yields are reproduced. Inaccuracies in simulating current yields tend to weaken the reliability of future climate impact assessments on rice production across different Indian states. They advocated applying bias correction methods to improve the reliability of projections while using GCMs.

9.6 ADAPTATION/MITIGATION STRATEGY FOR AROMATIC BASMATI RICE

Adaptation refers to the adjustment in response to existing or forecasted adverse effects of climate change (Iqbal et al., 2009a). Adaptation measures could be improvements in the genetic characteristics of a crop to withstand a warmer temperature regime or adopting appropriate crop management practices to negate or minimise the climate change effects. Follow-up of these adaptation measures needs policy and financial support from the governments. At the national level, the Government of India has launched a National Action Plan for Climate Change (NAPCC) in the year 2008, which provides a policy framework to address climate challenges, with the Ministry of Environment, Forests, and Climate Change providing Inter-Ministerial coordination between the sectoral ministries. NAPCC has outlined nine strategic National Missions in different sectors, including agriculture, named as the National Mission for Sustainable Agriculture (NMSA). The NMSA works to build agricultural resilience through: (i) Climate resilient crop varieties development, (ii) Risk management by expanding weather-based crop insurance system, (iii) Water efficiency by promoting drip and other water-saving technologies, and (iv) Soil health management, by enhancing carbon sequestration in agricultural soils. In Pakistan, the Ministry of Climate Change has been created to coordinate climate issues at the national level.

A national Climate Change Policy (NCCP) was launched in 2013 by the Ministry of Climate Change. NCCP prioritises adaptation measures over mitigation. Some of the mitigation and adaptation measures with specific reference to aromatic Basmati rice may be outlined as:

9.6.1 Breeding Climate-resilient Aromatic Basmati Rice Varieties

Developing crop varieties with enhanced tolerance to climatic extremes is a key adaptation strategy, aiming to ensure improved performance under future weather uncertainties compared to existing cultivars (GRiSP, 2013). Research efforts need to be augmented towards breeding aromatic rice varieties for high temperature tolerance, drought- and flood-tolerance, and pest and disease resistance without compromising the grain quality characteristics. Future breeding programs can benefit from exploring climate-resilient germplasm and incorporating adaptive traits into widely adopted cultivars using marker-assisted selection (MAS) and complementary advanced techniques (Nagarajan et al. 2010). IARI, New Delhi, released in 2021 three pest and disease resistant Basmati varieties by incorporating resistance genes through MAS in the popular, mega-varieties of Basmati rice (PIB, 2022). The newly developed varieties include Pusa Basmati-1847, enhanced for resistance to bacterial blight and blast while retaining the early maturity of the widely grown PB-1509; Pusa Basmati-1885, which offers improved disease resistance in the widely preferred extra-long grain PB-1121 background; and Pusa Basmati-1886, designed to combine bacterial blight and blast resistance with the aromatic and long-duration characteristics of PB-1401.

Some of the existing aromatic varieties can also be used in future breeding programs. Basmati rice varieties Taraori Basmati and Pusa Basmati-1 were found to be least affected at 3°C higher temperature compared to Basmati-370, Basmati-386, and Pusa Sugandh-2 in experiments conducted at IARI, New Delhi (Johnson et al., 2011). These high-temperature-tolerant varieties could be used in breeding climate-resilient Basmati varieties. Similarly, on the basis of observed morphological and physiological attributes such as chlorophyll content, relative water content, and relative cell injury at the seedling stage, Basmati-515, Super Basmati, and NIAB-16 showed tolerance to heat stress experiments conducted in Punjab, Pakistan, while Kisan and PK-1121 aromatic Basmati were graded as thermo-sensitive varieties (Baloch et al., 2024). In Pakistan, the adverse effects of rising temperatures on the phenology of the rice-wheat system were partially offset by introducing newer Basmati varieties with shorter durations and higher thermal time requirements for maturity (Ahmad et al., 2019). In India, some of the recently released Basmati varieties like Pusa Basmati-1509 have become popular among farmers due to their short duration and less irrigation requirement.

Mitigating crop residue burning helps not only in reducing GHG emissions, but also in providing local communities with a pollution-free and healthy environment. Despite a national ban on agricultural residue burning, several farmers in northern India are still practicing it. A study conducted in rural Punjab, India, between 2015 and 2017 showed a growing trend of adoption of no-burn farming technologies among farmers with large landholdings, but medium and small farmers largely remained non-adopters (Krishnapriya et al., 2024). These large farmers are using either no-till and no-burn machines like *Happy Seeder*, which uses crop residue as a mulch, or *Super Seeders*, which are no-burn tillage machines. Other measures like baling of residue for energy production, and the use of Pusa Bio-decomposers are also being practiced. The cumulative effect of these technologies might have reduced the number of fire counts (Haq, 2023).

A study carried out by the authors of this book (unpublished) to observe the status of rice residue burning and its impact on CO columnar concentration in the state of Haryana, India, during 2019 and 2023 also showed a reduction in burning activities. Active fire locations from the VIIRS sensor onboard the Suomi NPP satellite were obtained via the NASA FIRMS portal for the post-harvest rice season from October to mid-November. Using Sentinel-5P TROPOMI data for the years 2019 and 2023, covering four weeks of October and two weeks of November, weekly carbon monoxide (CO) columnar concentrations were assessed. The analysis was conducted in Google Earth Engine (GEE), where daily Level 2 CO data were quality-filtered (qa_value >0.5) and aggregated into weekly mean composites. The CO column values (mol/m^2) were converted to mass concentration (mg/m^3) assuming a 5 km vertical column and using the molecular weight of CO (28.01 g/mol). To contextualise emission sources, fire point locations were overlaid on the CO maps. This spatial integration enabled correlation between residue burning activity and elevated CO concentrations, particularly during late October (Figure 9.1). A comparison of burning points (shown in black dots in the figure) on the weekly images of two years shows a significant reduction in 2023 compared to 2019. Higher CO columnar concentration in the areas located above or around the burning points indicates the influence of rice residue burning on it. However, consistently high CO concentrations observed over Faridabad and Gurgaon districts (in the south-eastern part of the image), where few fire events were recorded, suggest contributions from intense urban transportation and industrial activity prevailing in this part of Haryana. Additionally, elevated CO levels in certain non-burning zones indicate the atmospheric transport of CO by prevailing winds, underscoring the regional-scale impact of both agricultural and non-agricultural emission sources on the atmospheric air quality.

9.6.2 Management Practices to Counter Negative Effects of Climate Change

Farmers and other stakeholders can adapt and reduce the losses due to climate change (Aggarwal, 2008). Adjusting sowing schedules and adopting climate-resilient crop varieties can play a key role in lessening the negative effects of climate variability on rice and other crops. Other adaptations related to soils, like minimum tillage, residue mulching, and land levelling; and related to water use, like direct seeding of rice, alternately irrigating and drying the cultivated fields, water resource conservation measures, use of solar pumps, sprinklers, and so on, will help in increasing resource use efficiency, and also reduce emissions. These adaptation technologies are commonly referred to as climate-smart agriculture (CSA) technologies.

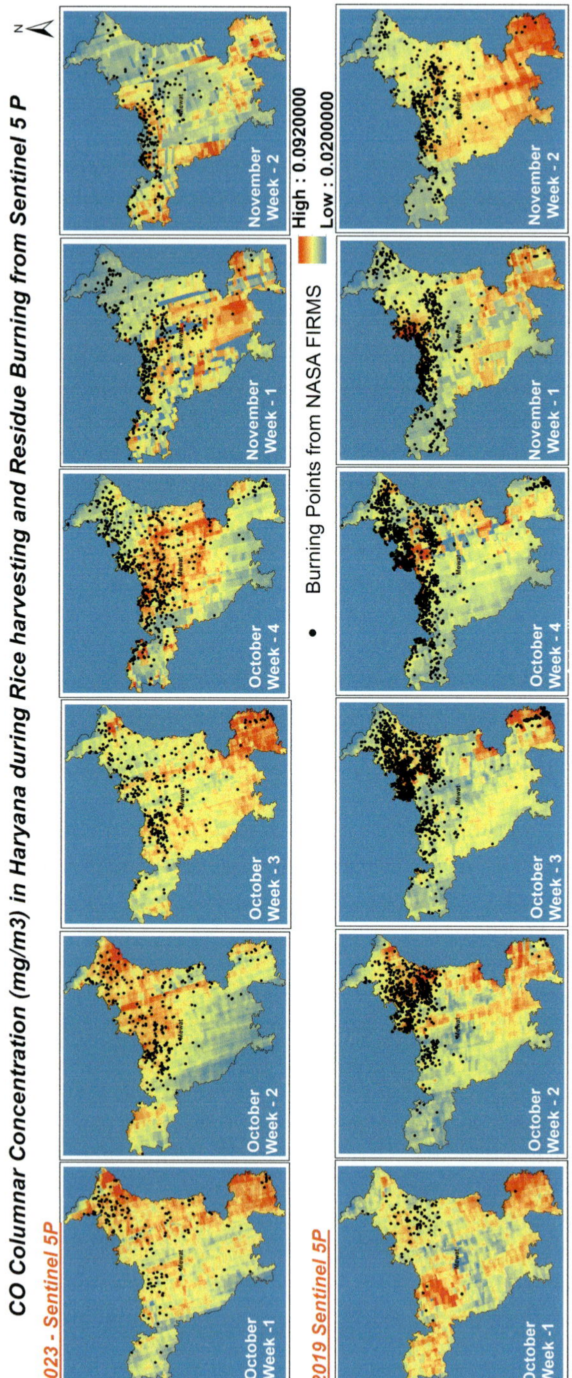

FIGURE 9.1 Comparison of weekly carbon monoxide (CO) columnar concentration (mg/m^3) using Sentinel 5 P Data and residue burning locations in Haryana state, India, during rice harvesting seasons of 2019 and 2023.

Optimisation of transplanting dates: One of the adaptation strategies to overcome the effect of rising temperature and CO_2 concentration can be the optimisation or shifting of transplanting dates in rice. The optimum planting dates for aromatic Basmati Super variety were determined in Pakistan, Punjab, under warmer temperatures and higher CO_2 concentration (Iqbal et al., 2009a). Simulation results indicated that the optimal window for transplanting rice lies between June 20 and July 20 under current temperature and CO_2 conditions. Delaying transplanting up to July 20 may serve as a potential adaptation strategy to maintain baseline yields, even with a temperature increase of up to 2°C and atmospheric CO_2 concentration reaching 550 ppm, assuming other agronomic practices remain unchanged.

Water-saving technologies: Water-saving technologies in rice, such as alternate wetting and drying, land levelling, improved tillage, bund preparation, and direct seeding, can help address the likely water shortage due to irregular rainfall associated with climate change (GRiSP, 2013). Replacing transplanting of rice (TPR) with dry-seeded rice (DSR) has been shown to save labour costs of transplanting and also reduce water requirement in puddling the field. In Pakistan Punjab, assessment of DSR technology using the CERES-Rice model showed the water use efficiency (WUE) to increase from 2.91 kg/m^3 in TPR to 3.68 kg/m^3 in DSR (Iqbal et al., 2009a). The irrigation requirement was also reduced from 17 in TPR to only 13 in DSR for similar yield performance. Another advantage of DSR was early anthesis in rice plant by two days, and reduction in total maturity period by seven days (Iqbal et al., 2009a).

However, in the DSR, farmers face the problem of weeds and need to use weedicides to control them. Recently, two herbicide-tolerant Basmati varieties, Pusa Basmati-1979 and Pusa Basmati-1985, were released by IARI, New Delhi, and seeds were provided to farmers during the *kharif* 2024 season. These two varieties are the MAS (marker-assisted selection)–derived, herbicide-tolerant, near-isogenic lines of Pusa Basmati-1121 and Pusa Basmati-1509, respectively, and are tolerant to Imazethapyr herbicide (IARI, 2024).

Alternate wetting and drying: Alternate wetting and drying rice fields is another climate-smart technology that not only saves on the total water required in rice cultivation but also reduces methane emissions. In this method, instead of continuously keeping the rice fields flooded, the fields are alternately irrigated and then dried.

Use of Biofertilisers: The use of biofertilisers, like blue-green algae (BGA), could be another potential option for reducing methane emissions from rice fields. The use of BGA has been found to reduce methane emissions from flooded rice fields, besides supplying nitrogen to the crop (ICAR Network Study, 2012).

Change in input use: The adoption of high-performing rice varieties with better input use efficiency, combined with a 25% increase in nitrogen application, has the potential to counteract the adverse effects of climate change. Under future climate conditions, this approach could lead to a production increase of 6–17% in irrigated systems and 20–35% in rainfed areas (ICAR Network Study, 2012).

Understanding pest dynamics: Raising farmer awareness about shifting pest patterns and the role of biodiversity in natural pest regulation is essential for reducing reliance on excessive pesticide use (GRiSP, 2013).

Mitigating crop residue burning: Mitigating crop residue burning helps not only in reducing GHG emissions, but also in local communities living in a pollution-free and healthy environment. Despite a national ban on agricultural residue burning, several farmers in northern India are still practicing it. A study conducted in rural Punjab, India, between 2015 to 2017 showed a growing trend of adoption of no-burn farming technologies among the farmers having large land holdings, but medium and small farmers largely remained non-adopters (Krishnapriya et al., 2024). These large farmers are using either no-till and no-burn machines like *Happy Seeders*, which use crop residue as a mulch, or *Super Seeders*, which are no-burn tillage machines. Other measures like baling of residue for energy production and the use of Pusa Bio-decomposers are also being practiced. The cumulative effect of these technologies might have reduced the number of fire counts (Haq, 2023). Apart from these measures, government regulations, provision of subsidies to farmers, information dissemination, training to farmers, and so on, have also played their role in reducing residue burning (Krishnapriya et al. 2024). However, using smallholders as field demonstration sites and focusing extension services on medium farmers might help in the large-scale conversion of small and medium farmers from non-adopters to adopters of no-burn technology in the future.

Climate risk management services: Providing value-added climate risk management services to farmers, like weather forecasts and related agro-advisory, would help them take urgent follow-up actions and reduce risk (Aggarwal, 2008). Similarly, early warning of extreme weather phenomena, like floods, droughts, cyclones, heat and cold waves, and so on, and pest and disease outbreaks would prove effective in reducing damage.

9.6.3 Role of Different Stakeholders in Managing Adverse Climate Effects

Managing the adverse impact of climate change through various mitigation and adaptation measures will require concerted and coordinated efforts from all stakeholders, from farmers to government agencies, research institutions, NGOs, and policymakers. Higher investment will be required in research, infrastructure, capacity building, and risk management. Policy changes may also be needed to facilitate adaptation to climate change.

9.6.4 Overcoming Adoption Barriers

Adoption of climate-smart agriculture, however, is faced with various challenges. Ishtiaque et al. (2024) highlighted several critical barriers hindering the uptake of climate-smart agriculture (CSA) practices in South Asia. They also proposed actionable strategies to promote wider adoption, many of which are applicable to the aromatic Basmati-producing regions:

Weak organisational capacities: Due to a large part of the rural population practicing agriculture in the Basmati growing region, it becomes difficult to reach all farmers with the new practices and technologies, in view of the limited government resources. Developing effective networks that can serve as a source of information on CSA technologies to the farmers, therefore, becomes imperative. Sharing of knowledge and practices through peer-to-peer farmer networks is more effective and needs promotion (Ishtiaque et al., 2024). Agricultural Extension Departments of the government, which are primarily responsible for the dissemination of information on climate-smart agriculture, often suffer from a lack of staff, inadequately trained personnel, and limited resources (transport, finance, etc.). The number of field schools, demonstration plots, and training programmes is not sufficient. Increased investment in extension services and incentives to the private sectors involved in information dissemination may prove helpful. Custom hiring centres and service providers have also facilitated the adoption of farm machine-related technologies to some extent (Ishtiaque et al., 2024). They can be more helpful if they consider local market demand and equity concerns. Enhancing their management skills may improve adoption.

Inadequate targeted incentives: Scaling up CSA practices and technologies requires their affordability for smallholder farmers. Government policies and targeted subsidies to marginal and small farmers can help in technology adoption, as for example in promoting water-saving technologies in Pakistan's Punjab, and zero-tillage technology in the Indo-Gangetic Plains of India (Ishtiaque et al., 2024).

Limited post-adoption follow-up: Post-adoption follow-up is often of low priority. Evaluating adoption over time can identify challenges that farmers face in the continued use of technology. Revisiting the technology adopters can help in understanding the bottlenecks and, in turn, taking steps to improve the situation.

Inequities in information dissemination: Climate-smart agriculture-related information and technology dissemination often do not reach small and marginal farmers (Ishtiaque et al., 2024). It is generally observed that the large farmers are preferred for field demonstrations and provisions. These groups have less access to resources, which could be overcome by creating institutions that also enhance economic empowerment, like self-help groups.

9.7 CONCLUSIONS

Climate change has already started exerting pressure on our agricultural systems and is likely to further constrain food production systems even more in the future in India unless we take appropriate measures to overcome it. Aromatic Basmati rice is sensitive to rising ambient temperature, which may affect not only its productivity but also the grain quality and aroma. Studies have shown that following appropriate adaptation measures may alleviate the situation by reducing crop production losses and also help in reducing GHG emissions. A key approach to climate adaptation involves developing resilient aromatic Basmati rice varieties that are better equipped to handle extreme weather events and outperform existing cultivars under changing climatic conditions. Strengthening research efforts towards breeding Basmati rice varieties not only for high-temperature tolerance, drought-, and flood-tolerance, but also for pest and disease resistance, would be of utmost necessity. Some efforts have been made in this direction, and three disease-resistant Basmati varieties developed recently in India have already reached farmers' fields.

Emissions of methane from the rice fields, nitrous oxide from agricultural soils, and carbon dioxide and nitrous oxide from crop residue burning are the three main GHG sources in rice cultivation regions. Emissions from the first two sources can only be reduced, whereas the third can be completely curbed. DSR and alternate wetting and drying (of rice fields) technologies can help reduce methane emission, and appropriate nutrient management practices can reduce fertiliser use, and in turn nitrous oxide emission, only to a smaller extent. DSR and resource-conserving zero-till technologies are being adopted by Basmati rice farmers. However, scaling it up further would require more support from the government and other agencies. Following appropriate adaptation options, resource-conserving and energy-saving technologies are likely to help farmers overcome adverse climate change effects in the future.

REFERENCES

Aggarwal, P. K. 2007. Climate change: Implications for Indian agriculture. *Jalvigyan Sameeksha* 22: 34–46.

Aggarwal, P. K. 2008. Global climate change and Indian agriculture: Impacts, adaptation and mitigation. *Indian Journal of Agricultural Sciences* 78: 911–919.

Ahmad, A., M. Ashfaq, G. Rasul et al. 2015. Impact of climate change on the rice-wheat cropping system of Pakistan. In *Handbook of Climate Change and Agro-Ecosystems*, eds., D. Hillel, C. Rosenzweig, Vol. 3. London: Imperial College Press, pp. 219–258.

Ahmad, A., G. Abbas, M. Ahmed et al. 2019. Climate warming and management impact on the change of phenology of the Rice-Wheat Cropping System in Punjab, Pakistan. *Journal of Field Crop Research* 230: 46–61.

Ahmed, A. and S. Suphachalasai. 2014. *Assessing the Cost of Climate Change and Adaptation in South Asia*. Manila: Asian Development Bank.

Ahmed, N., I. J. Tetlow, S. Nawaz et al. 2015. Effect of high temperature on grain filling period, yield, amylose content, and activity of starch biosynthesis enzymes in endosperm of Basmati rice. *Journal of Science of Food and Agriculture* 95: 2237–2243.

Baloch, R., M. F. Saleem, M. Shahbaz, and M. Sarwar. 2024. Evaluation of basmati rice (Oryza sativa L.) genotypes for seedling growth and leaf physiology under high temperature stress. *Journal of Agronomy and Crop Science* 210: e12777.

Bokhari, S. S. A., G. Rasul, A. C. Ruane, G. Hoogenboom, A. Ahmad. 2017. The past and future changes in climate of the rice-wheat cropping zone in Punjab, *Pakistan. Pakistan Journal of Meteorology.* 13: 9–23.

Chakrabarti, B., P. K. Aggarwal, S. D. Singh, S. Nagarajan, and H. Pathak 2010. Impact of high temperature on pollen germination and spikelet sterility in rice: Comparison between Basmati and non-Basmati varieties. *Crop and Pasture Science* 61: 363–368.

Chaturvedi, A. K., R. N. Bahuguna, M. Pal, D. Shah, S. Maurya, and K. S. Jagdish. 2017. Elevated CO_2 and heat stress interaction affect grain yield, quality, and mineral nutrient composition in rice under field conditions. *Field Crops Research* 206: 149–157.

Chaudhry, Q. U. Z. 2017. *Climate Change Profile of Pakistan*. Manila: Asian Development Bank. 130p.

Chaudhry, Q. Z. et al. 2009. *Climate change indicators of Pakistan*. Tech. Report No. PMD 22/2009. Pakistan Meteorological Department, Islamabad.

Copernicus Global Climate Highlights. 2025. *Copernicus: 2024 is the First Year to Exceed 1.5 °C Above the Pre-Industrial Level*. Copernicus Press Release 10 January 2025.

Figueiredo, N., C. Carranca, H. Trindade et al. 2015. Elevated carbon dioxide and temperature effects on rice yield, leaf greenness, and phenological stages duration. *Paddy, Water, and Environment* 13: 313–324.

GRiSP (Global Rice Science Partnership) 2013. *Rice Almanac*, 4th ed. Los Banos (Philippines): International Rice Research Institute. 283p.

Guhey, A., K. C. Patel, R. Saxena, and S. B. Verulker. 2009. Influence of temperature and humidity on physiology, phenology, and yield traits of rice. *Journal of Rice Research* 2: 1–15.

Gummadi, S., T. Wheeler, T. Osborne, and A. Turner. 2016. Addressing the uncertainties associated in assessing the impacts of climate change on agricultural crop production using model simulations. *Academic Research Journal of Agricultural Science and Research* 4: 206–221.

Haq, Z. 2023. Farm fire cases on the decline, Parliament told *Hindustan Times*. (Available at www.hindustantimes.com/india-news/farm-fire-cases-on-the-decline-parliament-told-101702489750674.html).

IARI. 2024. Launch of seed sale of RobiNOweed Basmati rice varieties for direct-seeded rice cultivation. *Press Note 24052024*. Indian Agricultural Research Institute, New Delhi.

ICAR Network Project. 2012. *Climate Change and Indian Agriculture: Impact, Adaptation, and Vulnerability: Salient Achievements from ICAR Network Project*. New Delhi: Indian Agricultural Research Institute. 32p.

IPCC. 2000. *Emission Scenarios- A Special Report of IPCC Working Group III*. Summary for Policymakers. Intergovernmental Panel on Climate Change.

IPCC. 2007. *Climate Change 2007: The Physical Science Basis*. Summary for Policymakers. Inter-Governmental Panel on Climate Change.

IPCC. 2013. Summary for Policymakers. In *Climate Change 2013: Physical Science Basis. Contribution of Working Group I to the Fifth Assessment Report of the Intergovernmental Panel on Climate Change*. eds., (Stocker, T. F., D. Qin, G. K. Plattner, eds. Cambridge, UK, and New York, NY, USA: Cambridge University Press.

IPCC. 2014. *Climate Change 2014: Impacts, Adaptation and Vulnerability*. IPCC Working Group II Contribution to AR5. https://ipcc-wg2.gov/AR5/

IPCC. 2022. *Sixth Assessment Report on mitigation of climate change*. Working Group III. Summary for Policymakers. Intergovernmental Panel on Climate Change.

Iqbal, W. and M. Zahid. 2014. Historical and future trends of summer mean air temperature over South Asia. *Pakistan Journal of Meteorology* 10 (20): 67–74.

Iqbal, M. M., M. A. Goheer, H. Sultana et al. 2009a. *Climate change and agriculture in Pakistan: Adaptation strategies to cope with negative impacts*. Research Report GCISC-RR-16. Global Change Impact Studies Centre, Islamabad, Pakistan.

Iqbal, M. M., M. A. Goheer, and M. A. Khan. 2009b. Climate change aspersions on food security of Pakistan. *Science Vision* 15: 15–23.

Iqbal, M. M., M. A. Goheer, S. A. Noor, H. Sultana, K. M. Salik, and A. M. Khan. 2009c. *Climate change and rice production in Pakistan: Calibration, validation, and application of CERES-Rice model*. Research Report GCISC-RR-15. Global Change Impact Studies Centre, Islamabad, Pakistan.

Ishtiaque, A., T. J. Krupnik, V. Krishna et al. 2024. Overcoming barriers to climate-smart agriculture in South Asia. *Nature Climate Change* 1–3. https://doi.org/10.1038/s41558-023-01905-z

Johnson, T.A., S. Singh, S. Mishra, and S. Kalpana. 2011. Effect of elevated temperature and low radiation on growth and yield of Basmati rice (*Oryza sativa* L.). *Indian Journal of Plant Physiology* 16: 147.

Krishnapriya, P. P., S. K. Pattanayak, E. Somanathan et al. 2024. Mitigating residue burning challenges and solutions across land classes in Punjab. *Environmental Research in Food Systems* 1: 015001. https://doi.org/10.1088/2976-601X/ad2689

Long, S. P., E. A. Ainsworth, A. D. B. Leakey and P. B. Morgan. 2005. Global food insecurity: Treatment of major food crops with elevated carbon dioxide or ozone under large-scale fully open-air conditions suggests recent models may have overestimated future yields. *Philosophical Transactions of the Royal Society B*, 360: 2011–2020.

Mir, A. K. and M. Ijaz. 2015. *Greenhouse gas emission inventory of Pakistan for the year 2011-2012*. Islamabad: Global Change Impact Study Centre.

Mishra, A., B. B. Singh, N. A. Shakil et al. 2024. Effect of high temperature stress on metabolome and aroma in rice grains. *Plant Gene* 38: 2024, https://doi.org/10.1016/j.plgene.2024.100450

MOEFCC. 2023. *India: Third National Communication and Initial Adaptation Communication to United Nations Framework Convention on Climate Change*. New Delhi: Ministry of Environment, Forests, and Climate Change, Government of India.

Muruganandham, A. and J. M. Singh. 2024. Rainfall dynamics of agro-climatic zones in the green revolution belt of India. *Water Supply* 24: 3568.

Nagarajan, S., S. V. K. Jagdish, A. H. Prasad, A. K. Tomar, A. Anand, M. Pal, and P.K. Aggarwal. 2010. Local climate affects growth, yield, and grain quality of aromatic and non-aromatic rice in northwestern India. *Agriculture, Ecosystems, and Environment* 138: 274–281.

NATCOM. 2004. *India's Initial National Communication to the United Nations Framework Convention on Climate Change*. New Delhi: Ministry of Environment and Forests, Government of India. 268p.

PIB. 2022. *IARI organised Kisan Sampark Yatra to obtain farmers' feedback of newly released IARI Basmati varieties*. Posted on 30 September 2022 by PIB Delhi, Release ID: 1863885. Ministry of Agriculture & Farmers' Welfare, Government of India. https://pib.gov.in/PressReleasePage.aspx?PRID=1863885

Saseendran, A. S., K. K. Singh, L. S. Rathore, S. V. Singh, and S. K. Sinha. 2000. Effects of climate change on rice production in the tropical humid climate of Kerala, India. *Climate Change* 44: 495–514.

Sheikh, M. M., N. Manzoor, M. Adan, J. Ashraf, and A. M. Khan. 2009.*Climate Profile and Past Climate Changes in Pakistan*. GCISC-RR-01. Islamabad, Pakistan: Global Change Impact Studies Centre.

Singh, V. P. 2000. Basmati rice of India. In *Aromatic Rices*, ed. R. K. Singh, U. S. Singh, and G. S. Khush, (pp. 135–153. New Delhi: Oxford & IBH Publishing Co. Pvt. Ltd.

Singh, J., S. Sahany, K. K. Singh, A. Robock, L. Xia. 2024. Future climate change impacts on rice in Uttar Pradesh- India's most populous agrarian state. *Earth's Future* 12, e2023EF004009. https://doi.org/10.1029/2023EF004009

Srinivasa Rao, C., K. Baral, V. Manichandana, M. Jagadesh, and R. Karthik. 2024. Climate change adaptation and mitigation in agriculture. *Journal of Agrometeorology* 26: 137–148.

Tao, F., Z. Zhang, W. Shi et al. 2013. Single rice growth period was prolonged by cultivar shifts, but yield was damaged by climate change during 1981-2009 in China, and late rice was just opposite. *Global Change Biology* 19: 3200–3209.

Van Oort, P. A. J., T. Zhang, M. E. deVries, A. B. Heinemannd, and H. Meinkea. 2011 Correlation between temperature and phenology prediction error in rice (*Oryza sativa* L.). *Agriculture, Forests and Meteorology* 151: 1545–1555.

Zahid, M. and G. Rasul. 2012. Changing trends of thermal extremes in Pakistan. *Climate Change* 113: 883–896.

Index

Pages in *italics* refer to figures and pages in **bold** refer to tables.

B

C

D

E

H

I

L

M

N

O

P

S

T

U

V

For Product Safety Concerns and Information please contact our
EU representative GPSR@taylorandfrancis.com Taylor & Francis
Verlag GmbH, Kaufingerstraße 24, 80331 München, Germany